A **REVIEW** OF THE **DOSE RECONSTRUCTION PROGRAM**
OF THE DEFENSE THREAT REDUCTION AGENCY

Committee to Review the Dose Reconstruction Program
of the Defense Threat Reduction Agency

Board on Radiation Effects Research
Division on Earth and Life Studies

NATIONAL RESEARCH COUNCIL
OF THE NATIONAL ACADEMIES

THE NATIONAL ACADEMIES PRESS
Washington, D.C.
www.nap.edu

THE NATIONAL ACADEMIES PRESS • **500 Fifth Street, N.W.** • **Washington, DC 20001**

NOTICE: The project that is the subject of this report was approved by the Governing Board of the National Research Council, whose members are drawn from the councils of the National Academy of Sciences, the National Academy of Engineering, and the Institute of Medicine. The members of the committee responsible for the report were chosen for their special competences and with regard for appropriate balance.

This study was supported by contract DTRA01-01-C-0012 between the National Academy of Sciences and the Defense Threat Reduction Agency. Any opinions, findings, conclusions, or recommendations expressed in this publication are those of the author(s) and do not necessarily reflect the views of the organizations or agencies that provided support for the project.

International Standard Book Number 0-309-08902-6 (Book)
International Standard Book Number 0-309-51696-X (PDF)
Library of Congress Control Number 2003094456

Additional copies of this report are available from the National Academies Press, 500 Fifth Street, N.W., Lockbox 285, Washington, DC 20055; (800) 624-6242 or (202) 334-3313 (in the Washington metropolitan area); Internet, http://www.nap.edu

COVER PHOTO. Soldiers watching atomic-bomb detonation at Yucca Flat, Nevada. This photo is Shot DOG in which a 21.5-kiloton device was dropped from a B-50 bomber in the BUSTER-JANGLE series of tests on November 1, 1951. This photo was also published in the November 12, 1951, issue of Life Magazine.

THE NATIONAL ACADEMIES
Advisers to the Nation on Science, Engineering, and Medicine

The **National Academy of Sciences** is a private, nonprofit, self-perpetuating society of distinguished scholars engaged in scientific and engineering research, dedicated to the furtherance of science and technology and to their use for the general welfare. Upon the authority of the charter granted to it by the Congress in 1863, the Academy has a mandate that requires it to advise the federal government on scientific and technical matters. Dr. Bruce M. Alberts is president of the National Academy of Sciences.

The **National Academy of Engineering** was established in 1964, under the charter of the National Academy of Sciences, as a parallel organization of outstanding engineers. It is autonomous in its administration and in the selection of its members, sharing with the National Academy of Sciences the responsibility for advising the federal government. The National Academy of Engineering also sponsors engineering programs aimed at meeting national needs, encourages education and research, and recognizes the superior achievements of engineers. Dr. Wm. A. Wulf is president of the National Academy of Engineering.

The **Institute of Medicine** was established in 1970 by the National Academy of Sciences to secure the services of eminent members of appropriate professions in the examination of policy matters pertaining to the health of the public. The Institute acts under the responsibility given to the National Academy of Sciences by its congressional charter to be an adviser to the federal government and, upon its own initiative, to identify issues of medical care, research, and education. Dr. Harvey V. Fineberg is president of the Institute of Medicine.

The **National Research Council** was organized by the National Academy of Sciences in 1916 to associate the broad community of science and technology with the Academy's purposes of furthering knowledge and advising the federal government. Functioning in accordance with general policies determined by the Academy, the Council has become the principal operating agency of both the National Academy of Sciences and the National Academy of Engineering in providing services to the government, the public, and the scientific and engineering communities. The Council is administered jointly by both Academies and the Institute of Medicine. Dr. Bruce M. Alberts and Dr. Wm. A. Wulf are chair and vice chair, respectively, of the National Research Council.

www.national-academies.org

COMMITTEE TO REVIEW THE DOSE RECONSTRUCTION PROGRAM OF THE DEFENSE THREAT REDUCTION AGENCY (DTRA)

A Note on the Units of Measurement Used in this Report

It has been the custom of the Board on Radiation Effects Research (BRER) to use the International System of Units (SI) in its reports. In this report, however, exceptions are made in presenting data on radiation exposure, radiation dose, and activity of radionuclides. In all such cases, traditional non-SI units and their special names are used. Thus, exposure in air is given in roentgen (R), absorbed dose in rad, equivalent dose in body organs or tissues in rem,[1] and activity of radionuclides in curies (Ci). Decimal submultiples of the units also are used. For example, equivalent dose may be given in millirem (mrem), or one thousandth (10^{-3}) of a rem, and activity may be given in microcuries, or one millionth of a curie (μCi), or in nanocuries, or one billionth of a curie (nCi). The traditional units are used in this report because they have been used exclusively in all dose reconstructions for atomic veterans and in other documents of the dose reconstruction program and therefore are the units with which veterans are familiar.

The relationships between the units used in this report and the corresponding SI units and special names are given in the table below.

Quantity	Previous unit	SI unit	Special name of SI unit	Conversion
Exposure	roentgen (R)	coulomb per kilogram (C kg^{-1})		1 R = 2.58×10^{-4} C kg^{-1}
Absorbed dose	rad	joule per kilogram (J kg^{-1})	gray (Gy)	1 rad = 0.01 Gy
Equivalent dose	rem	joule per kilogram (J kg^{-1})	sievert (Sv)	1 rem = 0.01 Sv
Activity	curie (Ci)	disintegration per second (s^{-1})	becquerel (Bq)	1 Ci = 3.7×10^{10} Bq

[1] In dose reconstructions for atomic veterans, the biologically significant dose to body organs or tissues is called "dose equivalent." In the present report, however, this quantity is called "equivalent dose," as recommended by the International Commission on Radiological Protection (ICRP, 1991a).

Preface

From 1945 through 1962, the US atmospheric nuclear weapons testing program involved hundreds of thousands of military and civilian personnel, and some of them were exposed to ionizing radiation. Veterans' groups have since been concerned that their members' health was affected by radiation exposure associated with participation in nuclear tests and have pressured Congress for disability compensation. Several pieces of legislation have been passed to compensate both military and civilian personnel for such health effects. Veterans' concerns about the accuracy of reconstructed doses prompted Congress to have the General Accounting Office (GAO) review the dose reconstruction program used to estimate exposure. The GAO study concluded that dose reconstruction is a valid method of estimating radiation dose and could be used as the basis of compensation. It also recommended an independent review of the dose reconstruction program. The result of that recommendation was a congressional mandate that the Defense Threat Reduction Agency (DTRA), a part of the Department of Defense, ask the National Research Council to conduct an independent review of the dose reconstruction program. In response to that request, the National Research Council established the Committee to Review the Dose Reconstruction Program of the Defense Threat Reduction Agency in the Board on Radiation Effects Research (BRER).

The committee randomly selected sample records of doses that had been reconstructed by DTRA and carefully evaluated them. The committee's report describes its findings and provides responses to many of the questions that have been raised by the veterans.

Throughout the study, the committee's work was greatly aided by the efforts of D. Michael Schaeffer of DTRA, the DTRA contractor team, Bradley Flohr and Neil Otchin of the Department of Veterans Affairs, and Department staff; we thank them for providing valuable historical insights for the committee's study and providing feedback and materials for additional review. The committee and the BRER staff are grateful for the information provided by invited speakers who generously contributed their time and participated in the committee's information-gathering meetings: D. Michael Schaeffer, Steve Powell, W. Jeffrey Klemm, Julie Fisher, Cindy Bascetta, Neil Otchin, Bradley Flohr, Pat Broudy, Richard Conant, Andy Nelson, Barry Pass, and Alex Romanyukha. The committee thanks Tony E. Carter and Sandy Ford for redacting case files.

The committee is especially grateful for the assistance provided by "atomic veterans" throughout the course of its work. Veterans provided records, explained their concerns, and assisted us in understanding the conditions surrounding the nuclear-weapon tests. The committee and the BRER staff are appreciative of the information, feedback, and background materials for review provided by Gilbert Acciardo, James Avans, James Bradley, Robert Brenner, Frank Bushey, Thomas Caffarello, Boley Caldwell III, Sarah Comley, Joseph Ceonzo, Fred Clapp, Fred Clark, daughterrad (e-mail), William Duffy, Theodore Dvorak, Frank Fancieullo, William Fish, Walter Furbee, Richard Gilson, Glen Howard, Thomas Hughes, Jennifer Jones, Martin Kinney, Harold Kolb, David Lloyd, John Locke, Michael Lynch, James McDonald, Jack Nelson, James Robert Peden, Howard Pettet, Howard Pierson, Bernard Reynolds, Claude Richard, Keith Schwenk, James Warren Scott, Rodney Seidler, Delinda Sterling, R. Stockwell, Gerald Stone, Richard Stoyle, Herb Stradley, strawbry (e-mail), Arthur Templin, James Tokar, Paul Tutas, Lawrence Wagner, and Sidney Wolfeld. We hope that we have responded usefully to the veterans' questions about dose reconstruction and the claims process. We also hope that our work will help to generate changes in the dose reconstruction program that will make it more effective.

Finally, the committee thanks the National Research Council staff who worked directly with us, especially Study Director Isaf Al-Nabulsi for keeping the committee focused and assisting in the writing and preparation of our report. Dr. Al-Nabulsi was well assisted in the administration of the committee's work by Dianne Stare and Doris Taylor.

John E. Till
Chairman, Committee to Review the
Dose Reconstruction Program of the
Defense Threat Reduction Agency

Acknowledgments

This report has been reviewed in draft form by persons chosen for their diverse perspectives and technical expertise in accordance with procedures approved by the National Research Council's Report Review Committee. The purposes of this review are to provide candid and critical comments that will assist the institution in making the published report as sound as possible and to ensure that the report meets institutional standards of objectivity, evidence, and responsiveness to the study charge. The review comments and draft manuscript remain confidential to protect the integrity of the deliberative process. We wish to thank the following for their participation in the review of this report:

Harry M. Cullings, Hiroshima, Japan
Edward R. Epp, Weston, MA
Naomi H. Harley, New York, NY
Milton Levenson, Menlo Park, CA
Francis X. Masse, Middleton, MA
Bruce A. Napier, Richland, WA
Andrew M. Sessler, Berkeley, CA

Although the reviewers listed above have provided many constructive comments and suggestions, they were not asked to endorse the conclusions or recommendations, nor did they see the final draft of the report before its release. The review of this report was overseen by Richard B. Setlow, Brookhaven National Laboratory (Senior Biophysicist) and Maureen M. Henderson, University of Washington (Professor Emeritus). Appointed by the National Research Council, they were responsible for making certain that an independent examination of this report was carried out in accordance with institutional procedures and that all review comments were carefully considered. Responsibility for the final content of this report rests entirely with the authoring committee and the National Research Council.

Contents

Public Summary

This report examines the radiation dose reconstructions for military personnel who participated in various activities during atmospheric nuclear-weapons tests. The tests took place in New Mexico, Nevada, and the Pacific from 1945 through 1962. Other personnel included in the dose reconstructions are those who were prisoners of war in Japan or who were stationed in Hiroshima or Nagasaki, Japan, after the atomic bombings of 1945. Hundreds of thousands of personnel were involved. Most of the radiation doses were received from exposures to radioactive fallout, and not from the nuclear-weapons detonations themselves. The soldiers were mostly too far away from the shot locations to receive radiations directly from a detonation.

Dose reconstruction efforts began in the late 1970s, and a compensation program for atomic veterans whose diseases might have been caused by radiation exposure began in the early 1980s. The Defense Threat Reduction Agency (DTRA) is the Department of Defense agency responsible for assessing radiation exposures of atomic veterans. Science Applications International Corporation (SAIC) has performed the dose reconstructions under a contract from DTRA. SAIC works with JAYCOR, which is responsible for confirming that a veteran was a participant in the testing program and for developing information about the veteran's activities that will help in estimating a dose. The Department of Veterans Affairs (VA) is the main contact with the veteran and is ultimately responsible for determining eligibility for compensation.

After Senate hearings in April 1998, the General Accounting Office (GAO) was asked to review the reliability of the dose reconstruction program. In January 2000, GAO reported that dose reconstruction is a valid method for use in evaluating claims but noted that the program had no independent review process. In December 2000, the National Research Council formed a committee in response to a charge by Congress that directed it to evaluate randomly sampled dose reconstructions and address these four issues:

1. Whether or not the reconstruction of the sample doses is accurate.
2. Whether or not the reconstructed doses are accurately reported.
3. Whether or not the assumptions made regarding radiation exposure based on the sampled doses are credible.
4. Whether or not the data from nuclear tests used by DTRA as part of the reconstruction of the sampled doses are accurate.

The committee was also asked to recommend whether there should be a permanent system of review for the dose reconstruction program.

A number of laws and regulations apply to the dose reconstruction program. In particular, under 38 *CFR* 3.309, veterans who are confirmed participants and have any of 21 cancers are eligible for compensation regardless of their radiation exposures; this is often called the presumptive regulation. The list of cancers considered "presumptive" has been added to over the years. A different regulation, 38 *CFR* 3.311, applies to other diseases; this is the "nonpresumptive" regulation. For them, a dose assessment is used to help evaluate whether a veteran's disease is at least as likely as not to have been caused by radiation exposure during atomic testing. Furthermore, the veteran is to be given the benefit of the doubt in evaluating a claim for a nonpresumptive disease if his participation cannot be definitely confirmed. He is also to be given the benefit of the doubt in estimating his dose. Dose reconstruction involves estimating the most likely dose that a veteran received and also a higher number called an upper-bound dose, which is the dose to be considered in deciding compensation. For skin, eye, and inhalation exposures, only an upper bound is estimated. A stated goal of the dose reconstruction program is that there not be more than a 5% chance that the reported upper bound is lower than the actual dose the veteran received.

Radiation dose reconstruction can be tedious and complicated. Often historical information is lacking about individual activities that would help in estimating a dose. The committee recognized the difficulties that would face any agency or organization that took on this challenge. In addition, the science involved in dose reconstruction has changed in the last 25 years. For all those reasons, independent review of the process is important.

On the basis of its review of 99 individual dose reconstructions and other program documents, the committee reached these conclusions:

1. Although the methods used to estimate *average* doses to participants in various units are generally valid, many participants did not wear film badges all the times that they might have been exposed, so *individual* doses are often highly uncertain.

2. Upper bounds of doses from external exposure to gamma radiation are often underestimated because of questionable assumptions about a person's locations and durations of exposure.

3. Upper bounds of doses from external exposure to neutrons are always underestimated by a factor of about 3–5, but few participants received much neutron exposure.

4. Skin and eye doses from exposure to beta particles do not always seem to be credible upper bounds, and skin doses from radioactive particles on the skin do not seem to have been taken into account.

5. Methods used to estimate doses due to inhaled radioactive materials involve many assumptions that are subject to error because of a lack of data to monitor exposures. Nonetheless, in some exposure scenarios, estimates of inhalation dose appear to be credible upper bounds. In other cases, the estimates are too low but credible upper bounds would still be small doses. However, there were scenarios involving some maneuver troops and close-in observers at the Nevada Test Site in which upper bounds of inhalation dose were underestimated by large factors, and the doses in these cases often could be important. Large underestimates of inhalation dose were due mainly to neglecting the effects of the blast wave produced in a detonation, which could have caused resuspension of large amounts of radionuclides that had accumulated on the ground from previous tests.

6. Dose reconstruction has not routinely included exposure from ingestion of radioactive materials or contaminated food, but the committee does not believe this was an important source of radiation exposure for most participants.

7. In developing exposure scenarios and assessing film-badge data, veterans are not always given the benefit of the doubt and often were not contacted to verify their activities, so underestimates could have occurred in individual cases. The veterans themselves are a valuable resource that has been underused.

8. Because of problems of scenario development and estimation of external and internal doses, total doses do not always provide credible upper bounds, and the resulting underestimates often are substantial. Methods used to estimate doses and their uncertainties should be re-evaluated, and the requirement to give the veteran the benefit of the doubt should be applied more consistently in dose reconstructions.

9. Interaction and communication with the atomic veterans should be improved. For example, veterans should be allowed to review the scenario assumptions used in their dose reconstructions before the dose assessments are sent to the Department of Veterans Affairs for claim adjudication.

10. Dose reconstructions have been accurately reported to veterans, but uncertainty should also be reported and carefully explained to VA and the veterans. Also, since some changes in the dose reconstruction program could have made a substantial difference in some earlier dose estimates, veterans and their advisors should be notified when changes are made and that they can ask for updated dose assessments and re-evaluation of their prior claims.

11. More effective approaches should be established to communicate the meaning of doses to veterans in terms of their risk of disease and the probability that their disease was caused by radiation exposure from atomic testing.

12. A comprehensive manual of standard operating procedures for the conduct of dose reconstructions is needed. The lack of a procedures manual may have led to inconsistencies in dose reconstructions.

13. There was little evidence of quality control in dose reconstructions the committee reviewed. For example, many calculations are illegible or not explained. A comprehensive program of quality assurance and quality control of dose reconstructions is needed.

14. If the dose reconstruction program continues, the committee believes there should be an independent oversight system. For example, an advisory board could be established to include experts in the various parts of the program and at least one atomic veteran. Broad oversight would be desirable, including the roles of both DTRA and VA. The board should be able to conduct random audits, review methods and recommend changes, and meet with atomic veterans regularly and help DTRA and VA communicate with them.

About 70% of all dose reconstructions have been done in response to veterans' claims for compensation, but many of their diseases are now included in the presumptive category. Except for beta exposures and skin cancer, it appears to the committee that most future claims for nonpresumptive diseases would not qualify for compensation, even with revised upper-bound dose estimates.

The committee appreciates the sacrifices made by the veterans in the service of their country and their frustrations in dealing with the bureaucracy to obtain the compensation that they believe they are entitled to. Perhaps a few more veterans who filed claims in the past would have been compensated if the upper-bound dose estimates had been more credible. It is evident that only a very small number of awards have been granted for claims under the nonpresumptive regulation out of many thousand that have been filed. The exact number of successful claims is difficult to determine, but the committee has concluded that the number is probably on the order of 50, as has been previously reported. Obviously it is very unlikely that a claim will be granted when a veteran files under the nonpresumptive regulation.

Yet there are good reasons for the low rate of successful claims for nonpresumptive diseases. There is an extensive amount of information from radiation studies in humans which indicates that ionizing radiation is not a potent cause of cancer. Thus, although the committee believes that in many cases the veterans have legitimate complaints about their dose reconstructions, veterans also need to understand that in most cases their radiation exposure probably did not cause their cancer. Even if reasonable changes are made in the dose reconstruction program, it is not likely that the chance of a successful claim will increase very much when a dose reconstruction is needed, except possibly in cases of skin cancer.

Executive Summary

Radiation dose reconstruction is the process of estimating radiation doses that were received by individuals or populations at some time in the past as a result of particular exposure situations. This report is concerned with dose reconstructions for military personnel—atomic veterans—who participated in various activities during atmospheric testing of nuclear weapons at the Trinity site in New Mexico, at the Nevada Test Site (NTS), and in the Pacific in 1945-1962, or who were prisoners of war in Japan or were stationed in Hiroshima or Nagasaki, Japan, after the atomic bombings of 1945. The types of ionizing-radiation exposures received by military personnel depended on characteristics of the detonations, the roles of the participants, and the proximity of personnel to detonations and fallout of nuclear debris from each detonation. Few of the hundreds of thousands of military participants were close enough to the locations of shots to receive exposures from gamma rays or neutrons produced directly by a detonation. Most radiation doses to military personnel in the continental United States, the Pacific, and Japan were due to exposure to beta- and gamma-emitting fission and activation products produced by nuclear-weapon detonations and to plutonium that did not undergo fission.

Possible radiation exposures of military personnel during the atomic testing program have been of concern since the middle 1970s. Efforts to develop a program of dose reconstruction for atomic veterans began in the late 1970s, and a compensation program for atomic veterans whose diseases might have been caused by radiation exposure began in the early 1980s.

The Defense Nuclear Agency (now the Defense Threat Reduction Agency, DTRA) was designated as the responsible Department of Defense (DOD) agency to assess radiation exposures of atomic veterans. Science Applications Incorporated (now Science Applications International Corporation, SAIC) has held a contract to perform dose reconstructions for military personnel almost since the inception of the Nuclear Test Personnel Review (NTPR) program. SAIC eventually teamed with JAYCOR, which is responsible for confirming each veteran's status as a participant in the testing program and developing background information for estimating exposures to the veterans, such as detailed records of activities of veterans' units at the NTS or in the Pacific. From the inception of the dose reconstruction and compensation programs, the responsibilities of DTRA and the Department of Veterans Affairs (VA) have been different. DTRA is responsible for confirming service status, estimating doses to participants, and reporting doses to VA; VA is the primary avenue of contact for the veterans and is responsible for determining eligibility for compensation.

In April 1998, the Senate Committee on Veterans Affairs held a hearing that focused on radiation issues concerning the efficacy of current legislation governing compensation benefits for radiation-exposed veterans. The hearings highlighted the controversy about the use of dose reconstruction as a tool for determining veterans' eligibility for benefits. In August 1998, the Senate committee asked the General Accounting Office (GAO) to review available information related to dose reconstruction to determine its reliability for measuring veterans' radiation exposures and to assess the completeness of historical records that are used to assign radiation doses. GAO completed its review in January 2000 and found that although dose reconstruction is a valid method of estimating veterans' doses for compensation claims and no better alternative was identified, the program lacks an independent review process.

In December 2000, in response to GAO's findings, the National Research Council was asked to review the DTRA dose reconstruction program, and the present committee was formed for this purpose. The committee was charged by Congress to conduct a review that included the random selection of samples of doses reconstructed by DTRA to determine

1. Whether or not the reconstruction of the sample doses is accurate.
2. Whether or not the reconstructed doses are accurately reported.
3. Whether or not the assumptions made regarding radiation exposure based on the sampled doses are credible.
4. Whether or not the data from nuclear tests used by DTRA as part of the reconstruction of the sampled doses are accurate.

The committee was asked to make recommendations, if appropriate, on a permanent system of review of the DTRA dose reconstruction program.

To address the questions posed in the committee's statement of task summarized above, it is important to understand the capabilities and limits of historical dose reconstruction in general. Dose reconstruction can be a complex, tedious, and intensive undertaking, and there often is substantial uncertainty in estimates of dose to individuals. In the present context of adjudicating claims, it is also important to have some knowledge of the history of the atomic-veterans compensation program and the laws, regulations, and objectives that guide it. The public laws and regulations governing compensation for atomic veterans have changed over the last two decades. And, there have been advances in the science and tools available for the conduct of dose reconstruction and for administering the program. Those changes are reflected in changes to the dose reconstruction process noted by the committee during its review.

The various laws governing the dose reconstruction and compensation programs are implemented in Title 38, *Code of Federal Regulations*, Part 3 (38 *CFR* Part 3). Those regulations authorize the VA to provide medical care and pay compensation benefits to confirmed test participants and their dependents and to pay indemnity compensation to some survivors. Under 38 *CFR* 3.309, the so-called presumptive regulation, veterans who are confirmed participants and experience any of 21 specified cancers are eligible to receive compensation regardless of their radiation exposures. Under 38 *CFR* 3.311, the nonpresumptive regulation, a dose assessment is used to evaluate whether a veteran's disease was at least as likely as not to have been caused by radiation exposure during the atomic-testing program. In accordance with the policy set forth in 38 *CFR* 3.102, a veteran is to be given the benefit of the doubt if his[1] participation cannot be definitely confirmed when the nonpresumptive regulation is applied. Most important, veterans are to be given the benefit of the doubt in the estimation of their doses. That requirement led to the policy of the NTPR program that estimates of doses to atomic veterans should include an upper bound that is intended to represent at least a 95% confidence limit or that the dose estimates themselves should be sufficiently "high-sided" that they represent an upper bound. Thus, the goal of the NTPR program is that there should be only a small chance (no more than 5%) that an upper bound or "high-sided" estimate of dose to an atomic veteran is lower than the true dose.

The committee's report addresses all aspects of the process of dose reconstruction for atomic veterans, including how the estimated doses are used in the compensation program for the veterans. The committee is fully cognizant of the importance of the dose reconstruction program and of the controversies about the feasibility and value of dose reconstruction in the compensation program.

During the 25-year existence of the NTPR program, there have been significant improvements in the scientific foundations of dose reconstruction and in the tools that can be used to estimate doses and evaluate uncertainty. Many of these improvements are discussed in this report and are reflected in the committee's findings and recommendations on specific technical issues related to methods of dose reconstruction used in the NTPR program. The committee

---

[1]Masculine pronouns are used throughout this report, as needed; fewer than 1% of the participants were female.

recommends that the improvements be evaluated and incorporated into the NTPR program in a timely manner. That does not mean that methods of dose reconstruction for atomic veterans need to be changed each time a new piece of information becomes available, but there must be deliberate and periodic efforts to evaluate changes in data and methods of dose reconstruction and to incorporate improvements into the dose reconstruction process as warranted. The committee recognizes that many improvements have been made in the NTPR program since its inception, and it recognizes the challenge confronting DTRA and VA associated with the need to use records and data that are incomplete and often difficult to piece together to reconstruct historical doses and make decisions about compensation to thousands of veterans who were exposed decades ago. Peer review of the methods of dose reconstruction and the availability of a detailed procedures manual, with proper procedures for document control and updating, are important.

Most of the committee's effort in reviewing the program of dose reconstruction for atomic veterans was directed at the first part of the statement of task concerning whether doses to atomic veterans estimated by the NTPR program are "accurate." Because dose reconstruction is not an exact science, the committee has interpreted the question to be whether uncertainty in estimating dose has been appropriately addressed in dose reconstructions and whether credible upper bounds of doses to atomic veterans have been obtained. That interpretation is consistent with the policy of giving the veterans the benefit of the doubt in reconstructing their doses and with the intent of the NTPR program that the dose reports provided to VA for use in evaluating claims for compensation include upper bounds (95% confidence limits) of uncertain doses. In addressing the issue, the committee conducted a detailed review of 99 randomly selected dose reconstructions for individuals and many other documents of the NTPR program, including unit dose reconstructions for participant groups, documents describing methods of calculation and databases used in dose reconstructions, and documents describing participant activities and other conditions at various atomic tests.

On the basis of its review, the committee reached the following conclusions related to all aspects of the process of dose reconstruction for atomic veterans:

1. The methods used to estimate average doses to participants in various military units from external exposure to gamma rays, on the basis of exposures measured by film badges worn by participants or by field survey instruments, and from external exposure to neutrons, on the basis of established methods of calculation, are generally valid. However, because the specific exposure conditions for any individual often are not well known, many participants did not wear film badges during all possible times of exposure, and the available survey data used as input to the models often are sparse and highly variable, the resulting estimates of total dose for many participants are highly uncertain.

2. Upper bounds of doses from external exposure to photons often are underestimated, sometimes considerably (for example, by a factor of 2-3), particularly when reconstructed doses are based on field survey data and uncertain assumptions about an individual's locations and times of exposure, as opposed to being based on film-badge data.

3. Upper bounds of doses from external exposure to neutrons are always underestimated—by a factor of about 3-5, depending on the value of the neutron quality factor assumed in a dose reconstruction—because of neglect of the uncertainty in the biological effectiveness of neutrons relative to gamma rays in all calculations. However, few participants received significant doses from exposure to neutrons.

4. Doses to the skin and lens of the eye from external exposure to beta particles are claimed by the NTPR program to be upper bounds ("high-sided") because they are based on multiplying a presumed upper-bound external gamma dose by a calculated beta-to-gamma dose ratio, which also is presumed to be "high-sided." However, upper-bound gamma doses based on a reconstruction are often too low, as noted above, and the beta-to-gamma dose ratios are not evidently "high-sided" in all cases. In addition, the committee found no evidence in the 99 reviewed files that estimates of beta dose to skin include the dose due to contamination of the skin, for example, by means of adhering dirt particles. That probably was an important exposure pathway for many participants at the NTS because of the substantial dust in areas of participant activity at many shots.

5. Methods used to estimate inhalation doses are highly uncertain and subject to potentially important sources of error because of the lack of relevant air monitoring or bioassay data, and most uncertainties and sources of error have not been evaluated by the NTPR program. Nonetheless, in some exposure scenarios, the committee believes that inhalation doses assigned to atomic veterans are credible upper bounds. That is probably the case, for example, when veterans received inhalation exposures mainly from descending fallout at the NTS or in the Pacific or from resuspension of activation products in soil at the NTS. In other scenarios, such as exposure to resuspended fallout caused by walking or other light activity, upper bounds may have been underestimated but the doses still were apparently low. However, the committee has concluded that there are important scenarios in which credible upper bounds of inhalation doses exceed alleged "high-sided" doses estimated by the NTPR program by large factors. Large underestimates of upper bounds occurred in scenarios in which participants (including maneuver troops and close-in observers) were exposed in forward areas shortly after a detonation at the NTS, especially late in the period of atomic testing, and were due mainly to neglect of the effects of the blast wave produced in a detonation on resuspension of previously deposited fallout, the frequent neglect of aged fallout that accumulated at the NTS throughout the period of atomic testing, and the general neglect of fractionation of radionuclides, especially plutonium, in fallout. Furthermore, in scenarios in which inhalation

doses were underestimated by large factors, credible upper bounds of organ equivalent doses could be high enough to be important to a decision about compensation. Thus, the committee has concluded that the methods that have been used to estimate inhalation doses to atomic veterans do not consistently provide credible upper bounds of possible doses and that this could be an important deficiency in some exposure scenarios.

6. The possibility of ingestion exposures apparently is not considered routinely in dose reconstructions for atomic veterans. However, except in rare situations, the committee has concluded that potential ingestion doses were not significant. Therefore, in nearly all cases, neglect of ingestion exposures should not have important consequences with regard to estimating credible upper bounds of total doses to the veterans.

7. Veterans are not always given the benefit of the doubt in developing exposure scenarios and assessing film-badge data. Veterans often were not contacted to verify their exposure scenarios even when such contact was feasible and could have been helpful. In some cases, there was inadequate follow-up with other participants who might have been able to clarify scenario assumptions. As a result of inconsistent application of the policy of benefit of the doubt, the committee has concluded that upper bounds of dose have been underestimated substantially in a number of dose reconstructions for individual veterans.

8. As a result of problems identified by the committee in scenario development and estimation of external and inhalation doses, as summarized above, total doses reported by the NTPR program do not consistently provide credible upper bounds, and the degree of underestimation of upper bounds is substantial in many cases.

9. In response to the second part of the statement of task, the committee has concluded that doses, as they have been calculated by the NTPR program, have been accurately reported to VA and the veterans. However, the committee also believes that uncertainty in assigned doses should be reported and carefully explained to VA and the veterans. A broader communications issue is related to how changes within the program are communicated to the community of atomic veterans. The committee found that some of the changes that have been made in the dose reconstruction program over time, if adopted retroactively, would have changed a veteran's reconstructed dose and, in perhaps a few cases, even the result of the adjudication of a claim for compensation. There is no mechanism within the present system for revisiting these decisions when changes in methods of dose reconstruction are made. The committee found that veterans are not always aware of these changes or of the fact that they can request a re-evaluation of their dose reconstruction.

10. In response to the third part of the statement of task, the committee has concluded for reasons described in earlier statements that assumptions about input parameter values and exposure scenarios often were not credible (that is, reasonable and appropriate) and led to reported upper bounds of

external and internal doses that are less than the 95th percentile goal in many cases.

11. The committee found that existing documentation of individual dose reconstructions is unsatisfactory in a large majority of the 99 sampled cases reviewed by the committee. There is little evidence of quality control over the work, and many calculations in dose reconstructions are illegible or lack an explanation of their meaning or use. The committee also noted that information presented in dose reconstructions should be sufficiently complete and understandable to allow a knowledgeable individual to reproduce the calculations, but the committee found too few instances where this expectation reasonably could be met. The committee also believes that lack of a comprehensive manual of standard operating procedures is an important problem that has led to inconsistencies in dose reconstructions.

12. In response to the fourth part of the statement of task, the committee has concluded that the radiological and historical information compiled by the NTPR program is suitable and sufficient for use in historical dose reconstruction for the atomic veterans. All in all, the committee is impressed with the large amount of information that has been brought together by the NTPR program. There is a large repository of information from which to draw data about exposures. In addition, the committee believes that the veterans themselves are a valuable source of information about their own exposures. Although some attempts have been made to contact them and seek their input about scenarios of exposure, this source of information seems to be underused.

If the program of dose reconstruction continues, the committee recommends that an external system of review and oversight be established. The degree of review and oversight should be commensurate with the anticipated scope of the compensation program in the future. Although the responsibility for a permanent system of review rests with DTRA and VA, the committee provides some guidelines that may be helpful in its design and implementation.

One approach to continuing review and oversight among possible alternatives is to create an advisory board that consists of persons who can evaluate the many aspects of the program, such as historical dose reconstruction, radiation risk and probability of causation, communication with the veterans and between VA and DTRA, quality assurance and quality control, and historical research related to service experience. In addition to review and oversight of the dose reconstruction program of DTRA, review and oversight of the program as a whole, including the responsibilities of DTRA and VA in the administration of the atomic veterans' program, is desirable. If such an advisory committee is created, it should

- include at least one representative of the atomic veterans;
- meet frequently enough to understand the program fully, to conduct ran-

dom audits of doses being reconstructed and decisions regarding claims, to review methods, and to recommend changes when needed;

• meet with atomic veterans regularly, listen to their concerns, and ensure
that their concerns are addressed; and

• help DTRA and VA to provide information to veterans that effectively
communicates the program's mission and process and the health risks posed by
radiation exposures.

About 70% of all dose reconstructions have been in response to veterans'
claims for compensation, but many of the diseases that have been claimed by
veterans are now included in the presumptive regulation (38 *CFR* 3.309), and a
dose reconstruction is no longer required unless a veteran's participation cannot
be established. With the exception of dose reconstructions for beta exposures and
skin cancer, it is clear to the committee that in most future cases, even revised
upper-bound dose estimates, taking into account the committee's findings on
deficiencies in methods of dose reconstruction, would be too low for the VA to
conclude that the veteran's disease was at least as likely as not caused by his
radiation exposure and thus qualify the veteran for compensation.

The committee appreciates the frustrations of the veterans who willingly
performed their duties under extraordinary circumstances and who are confronted
with the burden of seeking compensation for diseases that they believe are related
to the service they performed for their country. Although the number is probably
small, the committee has concluded that some veterans would have been compensated if more-credible upper bounds of dose had been estimated in their dose
reconstructions. The committee's belief applies, for example, in cases of participants who could have received a much higher inhalation dose at the NTS than
was assigned in dose reconstructions, were nonsmokers, and later experienced
lung cancer.

One of the veterans' many concerns about the program of dose reconstruction and compensation is that very few claims have been granted for nonpresumptive diseases under 38 *CFR* 3.311. VA reported to the veterans in 1996 that
this number was on the order of 50. Confirmation of the number of claims
awarded under the nonpresumptive regulation from the beginning of the program
to the present time is difficult to obtain because the information needed to determine this number is not available in the VA database. In an effort to address the
veterans' concern, the committee looked at about 300 records of claims filed
and the disposition regarding awards. We concluded that the number of claims
awarded under the nonpresumptive regulation, excluding recent awards for skin
cancer, is indeed very small and likely to be on the order of 50, as previously
reported to the veterans. This indicates that when a veteran files a claim for a
disease other than skin cancer under the nonpresumptive regulation, the probability is very low (less than 1%) that the claim will be granted. If claims granted for
skin cancer since 1998 are included, the current rate of granting claims for all

nonpresumptive disease may be as high as 10%. Before 1998, few, if any, claims for skin cancer were granted.

However, it is important for the veterans to understand that there are legitimate reasons for the low number of successful claims for nonpresumptive diseases, and that these reasons are unrelated to any deficiencies in the methods of dose reconstruction used in the NTPR program. On the basis of studies of radiation dose and risk in human populations, it is evident that ionizing radiation is not a potent cause of cancer. That is indicated, for example, by the small number of excess cancers that have been observed in the Japanese atomic-bomb survivors, even though many in this population received doses much higher than the doses received by most atomic veterans. That conclusion is also indicated by the screening doses based on current radioepidemiological tables. For a given cancer type, the screening dose gives the 99% lower confidence limit of the dose associated with a probability of causation of 50%, taking the uncertainty in the cancer risk per unit dose into account. The screening doses are used to judge whether, given the current uncertain state of knowledge, it is at least as likely as not that a veteran's cancer was caused by radiation exposure, giving the veteran the benefit of the doubt. New screening doses that will be used in the future are 10 rem or greater for most cancers, and this indicates that high doses will be required to give an appreciable probability that a veteran's cancer was caused by the radiation exposure. Screening doses that have been used until now also are high for most cancers, although compensation could be awarded at doses of less than 10 rem in a few cases (such as 1 rem for liver cancer and 4 rem for lung cancer in a nonsmoker).

The committee emphasizes that the established policy of using upper-bound estimates of dose (95th percentiles) with the more extreme lower-bound estimates of doses associated with a 50% probability of causation of various cancers is highly favorable to the veterans' interests. If credible upper bounds of dose are obtained in dose reconstructions, atomic veterans can be compensated for nonpresumptive diseases even if the true probability that radiation exposure caused the diseases is substantially less than 50%.

None of that is to say that the veterans do not have legitimate complaints about their dose reconstructions; in many cases, they do. Rather, the committee hopes that veterans will understand that their radiation exposure probably did not cause their cancers in most cases and that reasonable changes in methods of dose reconstruction in response to this report are not likely to greatly increase their chance of a successful claim for compensation in most cases when a dose reconstruction is required.

The committee offers a number of recommendations that would, if implemented, improve the dose reconstruction process of DTRA and the atomic-veterans compensation program in general:

1. If the program of dose reconstruction continues, there should be ongoing external review and oversight of the dose reconstruction and compensation pro-

grams for atomic veterans. One way to implement this recommendation would be to establish an independent advisory board.

2. There should be a comprehensive re-evaluation of the methods being used to estimate doses and their uncertainties to establish more credible upper-bound doses to atomic veterans.

3. A comprehensive manual of standard operating procedures for the conduct of dose reconstructions should be developed and maintained.

4. A state-of-the-art program of quality assurance and quality control for dose reconstructions should be developed and implemented.

5. The principle of benefit of the doubt should be consistently applied in all dose reconstructions in accordance with applicable federal regulations.

6. Interaction and communication with the atomic veterans should be improved. For example, veterans should be allowed to review the scenario assumptions used in their dose reconstructions before the dose assessments are sent to the Department of Veterans Affairs for claim adjudication.

7. More effective approaches should be established to communicate the meaning of information on radiation risk to the veterans. In addition to presenting general information on radiation risk, information should be communicated to veterans who file claims regarding the significance of their doses in relation to their diseases.

8. The community of atomic veterans and their survivors should be notified when the methods for calculating doses have changed so that they can ask for updated dose assessments and re-evaluation of their prior claims.

I Introduction

I.A STUDY RATIONALE AND SCOPE

From 1945 through 1962, the United States conducted atmospheric tests of nuclear weapons. Hundreds of thousands of military personnel participated in the conduct of those tests, and many of the participants were exposed to ionizing radiation. The effects of that radiation on military personnel first became of interest in the middle 1970s, and since 1981 Congress has passed numerous laws concerning compensation of veterans who were exposed to radiation and later claimed health effects. In recent years, there has been renewed concern regarding the efficacy of those laws and, in particular, the use of dose reconstruction to assess radiation doses received by participants. In April 1998, the Senate Committee on Veterans Affairs held a hearing that focused on dose reconstruction and on issues regarding compensation for "atomic veterans."

In August 1998, the committee released its hearing report and asked the General Accounting Office (GAO) to review all available information related to dose reconstruction to determine its reliability for measuring veterans' radiation exposures and to assess the completeness of historical records used to assign radiation doses. GAO completed its review in January 2000 and found that dose reconstruction was a valid method of estimating veterans' doses for compensation claims and no better method was known, but that the program lacks an independent review process. GAO recommended that the Department of Defense (DOD) establish an independent review process for its dose reconstruction program. Section 305 of Public Law 106-419, which implemented the Senate Committee on Veterans Affairs legislation, called for the DOD to enter into a contract with the National Academy of Sciences for a review of the dose reconstruction program of the Defense Threat Reduction Agency (DTRA).

In response to the congressional mandate, the Academy formed a committee in the Board on Radiation Effects Research of the Division on Earth and Life Studies to conduct a review of the DTRA dose reconstruction program. Because dose reconstruction is a multidisciplinary science, the committee consisted of members with expertise in radiation physics, pathway analysis, biomedical ethics, health physics, biostatistics, and epidemiology. The study began in December 2000.

The task set before the committee is described in the following scope of work:

> The committee will conduct a review which will consist of the selection of random samples of doses reconstructed by the Defense Threat Reduction Agency (DTRA) in order to determine: 1) whether or not the reconstruction of the sample doses is accurate; 2) whether or not the reconstructed doses are accurately reported; 3) whether or not the assumptions made regarding radiation exposure based on the sampled doses are credible; and 4) whether or not the data from nuclear tests used by DTRA as part of the reconstruction of the sampled doses are accurate. The committee will produce a report that will include a detailed description of the activities of the committee. If appropriate, the committee will make recommendations regarding a permanent system of review of the dose reconstruction program of DTRA. If after a year the committee has concluded that its findings differ from the previously published and congressionally directed studies or that significant changes are required to the existing dose reconstruction procedures and methodology, it will issue an interim letter report summarizing its findings and will make appropriate recommendations for any changes warranted.

In this report, the committee is transmitting the results of the review of the DTRA dose reconstruction program that the committee conducted in fulfillment of its task.

I.B BACKGROUND AND HISTORY

I.B.1 The US Nuclear-Weapons Testing Program

The US nuclear-weapons testing program began during World War II with Shot TRINITY, the first test of an atomic bomb. TRINITY, a plutonium implosion device, was detonated at 5:30 a.m. on July 16, 1945, from a 100-ft tower in the Journada del Muerto (Journey of Death) Desert, in the Alamogordo bombing range in New Mexico (Malik, 1985). It had a yield equivalent to 21 kilotons of TNT.

Thunderstorms and rain squalls had threatened to postpone the test, but the weather improved and the test was allowed. Radiation monitoring of fallout material was accomplished by joint teams of Los Alamos Scientific Laboratory and military personnel, who searched the most probable fallout areas with radiation-detection instruments, questioned residents, and took soil samples for laboratory analyses and comparison with original radiation-monitor readings.

Residents of the village of Bingham, New Mexico, some 18 miles from ground zero, were startled by a brilliant flash of light that awakened them and their families, as recounted to monitoring teams in interviews a few hours after the detonation (Hoffman, 1947). Those interviews were the first of civilians who had witnessed an atomic-bomb explosion. A previous story had been arranged regarding an explosion of store ammunition at the Alamogordo bombing range to answer questions for national security purposes.

The first test was followed by the detonation over Hiroshima, Japan, on August 6, 1945, of the bomb named LITTLE BOY because it was small (10 ft long, 28 in. in diameter, and weighing 9,000 lb) compared with FAT MAN (12 ft long, 60 in. in diameter, weighing 10,800 lb, and having fins), which was detonated over Nagasaki, Japan, on August 9, 1945, and resulted in ending the war with Japan. Unlike TRINITY, LITTLE BOY was a gun-type device containing uranium-235 (^{235}U) and had not been previously tested; FAT MAN was identical with TRINITY except for added tail fins and associated hardware to convert it from a test device to a weapon.

The end of World War II was followed by two nuclear tests at Bikini Atoll in the Pacific Ocean during Operation CROSSROADS, in July 1946. The devices detonated, Shots ABLE and BAKER, each had a yield of 21 kilotons; they were essentially the same as the device detonated at Shot TRINITY and the bomb dropped over Nagasaki. The mostly military participants in CROSSROADS numbered about 43,000.

Operations SANDSTONE in 1948, GREENHOUSE in 1951, IVY in 1952, and CASTLE in 1954—all in the Pacific—were interspersed with Operations RANGER in January-February 1951, BUSTER-JANGLE in October-November 1951, TUMBLER-SNAPPER in 1952, and UPSHOT-KNOTHOLE in 1953 at what came to be called the Nevada Test Site (NTS) (it was called the Nevada Proving Ground, or NPG, until December 1954). Test yields at the Pacific tests increased to 15.3 megatons (MT) of TNT for CASTLE BRAVO in 1954, and yields at NTS tests increased to 74 kilotons with Shot HOOD during Operation PLUMBBOB in 1957, which was the highest atmospheric yield on the continent.

Additional atmospheric test series in the Pacific were WIGWAM in 1955, HARDTACK Phase I in 1958, and DOMINIC Phase I in 1962. Additional atmospheric test series in Nevada were TEAPOT in 1955, HARDTACK Phase II in 1958, and DOMINIC Phase II in 1962. Another test operation was ARGUS, consisting of three nuclear detonations on rockets, each 1-2 kilotons in yield, hundreds of miles above the Atlantic Ocean—far enough above the ocean surface that no detectable exposures of test participants occurred. Overall, the United

TABLE I.B.1   US Atmospheric Nuclear-Weapons Test Series[a]

Test Series	Dates	Location	No. Tests
TRINITY	July 1945	Alamogordo, NM	1
CROSSROADS	June-July 1946	Bikini	2
SANDSTONE	April-May 1948	Enewetak	3
RANGER	January-February 1951	NTS	5
GREENHOUSE	April-May 1951	Enewetak	4
BUSTER	October-November 1951	NTS	5
JANGLE	November 1951	NTS	2
TUMBLER-SNAPPER	April-June 1952	NTS	8
IVY	October-November 1952	Enewetak	2
UPSHOT-KNOTHOLE	March-June 1953	NTS	11
CASTLE	February-May 1954	Bikini, Enewetak	6
TEAPOT	February-May 1955	NTS	14
WIGWAM	May 1955	Pacific	1
REDWING	May-July 1956	Bikini, Enewetak	17
PLUMBBOB	May-October 1957	NTS	25
HARDTACK-I	April-August 1958	Enewetak, Bikini	31
HARDTACK-II	September-October 1958	NTS	19
DOMINIC-I	April-November 1962	Christmas Island, Johnston Island	31
DOMINIC-II	July 1962	NTS	4

[a]High-altitude rocket tests that did not result in exposure to veterans are not included. Details of individual tests (names, dates, types, yields) can be found in DOE (2000), which is available at http://www.nv.doe.gov/news&pubs/publications/historyreports/default.htm.

States conducted about 200 atmospheric weapons tests with a total fission yield of about 80 MT.

Table I.B.1 lists the atmospheric nuclear-weapons test series that have been conducted by the United States. The number of tests in each series does not include safety tests and some high-altitude rocket tests. Figure I.B.1 shows the locations of US atmospheric nuclear-weapons detonations where veterans may have been exposed to radiation. Hundreds of underground tests of different weapon designs have since been conducted. There have also been high-altitude, rocket-launched tests and tests of nuclear devices for peaceful uses in a program called PLOWSHARE.

I.B.2 Radiation Exposures of Military Personnel

In understanding the significance of radiation exposures of military personnel that resulted from their participation in the atmospheric nuclear-weapons testing program, it can be helpful to consider that these exposures were in addition to unavoidable exposures to natural background radiation that have been experienced by all personnel throughout their lives. As summarized by the Na-

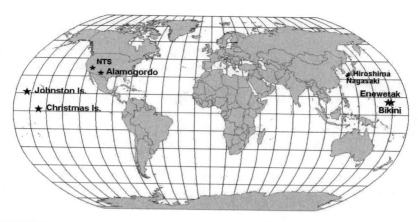

FIGURE I.B.1   Locations of atmospheric nuclear-weapons detonations where military personnel may have been exposed to radiation. A few additional high-altitude air blasts were conducted at sites in the south Atlantic and Pacific oceans, and two underwater tests that resulted in exposures to naval personnel were conducted in the Pacific off the southern California coast (DOE, 2000).

tional Council on Radiation Protection and Measurements (NCRP, 1987a), an average member of the US population receives in each year a total dose from natural background radiation of about 0.3 rem. Exposure to cosmic rays, naturally occurring radionuclides (potassium-40, radium, thorium, and uranium) in rock and soil, and naturally occurring radionuclides incorporated in body tissues (mainly by ingestion) gives a dose in each year of about 0.1 rem, and the remaining 0.2 rem in each year is due to indoor radon. Thus, for example, an average individual who lives for 70 years receives a total lifetime dose from natural background radiation, excluding the contribution from indoor radon, of about 7 rem, and the total lifetime dose, including the contribution from indoor radon, is about 21 rem.

A presentation of information on doses from natural background radiation is not intended to trivialize exposures that military personnel received during their participation in the weapons testing program or to convince individuals that their exposures are of no concern. However, such information can be helpful as each individual judges for himself the significance of his exposures during the testing program.

The types and amounts of ionizing-radiation exposures received by military personnel participating in atmospheric nuclear-weapons test detonations depended on the characteristics of the detonation, the role of the participants, and the proximity of personnel to detonations and fallout of nuclear debris.

When a nuclear explosion occurs, penetrating gamma rays and neutrons are emitted from fission of the fuel (usually ^{239}Pu or ^{235}U or a combination of the two). The gamma rays and neutrons can result in external radiation doses to unshielded participants at distances up to many kilometers depending on the yield and height of the burst. Emitted neutrons are also absorbed by elements in the soil and air, and this results in radioactive activation products. If the device has a fusion component, as well as a fission component, higher-energy neutrons are emitted that can travel even farther in air before they are absorbed. After the blast, a rapidly rising fireball forms; it contains vaporized fission products and other material, such as unfissioned fuel, debris from the device, and other surrounding material. Depending on the height of the burst, the fireball may also contain soil particles. As the fireball cools, a cloud of radioactive debris is formed. The cloud can extend from a few kilometers above the ground to the stratosphere (in the case of very-high-yield tests), and it expands and diffuses as it is transported away from the test site by the prevailing winds. As the cloud travels through the atmosphere, the radioactive debris begins to fall out, heavier particles first, because of gravitational settling.

Radioactive "fallout" includes hundreds of radioactive fission products with half-lives ranging from fractions of a second to millions of years and unfissioned plutonium and ^{235}U. This debris emits penetrating gamma radiation and less-penetrating beta and alpha particles. Test participants were exposed externally to those kinds of radiation as the debris descended through the atmosphere and after the debris was deposited on the ground and other surfaces. They were also exposed internally by inhaling particles as the fallout descended or was resuspended after being deposited. Because fairly heavy fallout can occur for many hours after a test, even people on ships and islands hundreds of kilometers downwind of a test were exposed. In fact, because the smaller particles can remain in the atmosphere for days, weeks, or even months, almost all the population of the United States was exposed to low levels of fallout from Nevada weapons tests. Debris injected into the stratosphere by the high-yield thermonuclear tests conducted in the Pacific resulted in low levels of fallout over the entire globe (Beck and Bennett, 2002).

Some members of the military participated directly in tests as observers or by conducting maneuvers close to ground zero shortly after a blast. Only a few of the hundreds of thousands of military participants were close enough to be exposed to the initial gamma rays and neutrons emitted in a detonation. The few exposed to neutrons from nuclear-weapon detonations, who were close to the blast, included pilots flying close to detonations to collect samples and volunteer officers in protective trenches at short distances selected to determine safety measures for other military personnel at different distances from detonations during a nuclear confrontation. The number of volunteer officers exposed during atmospheric nuclear-weapons tests was probably less than a few hundred. Most of the exposed military personnel were performing their duties in areas contaminated by fallout from the blasts, either during the fallout period itself or thereafter.

The total number of military and civilian personnel participating in atmospheric weapons tests was about 210,000. In addition to those exposed during the atmospheric-testing period in the Pacific and within the United States, some 195,000 military personnel were potentially exposed to residual fission products and fissile weapons material while on duty in Japan after and in the area of the bombings of Hiroshima and Nagasaki.

Most radiation exposures of military personnel in the Pacific or within the United States were external exposures to gamma- and beta-emitting fission products deposited on the ground and other surfaces and to activation products produced in soil, water, and other materials by neutrons from nuclear-weapon detonations. Participants were exposed externally to gamma and beta radiation and internally by inhaling descending fallout or resuspended debris. Exposures also resulted from inhalation of fission products and unfissioned plutonium and other fissile material from previous tests; these materials could be resuspended by wind and human activity, as well as by nuclear-blast shock waves, and could have exposed participants to radiation in addition to what they may have received from the tests they were participating in.

In addition to maneuver troops and observers, military personnel were exposed during performance of their support functions and monitoring activities. For example, pilots who took gaseous and particulate samples of clouds from nuclear detonations to be analyzed in determining yields were exposed externally to high levels of gamma radiation from the time they collected the samples until they left their aircraft. These pilots were authorized to receive some of the highest radiation exposures to accomplish their missions. Personnel who removed samples from the aircraft and decontaminated the aircraft were also exposed to radiation. Others were exposed while recovering equipment or decontaminating ships or aircraft. Many participants were exposed to fallout on residence islands or on support ships during tests in the Pacific. Thus, thousands of military participants were exposed to radiation by direct participation in tests, such as during maneuvers or as observers, and thousands more were exposed while performing their routine duties in support of nuclear tests.

From the beginning of the testing period, the primary means of measuring radiation dose to individuals was the film-badge dosimeter. The badges were effective and usually provide reasonable estimates of external gamma-radiation exposure. The accuracy of exposures recorded by film-badge dosimetry has previously been reviewed by the National Research Council (NRC, 1989). However, not all participants wore film badges, especially in the early period of testing, and exposures of most veterans could only be inferred from badges worn by other members of their cohorts or by making assumptions regarding exposure on the basis of radiation readings from the area and the veterans' locations and times of activities.

Exposures during atmospheric nuclear-weapons tests varied substantially from one veteran to another and depended on many factors. Furthermore, the reconstruction of doses to veterans for the purpose of compensation up to 50 years after the exposures occurred can be complex, tedious, and labor-intensive.

That is the challenge that has confronted DTRA and the Department of Veterans Affairs[1] for nearly two decades. The process of dose reconstruction and its validity for the purpose of awarding claims has been questioned by the veterans and by others since its beginning. This report evaluates the process of dose reconstruction for atomic veterans and considers some of the questions that have been raised.

To begin, it is important to have some knowledge of the history of the atomic-veterans compensation program and the laws and objectives that guide it. It is also important to have an understanding of the limits and capabilities of historical dose reconstruction in general. This background information is provided in the sections that follow.

I.B.3 Development of the Nuclear Test Personnel Review Program

Possible radiation exposures of military personnel during observer and maneuver programs at the NTS and during participation in support of testing in the Pacific and at the NTS have been of concern since 1977, when it was first reported that there might be an increase in leukemia among military personnel who participated in Shot SMOKY of Operation PLUMBBOB at the NTS that could be attributed to ionizing-radiation exposure (Caldwell et al., 1980). At that time, exposure to radiation was known to increase the risk of some types of cancer. Additional results reported by Caldwell et al. (1983) expanded observations on the same cohort to incidence of other types of cancer, in addition to mortality from cancers and other causes, and covered a period of 22 years, through 1979.[2] However, in this analysis of Shot SMOKY, it was assumed that 3,200 military personnel were exposed during the exercise when in fact only 572 participants were close enough to ground zero to receive exposure.[3] The other troops were either at News Nob, an observation point about 12 miles south of Shot SMOKY, or at Camp Desert Rock, about 40 miles south of Shot SMOKY. Because of the incorrect number of participants in the cohort exposed to radiation in Shot SMOKY, the estimate of the number of cases of leukemia might be in error. However, that error is understandable, considering the scarcity of data

---

[1]The Department of Veterans Affairs (VA) became a cabinet-level agency in 1989. It was formerly the Veterans Administration. Throughout this report, we refer to this organization as the Department of Veterans Affairs although it is recognized that for the early period of the atomic-veterans compensation program, Veterans Administration was the name of the agency.

[2]Although the dose reconstruction program for atomic veterans was initiated as a result of concerns that radiation exposure could have caused the unexpected increase in leukemia among participants at Shot SMOKY, the number of cases and the study population were both small, and the analysis by Caldwell et al. (1983) attributed the increase primarily to chance. The question of whether leukemia among SMOKY participants was caused by exposure to radiation or another agent during the atomic-testing program remains unanswered to this day.

[3]Personnel communication from Jay Brady.

available at the time on the activities of participants and the lack of dose information related to atomic tests. It is also important to note that these early studies were the first attempts to investigate effects among the veterans and set the stage for investigations that were to follow.

Concern among military personnel who participated in the testing program continued to grow during the late 1970s. By late 1977, funding was made available by the Energy Research and Development Administration (ERDA; now the Department of Energy, DOE) to begin reorganizing the master file of radiation-exposure records for the US nuclear-testing program. The Defense Nuclear Agency (DNA; now DTRA) had been designated as the responsible DOD agency to address radiation exposure of atomic veterans.

Effects of radiation exposure on military personnel participating in atmospheric nuclear-weapons testing soon became of interest to Congress, which held hearings on the matter. Congress played an important role not only in opening a forum and making sure funding was available to estimate personnel exposure, but also in opening archives to make available documents, many of which had to be declassified. In January 1978, the Nuclear Test Personnel Review (NTPR) program[4] was officially initiated as a coordinated effort of the DNA and the Energy Research and Development Administration. Science Applications Incorporated (now Science Applications International Corporation, SAIC) has held a contract to perform dose reconstructions on military personnel almost since the inception of the NTPR program.

One of the first attempts to gather information was a program where veterans were encouraged to call a toll-free number and register data related to their participation. This program was advertised in various military publications. Once a veteran called in, forms were sent to him to provide a written account of his experience in the atmospheric testing program. These data were collected in a database called "File A" that is still retained as part of a veteran's record. Subsequently, a "File B" was established to collect data from historical documents.

Originally, each branch of the service had an NTPR team to handle its own members' dose reconstructions. However, that led to disparities between methods and assumptions in estimating personnel doses. In 1983, it was decided to consolidate the teams at DNA and make procedures for dose reconstruction more consistent across the services.

SAIC continued to perform dose reconstructions for DNA and eventually teamed with JAYCOR, which is responsible for confirming a veteran's status as a participant in the testing program and for developing historical background

---

[4] The beginning effort to evaluate effects of atmospheric nuclear-weapons tests on atomic veterans was known as the Nuclear Test Personnel Review. Later, as more agencies were brought into the effort, it became known as the Nuclear Test Personnel Review program. Although the committee makes an effort to distinguish between the two names on the basis of the period being addressed, the reader should consider the two names as synonyms.

information for use by SAIC in estimating doses to the veterans. As of September 30, 2002, 4,048 partial or total dose reconstructions had been performed for specific veterans (Schaeffer, 2002a).

Since the inception of the program, the responsibilities of DTRA (then DNA) and the Department of Veterans Affairs (then the Veterans Administration) have been different. DTRA is responsible for confirming service status and reconstructing doses, and VA is the primary avenue of contact for a veteran and is responsible for determining eligibility for compensation. Each fulfills its role independently of the other, although close interaction is maintained.

I.B.4 Key Laws and Regulations Governing the NTPR Program

About 15 public laws form the basis of regulations that govern the administration of the NTPR program and determine the eligibility of veterans to receive service-connected disability compensation based on their radiation exposure during the nuclear-weapons testing program. Several of the key laws are described below, and then the regulations implementing them are discussed.

The primary law was enacted in 1981: Public Law (PL) 97-72, the Veterans' Health Care, Training, and Small Business Loan Act. It specified that atomic veterans were entitled to medical care if they could prove that their disease was service-connected, which few could do. The next law was PL 98-542, the Veterans' Dioxin and Radiation Exposure Compensation Standards Act of 1984, which listed in greater detail how such service connection was to be established. It also listed radiogenic diseases,[5] their latent periods,[6] and appropriate compensation. When a claim was filed under this law, the veteran had to obtain a dose estimate from the DNA and the estimated dose had to be above a certain level for an award to be made (usually greater than 5 rem).

In 1988, Congress passed PL 100-321, The Radiation-Exposed Veterans Compensation Act. Under this law, referred to as the presumptive law, it is presumed that a veteran's disease was caused by radiation if the veteran was present at a nuclear detonation, or some associated activities, and if the veteran developed one of the presumptive diseases, regardless of the veteran's dose. The original presumptive law listed 13 cancers as radiogenic: leukemia, except chronic lymphocytic leukemia; multiple myeloma; lymphoma, except Hodgkin's disease; primary liver cancer; and cancer of the thyroid, breast, pharynx, esophagus, stomach, small intestine, pancreas, bile duct, and gall bladder. Cancers of the salivary

---

[5]A radiogenic disease is a type of disease assumed on the basis of scientific studies to have an association with radiation exposure. A statement that a cancer is radiogenic does not imply that radiation is the only cause of the cancer but, rather, that radiation has been shown to be one of its causes. Exposure to other environmental substances could also cause the same type of cancer.

[6]The latent period is the time after exposure that it takes for a radiation-induced cancer to be manifested. Latent periods may vary widely between different types of cancer and within subgroups of one type of cancer.

gland and urinary tract were added in 1992 and bronchiolo-alveolar carcinoma was included in 1999. Finally, in 2002, cancer of the bone, brain, colon, lung, and ovary were added, bringing the total number of cancers considered radiogenic under the presumptive law to 21 (VA, 2002).

The various laws are implemented in Title 38, *Code of Federal Regulations*, Part 3 (38 *CFR* Part 3). The regulations authorize the VA to provide medical care and pay compensation benefits to confirmed test participants and dependence and indemnity compensation to certain survivors. A veteran seeking compensation can file a claim with VA. To resolve claims, VA uses one of two processes, depending on the specific type of disease being claimed. Under 38 *CFR* 3.309, if the veteran was a confirmed participant and has one of the 21 cancers presumed to be radiogenic, the veteran is eligible to receive compensation regardless of dose. Throughout this report, we refer to that situation as "presumptive."

The second regulation governing the claims process is described in 38 *CFR* 3.311. It applies to diseases that are not presumed to be caused by radiation exposure but could be linked to radiation if the veteran's dose was high enough. Alternatively, the veteran can supply evidence that the condition can be caused by radiation, that is, is a radiogenic disease. We refer to that type of claim as "nonpresumptive." Such a claim is used for veterans with all other forms of cancer and some nonmalignant ailments: tumors of the central nervous system, nonmalignant thyroid disease, posterior subcapsular cataract, and parathyroid adenoma. The process relies on estimating the radiation dose received and evaluating the probability that the disease was caused by the exposure. Under the nonpresumptive regulation, if a veteran's claim of presence at a nuclear test cannot be verified but the government cannot prove that the veteran was elsewhere at the time, it must be assumed that the veteran was present. The regulation also requires that in assessing a dose, the veteran must be given the benefit of the doubt if information needed to determine the dose is inconclusive or unavailable. The principle of benefit of the doubt is discussed in Section I.C.3.2.

It has happened that a claim was filed under the presumptive regulation but the veteran's presence at a nuclear test could not be verified, so the veteran was not eligible for compensation under that regulation. In such a case, however, the veteran's claim can be evaluated under the nonpresumptive regulation.

The laws governing compensation for atomic veterans have continued to change over nearly two decades. There have been advances in the science and tools available for administering the program and for estimating doses. Additional records have also come to light over the years. These changes are reflected in changes to the dose reconstruction process noted by the committee during its review.

I.B.5 Objectives of the NTPR Program

DTRA continues to administer the NTPR program for DOD. VA is responsible for making decisions about awarding claims. The primary function of the

NTPR program is to provide participation data and radiation-dose information to VA and to the veterans.

As described on its Web site (http://www.dtra.mil/news/fact/nw_ntprfact. html), the NTPR program has four primary objectives, which are summarized below:

- Providing participant and radiation-dose information to support medical and compensation programs administered by VA and the Department of Justice. The NTPR program also ensures that veterans can obtain access to relevant documents and records about their involvement in US atmospheric nuclear tests or in the occupation forces of Hiroshima and Nagasaki.
- Conducting historical-records research. Over 100 archives nationwide have been researched for relevant information. More than 40 historical volumes and 25 analytical reports have been developed to provide details on each test and operation. The program has located, retrieved, declassified as necessary, and preserved records pertaining to US atmospheric nuclear tests. The documentation includes service and medical records, film-badge records, pocket-dosimeter logs, special orders, muster rolls, unit memoranda, ship logs, morning reports, flight logs, personal accounts, diaries, and papers.
- Performing outreach service to veterans and their families and appointed representatives. The outreach includes personal contact with veterans and mass-media announcements to find veterans and publicize the availability of services and of VA's health-care and entitlement programs.
- Supporting independent scientific studies to determine whether US atmospheric nuclear-test participants have adverse health effects as a result of their participation. Some of the studies are described in the next section.

I.B.6 Previous National Research Council Studies on Military Personnel Exposed to Radiation in Atmospheric Nuclear-Weapons Tests

The National Research Council has conducted studies related to exposures of participants in atmospheric nuclear-weapons tests. The first report, *Mortality of Nuclear Weapons Test Participants* (NRC, 1985a), selected participants in five nuclear test series. Numbers of actual test participants were not well known during that study with the result that thousands of test participants were inadvertently omitted and thousands more military personnel who were not participants were included. An additional problem was the illness occurrences in the general populations as a comparison cohort, causing a "healthy soldier effect" that may have obscured illness in atomic veterans. As a consequence, the results of that study were later questioned and a second five-series study of atomic veterans was conducted (IOM, 2000).

The second report, *Review of the Methods Used to Assign Radiation Doses to Service Personnel at Nuclear Weapons Tests* (NRC, 1985b), was prepared to advise DNA on whether the methods used in the NTPR program to assign radiation doses were comprehensive and scientifically sound and to recommend improvements if needed. The committee was not charged to conduct audits of the dose assignments or reconstructions for specific veterans. The committee concluded that "the procedures used to estimate external radiation doses were reasonably sound." It also concluded that "if bias exists in the dose estimates, it is probably a tendency to overestimate the most likely dose, especially for internal emitters." Further discussion of the findings on methods of estimating dose from internal emitters is given in Section V.C.2. However, the National Research Council committee had difficulty in finding information that summarized the procedures being followed in the dose reconstruction process and recommended that DTRA develop a comprehensive report addressing the methods and procedures being used.

The third report, *Film Badge Dosimetry in Atmospheric Nuclear Tests* (NRC, 1989), was an in-depth evaluation of film-badge practices used during the weapons-testing period, of recording of dosimetric data, and of overall uncertainties associated with the film-badge dosimeter readings. The report concluded that "it is feasible to estimate dose for participants with reasonable certainty." The report included methods of addressing uncertainties in film-badge readings and of converting film-badge readings to doses received by organs that are important in assessing the biological significance of exposure.

In 1996, the Institute of Medicine published *Mortality of Veteran Participants in the CROSSROADS Nuclear Test* (IOM, 1996). The report described a cohort epidemiological study that investigated mortality in US Navy personnel who participated in Operation CROSSROADS, a 1946 atmospheric nuclear-test series in the Pacific. The study did not use dosimetry information for cases, because this information was not considered suitable to support an epidemiological analysis. The report stated that "the findings do not support a hypothesis that exposure to ionizing radiation was the cause of increased mortality among CROSSROADS participants."

As discussed above, a second five-series study (IOM, 2000) was undertaken to address problems associated with the 1985 report (NRC, 1985a). That project was a mortality study of about 70,000 military personnel who participated in at least one of five selected atmospheric nuclear-weapons test series. The study carefully confirmed the status of each participant and used nonparticipant military personnel as a comparison group. The analysis considered effects among participants for all tests and within each specific test series. The study found that, overall, participants and controls had similar risks of death from all causes. It originally intended to use specific dosimetry information on veterans to determine whether a dose-response analysis could be carried out, and a special working group was formed to investigate the feasibility of using doses generated by the NTPR program. The working group was specifically charged to determine whether the doses recorded for veterans could be used for epidemiological purposes and it considered four factors in forming its evaluation: consistency in technical approach, nondifferential methods in dose assignments, quality assurance, and consideration of uncertainties. The working group concluded that the available dose information did not meet the criteria for epidemiology, so dose-response analyses were not performed.

I.B.7 General Accounting Office Reports on the NTPR Program

GAO published two reports related to exposures of veterans in the atmospheric nuclear-weapons testing program. The first, *Nuclear Health and Safety: Radiation Exposures for Some Cloud-Sampling Personnel Need to be Reexamined* (GAO, 1987), was undertaken at the request of the chairman of the Senate Committee on Veterans Affairs. It addressed the concern expressed by a public-interest group (the Environmental Policy Institute) that radiation exposures of about 300 Air Force personnel associated with flying and decontaminating aircraft had been substantially underestimated in the NTPR program. The report concluded that the doses were indeed underestimated and needed to be re-examined. The NTPR program took actions to address the report's conclusions.

A second GAO report, *Veteran's Benefits: Independent Review Could Improve Credibility of Radiation Exposure Estimates* (GAO, 2000), responded to veterans' concerns about the methods being used in dose reconstructions performed under the nonpresumptive regulation. GAO's conclusion was that although studies appeared to validate DOD's dose reconstruction program for deciding claims, "the agency is not providing independent oversight of the program." The report noted that VA did not believe that it was responsible for overseeing the DOD program. The GAO report recommended that the dose reconstruction program be continued as a means for deciding claims but also recommended that independent oversight of the NTPR program be considered. The present National Research Council report responds to the issue of independent oversight and to other questions raised by Congress.

I.C PRINCIPLES AND PROCESS OF DOSE RECONSTRUCTION

I.C.1 Introduction to the Process of Dose Reconstruction

Dose reconstruction refers to the process of estimating radiation doses that were received by individuals or populations in the past as a result of particular exposure situations of concern. For example, this report is concerned with radiation exposure of military personnel (the atomic veterans) who were prisoners of war in Japan or were stationed in Hiroshima or Nagasaki after the atomic bombings of 1945 or who participated in various activities during atmospheric testing of nuclear weapons at the Trinity site in New Mexico, at the Nevada Test Site (NTS), and in the Pacific in 1945-1962.

In many respects, the process of dose reconstruction is similar to the process of estimating radiation doses to workers at an operating nuclear facility while they are on the job or doses to members of the public who are exposed to continuing releases of radionuclides from a nuclear facility. The principal distinction is that dose reconstruction generally is concerned with estimating doses that resulted from exposures in the past. The terms *historical* and *retrospective* often are used to indicate that characteristic of a dose reconstruction. The distinction is made even though the types of information that can be used to estimate doses may be similar. For example, some external doses received by atomic veterans were monitored at the time with film badges (NRC, 1989) in the same way that external doses to workers at an operating nuclear facility are monitored. However, the quantity and quality of historical data used to support a dose reconstruction may be inferior to data that can be used to monitor exposures of workers and the public today.

Many dose reconstructions have been undertaken over the last two decades. In the United States, dose reconstructions have been performed for members of the public who were exposed to fallout from atmospheric testing of nuclear weapons at the NTS, both in regions near the NTS (Anspaugh and Church, 1986; Anspaugh et al., 1990; Simon et al., 1990; Till et al., 1995; Kirchner et al., 1996; Whicker et al., 1996) and throughout the country (NCI, 1997; IOM/NRC, 1999). Dose reconstructions also have been performed for members of the public who were exposed to releases of radionuclides from nuclear facilities of the DOE and its predecessor agencies (Brorby et al., 1994; Farris et al., 1994a; Farris et al., 1994b; Killough et al., 1998; Apostoaei et al., 1999a; Apostoaei et al., 1999b; Grogan et al., 1999; Rood and Grogan, 1999; Rood et al., 2002), and a program of dose reconstruction to address historical exposures of workers at DOE facilities is under way (DHHS, 2002).

The process of dose reconstruction for atomic veterans generally is similar to the process that has been used in other cases. Dose reconstructions may differ substantially in some respects, such as in the radionuclides, radiation types, and exposure pathways of concern; the types, quality, and quantity of information available for estimating doses; the degree to which modeling, rather than relevant measurements, must be used to estimate doses; and the importance of subjective judgment, both scientific and nonscientific, in estimating doses. Nonetheless, all dose reconstructions, if conducted properly, incorporate a few basic principles. The main purpose of this section is to identify and briefly discuss those basic principles and to present examples of how they have been applied in dose reconstructions for atomic veterans. The dose reconstruction process for atomic veterans is discussed in more detail in Chapter IV.

This section also discusses three other aspects of the dose reconstruction process for atomic veterans that are particularly important: dose reconstructions for atomic veterans are used to evaluate claims for compensation by veterans who developed cancers or other diseases that could have been caused by radiation exposure; regulations governing dose reconstruction specify that atomic veterans will be given the benefit of the doubt in evaluating claims for compensation; and dose reconstructions for atomic veterans have been performed for many years. Proper consideration of those aspects places unusual demands on the dose reconstruction process for atomic veterans.

I.C.2 Elements of Dose Reconstruction Process

Regardless of the purpose of a dose reconstruction, the process has several basic elements, which the committee identifies as follows:

- Definition of exposure scenarios,
- Identification of exposure pathways,
- Development and implementation of methods of estimating dose,
- Evaluation of uncertainties in estimates of dose,
- Presentation and interpretation of results,
- Quality assurance and quality control.

Those elements constitute the basic principles of dose reconstruction. They are summarized in Table I.C.1 and are briefly described on the following pages.[7]

I.C.2.1 *Definition of Exposure Scenarios*

As used in this report, the term *exposure scenario* (or *scenario*) refers to a set of assumptions about the conditions of exposure of an individual or group of individuals for whom a dose reconstruction is being performed. A properly defined exposure scenario incorporates two kinds of information:

- A description of assumed locations of the individuals of concern, their activities at those locations, and the time spent at each location,
- A description of the radiation environment at assumed locations of the individuals during the time spent at those locations.

The dose received is determined by combining information on the radiation environment with information on the locations and activities of an individual or group of individuals.

For example, consider an atomic veteran who was a member of a maneuver unit at a nuclear-weapons test at the NTS. Information about the veteran that would be needed to develop an exposure scenario includes the location of the maneuver unit at the time of detonation (the distance from ground zero), whether

---

[7]A previous report of the National Research Council describes the dose reconstruction process for the specific purpose of supporting epidemiological studies (NRC, 1985c). Some elements of the process discussed in that report may be applied differently when the purpose is to evaluate claims for compensation.

TABLE I.C.1 Summary of Basic Elements of Dose Reconstruction Process

Basic element	Summary description
Definition of exposure scenarios	Activities of individuals in areas where radiation exposure could occur and characteristics of radiation environment in those areas
Identification of exposure pathways	Relevant pathways of external and internal exposure
Development and implementation of methods of estimating dose	Data, assumptions, and methods of calculation used to estimate dose from relevant exposure pathways in assumed scenarios
Evaluation of uncertainties in estimates of dose	Evaluation of effects on estimated dose of uncertainties in assumed exposure scenarios and uncertainties in models and data used to estimate dose in assumed scenarios, to obtain expression of confidence in estimated dose
Presentation and interpretation of results	Documentation of assumptions and methods of estimating dose and discussion of results in context of purpose of dose reconstruction
Quality assurance and quality control	Systematic and auditable documentation of dose reconstruction process and results

the unit was huddled in trenches at the time of detonation or standing or sitting in an unshielded position, where and for how long the unit marched or was transported after a detonation, the time spent at the objective of the maneuver or at other locations if the objective was not reached, as well as activities undertaken at the objective and whether the veteran used respiratory-protection equipment during any part of the maneuver. A description of the radiation environment could include the height of the weapon above ground at the time of detonation, the amounts of different kinds of radiations and radionuclides produced by and after the detonation, whether radiation emitted in the blast reached the location of the maneuver unit at the time of detonation, whether the maneuver unit was exposed to airborne radionuclides after detonation, whether fallout occurred at locations of the maneuver unit, whether there was fallout from previous tests at those locations, and available data on external exposure rates or concentrations of radionuclides in the air or on the ground at various locations and times after a detonation.

The committee emphasizes that no single approach to defining exposure scenarios is suitable in all cases. If activities of an atomic veteran were simple and indisputable and the radiation environment was well characterized, defining an adequate exposure scenario is usually straightforward. But defining an adequate scenario can be challenging, especially if a veteran engaged in unusual or complex activities or if important information on the veteran's activities or the radiation environment is lacking, is quite uncertain, or is a matter of dispute. In its review of dose reconstructions for individual atomic veterans, the committee encountered both extremes of difficulty in scenario development.

The task of defining exposure scenarios is the most important part of the dose reconstruction process for atomic veterans. An assumed exposure scenario provides the basis of an estimate of dose, so the adequacy of an estimated dose for purposes of dose reconstruction can be no better than the validity of the assumed scenario. The validity of an assumed scenario often cannot be determined by objective means alone, such as film-badge measurements of dose or a complete and indisputable record of a veteran's activities. Rather, a considerable amount of subjective judgment is often required in defining exposure scenarios, and it is often the case that more than one scenario is plausible. Thus, it is critical that all plausible scenarios be investigated, especially when plausible alternatives would result in higher estimates of dose. Veterans themselves can often provide information that can be used to develop plausible scenarios.

I.C.2.2 *Identification of Exposure Pathways*

Radiation doses can be received as a result of external or internal exposure. The term *external exposure* refers to irradiation of the body by sources outside the body. Only radiation that can penetrate the body surface and irradiate radiosensitive tissues of the skin and deeper organs is of concern with respect to external exposure. For atomic veterans, that radiation includes neutrons and photons (gamma rays and X rays) of any energy produced in a detonation or by decay of radionuclides and higher-energy electrons (beta particles) produced in radioactive decay. Nonpenetrating radiation, such as lower-energy electrons and alpha particles produced in radioactive decay, generally is not of concern in estimating dose from external exposure. External exposure to neutrons and photons usually is assumed to result in nearly the same dose to all body organs and tissues, whereas external exposure to higher-energy electrons results in a dose only to tissues near the body surface, including the skin and lens of the eye.

The term *internal exposure* refers to irradiation of the body by sources inside the body. Internal exposure can occur as a result of intakes of radionuclides by inhalation, ingestion, or absorption through the skin or an open wound. In cases of internal exposure, all radiation (photons, electrons, and alpha particles) emitted by the radionuclides of concern is important and is taken into account in estimating dose. Internal exposure can result in doses that are nearly the same in all organs or tissues, as when a radionuclide is distributed throughout the body (for example, ^{3}H, ^{14}C, and ^{137}Cs), or doses that are highly nonuniform and occur mainly at sites of radionuclide deposition in the body (for example, irradiation of the thyroid due to intakes of ^{131}I, irradiation of bone surfaces and bone marrow due to intakes of ^{90}Sr, and irradiation of the lungs, bone surfaces, bone marrow, and liver due to inhalation of plutonium).

I.C.2.2.1 *Pathways of external exposure*

Pathways of external exposure that could be of concern in dose reconstructions for atomic veterans include the following:

- Direct exposure to radiation emitted in a nuclear detonation,
- Exposure due to immersion in contaminated air,
- Exposure due to immersion in contaminated water,
- Exposure to radionuclides deposited on the ground or other surfaces or to radionuclides distributed in surface soil or water,
- Exposure to radionuclides deposited on the surface of the body.

The importance of different pathways of external exposure depends on the exposure scenario. For example, external exposure to radionuclides deposited on the ground or other surfaces often is the most important pathway of external exposure at the NTS and in the Pacific.

Direct exposure to radiation (neutrons and higher-energy photons) produced in a detonation occurred only when a veteran was relatively close to ground zero and the radiation was not completely absorbed during transport from the source to the receptor location. This pathway is relevant, for example, to participants at the NTS who observed shots from trenches that were within a few kilometers of ground zero.

Exposure due to immersion in contaminated air occurred in a number of circumstances, such as: when a veteran was at a location of descending fallout; when radioactive materials that had been deposited on the ground, on surfaces of a ship, or on surfaces of equipment were resuspended in the air; or in an aircraft that flew through a contaminated atmospheric cloud. Radioactive material on a surface can be resuspended by natural processes (such as wind), by the blast wave of a detonation, or by such activities as marching or transport through an area where fallout occurred, digging trenches in a contaminated area, handling contaminated equipment, and decontaminating ships with water hoses.

Exposure due to immersion in contaminated water generally is of concern only for nuclear tests in the Pacific. This pathway is important mainly for veterans who undertook diving activities after an underwater test.

Exposure to radionuclides deposited on the ground or other surfaces or to radionuclides distributed in surface soil or water is important in a number of circumstances. At the NTS and in the Pacific, external exposure to a contaminated ground surface is important if a veteran was in an area where fallout occurred. Exposure to other contaminated surfaces occurred, for example, on ships that experienced fallout or in the handling of contaminated equipment. Exposure to radionuclides distributed in surface soil occurred, for example, when a veteran was in an area of the NTS where irradiation by neutrons from a detonation resulted in activation of the nuclei of various elements in soil. Exposure to radionuclides distributed in water occurred in some circumstances in the Pacific, such as when a veteran spent time on a ship or in a small boat in a contaminated lagoon. This exposure pathway is different from that involving immersion in contaminated water in that the exposed person is above the body of contaminated water.

Exposure to radionuclides deposited on the surface of the body is of concern, for example, if a veteran was in an area of descending fallout or settling of resuspended material or handled contaminated soil or equipment. This pathway was especially important for maneuver troops and close-in observers at the NTS. The primary concern in cases of contamination of the body surface is irradiation of radiosensitive tissues of the skin and lens of the eye by higher-energy beta particles emitted by radionuclides.

I.C.2.2.2 *Pathways of internal exposure*

Inhalation and ingestion of radionuclides can occur by several pathways. In most cases of internal exposure of atomic veterans, intakes by inhalation probably were the most important. Inhalation exposure occurred, for example, when a person was in descending fallout or in a cloud of airborne radionuclides that were resuspended from a contaminated surface, such as the ground surface, surfaces of ships, or surfaces of equipment. The most likely pathway of ingestion exposure of atomic veterans involved inadvertent ingestion of contaminated material that originated in soil or on surfaces. However, ingestion may have occurred otherwise in unusual circumstances, for example, if a person consumed food or water that had been directly contaminated by fallout, or if a person on a residence island in the Pacific consumed local terrestrial foodstuffs that were contaminated by fallout or root uptake of fallout radionuclides from soil or consumed contaminated seafood obtained from local waters.

Absorption of radionuclides through the skin or an open wound probably is relatively unimportant in exposures of atomic veterans. Skin absorption would be important only if a veteran were in an atmospheric cloud that contained substantial amounts of ^{3}H in the form of tritiated water vapor (ICRP, 1979a). Absorption through an open wound could occur if radioactive materials were deposited on the body surface.

I.C.2.3 *Development and Implementation of Methods of Estimating Dose*

Once an exposure scenario is defined, including the assumed locations and activities of an atomic veteran at various times and the radiation environment at those locations and times, and relevant exposure pathways are identified, the radiation dose received by the veteran by all pathways can be estimated.

Estimation of dose is based on a combination of available data and modeling. Important data that can be used to estimate dose to an atomic veteran directly include film-badge readings and measured external exposure rates at various locations and times in the veteran's exposure environment. Data on humans also are used to estimate dose. For example, models used to estimate dose due to inhalation on the basis of estimates of concentrations of radionuclides in air include an assumption about the breathing rate.

When data that could be used to estimate dose directly are not available, mathematical models that incorporate available information and other assumptions must be used. The extent to which models must be used depends on the particular exposure situation. All estimates of internal dose must be based on models because internal dose cannot be measured directly. The availability of reliable and complete film-badge readings can minimize the need to use models to estimate external dose from exposure to photons. However, models to describe the time and spatial dependence of external dose rates must be used whenever film-badge or other radiation survey data are unavailable or inadequate, and complex models generally must be used to estimate dose from external exposure to neutrons and higher-energy electrons (beta particles).

One simplification in dose reconstructions for atomic veterans, compared with dose reconstructions for members of the public exposed to releases from the NTS or DOE facilities, is that models of radionuclide transport in air or water are not needed. Transport of radionuclides in the environment after a detonation generally was tracked by using cloud sampling and measurements of fallout on the ground or on ships. However, models of the environmental behavior of radionuclides are needed in some cases. For example, estimates of dose to atomic veterans due to inhalation of fallout that was resuspended in the air from the ground or other surfaces must be based on models to estimate concentrations of specific radionuclides on the surface and the extent of resuspension, because radionuclide concentrations on surfaces or in air generally were not measured during periods of exposure.

The dosimetric quantity calculated in dose reconstructions generally is the biologically significant radiation dose to organs and tissues of humans. In the NTPR program, that quantity is called the *dose equivalent* and is expressed in rem. Dose equivalent is calculated as the average absorbed dose in an organ or tissue, given in rad, modified by a factor that represents the biological effectiveness of the type of radiation that delivers the dose. The modifying factor takes into account that for a given absorbed dose in an organ or tissue of humans, the probability that a cancer or other stochastic radiation effect will result depends on the radiation type and the absorbed dose.[8] In this report, however, the biologically significant dose to organs and tissues is called the *equivalent dose* to conform to the terminology currently used by the International Commission on Radiological Protection (ICRP, 1991a).

---

[8]In the NTPR program, the modifying factor representing the biological effectiveness of different radiation types is called the *quality factor*. Quality factors greater than unity are used only in cases of exposure to neutrons and alpha particles.

I.C.2.4 *Evaluation of Uncertainties in Estimates of Dose*

Any estimate of dose obtained in a dose reconstruction has some uncertainty because of the variability in relevant measurements or a lack of knowledge of relevant processes or an individual's exposure scenario. All uncertainties, including uncertainties in exposure scenarios that form the basis of estimates of dose and uncertainties in the data and models used to estimate dose, must be considered and taken into account in a dose reconstruction. The essential purpose of an uncertainty analysis is to provide a credible range within which one can be reasonably confident that the true dose lies. Without proper consideration of uncertainty, the results of a dose reconstruction cannot be regarded as credible.

There are two ways of accounting for uncertainty in a dose reconstruction. An approach that can be used in any dose reconstruction, regardless of its purpose, is to obtain a best (central) estimate of dose and quantify its uncertainty due to uncertainties in the assumptions, data, models, and parameter values used. For example, the uncertainty in an estimate of dose can be represented by a 90% confidence interval giving a range of plausible values.[9] Such a confidence interval can be based on a combination of rigorous methods of statistical analysis and the use of subjective scientific judgment, depending on the quantity and quality of information used to estimate dose.

An alternative approach to addressing uncertainty that can be useful, depending on the purpose of a dose reconstruction, is to provide a credible upper bound of the dose (rather than a best estimate and confidence interval) on the basis of an argument that such a value overestimates the actual dose in almost all cases (for example, at least 95% of the time). This approach is particularly appropriate in dose reconstructions for atomic veterans because veterans are to be given the benefit of the doubt in estimating doses used to evaluate claims for compensation for radiation-related diseases (see Section I.C.3.2).

Both approaches to accounting for uncertainty have been used in dose reconstructions for atomic veterans. It is important to recognize that either can involve substantial subjective judgment (in addition to more rigorous methods of statistical analysis) depending on the importance of judgment in developing the assumptions, methods, and data used to estimate dose.

The importance of uncertainty in estimated doses to atomic veterans can depend on the magnitude of the dose. If the estimated dose is very low, the uncertainty can be large and still have no effect on a decision regarding compensation for a radiation-related disease. At higher doses, however, smaller uncertainties can be important to a decision about compensation.

---

[9]A 90% confidence interval of an estimated dose is a range within which it is believed that there is a 90% probability that the true but unknown dose lies; that is, there is only a 10% chance that the true dose lies outside the range. The upper bound of this range is referred to as the 95% confidence limit, meaning that there is only a 5% probability that the true dose is greater than the upper bound.

A wide range of plausible estimates of uncertainty may be encountered in dose reconstructions for atomic veterans. For example, uncertainties should be relatively low in scenarios where the dose was due mainly to external exposure and a film badge was worn at all times of exposure, but they can be much higher in scenarios where a veteran might have received a high dose by inhalation of resuspended radionuclides.

I.C.2.5 *Presentation and Interpretation of Results*

After estimates of dose have been obtained, it is important that the results of a dose reconstruction be presented in a way that can be understood by the veteran in question and others. For example, the assumptions, data, and models used in the dose reconstruction must be clearly identified and explained, and uncertainty in the results must be addressed. Key assumptions and conclusions should be provided, and the work should be signed and dated by the analyst. The analysis must be documented in such a way that other scientists can understand and verify the calculations. Especially in the dose reconstruction program for atomic veterans, proper communication of the analysis and results to the veterans themselves is essential.

It also is important that the results of a dose reconstruction be interpreted properly by discussing the results in the context of the purpose of the analysis. For example, the interpretation of results generally would depend on whether the purpose of the dose reconstruction is to provide best estimates of the dose to a specific person, best estimates of dose to a representative person in a group, upper-bound estimates of dose to persons or groups, or assurance that the dose received by a specific person or representative person did not exceed a specified value. As discussed throughout this report, the proper interpretation of results of dose reconstructions for atomic veterans is in terms of obtaining credible upper-bound estimates of dose to individuals.

Thus, in general, presentation and interpretation of results of a dose reconstruction should provide a reasonably complete, coherent, and understandable picture of the analysis that would allow others to judge the adequacy of the dose reconstruction for its intended purpose. Knowledgeable scientists should be able to reproduce the calculations on the basis of the information documented in the dose reconstruction, and persons or groups whose doses have been estimated should be able to understand the assumptions used in the analysis, especially assumptions about exposure scenarios and exposure pathways, and the results of the analysis.

I.C.2.6 *Quality Assurance and Quality Control*

Use of proper quality assurance and quality control procedures is an essential aspect of all the other elements of the dose reconstruction process described above. The essential function of quality assurance and quality control is to ensure that there is a systematic and auditable documentation of the procedures used in dose reconstructions and that the methods of analysis and the calculations themselves are free of important error. Proper documentation must be provided for all data, interpretations of data and other assumptions, and computer codes or other methods of calculation used to estimate dose. Procedures to be used in dose reconstructions and changes in the procedures, including when they occurred, must be properly documented. If complex computer codes are used, they should be verified to ensure that they do not introduce important error. External peer review is an important means of achieving quality assurance.

In general, proper quality assurance and quality control are essential to developing confidence in the dose reconstruction process and the resulting estimates of dose. The issue of quality assurance and quality control is discussed in more detail in Section IV.G.

I.C.3 Special Aspects of Dose Reconstructions for Atomic Veterans

Three other aspects of dose reconstructions for atomic veterans warrant special consideration. First, dose reconstructions have been used to evaluate claims for compensation by specific persons who incurred a disease that could have been caused by exposure to ionizing radiation during the atomic-testing program; in many cases, estimates of dose to a particular organ or tissue in which a cancer or other disease has occurred are compared with a specified dose as part of the process of deciding whether the disease was at least as likely as not to have been caused by radiation exposure (see Section III.E). Second, the claimant is to be given the benefit of the doubt in estimating dose. Third, the NTPR program has been going on for more than two decades, and there have been many advances in the science of dose reconstruction over that time. As discussed below, these considerations have important implications for the dose reconstruction process for atomic veterans.

I.C.3.1 *Focus on Specific Persons*

Focusing on reconstruction of doses received by specific persons can place considerable demands on the dose reconstruction process, especially with regard to defining exposure scenarios, selection of parameter values for use in models to estimate dose, and treatment of uncertainty. For example, depending on the particular exposure situation, it could be inappropriate to ascribe average exposure conditions in a participant group to a specific person in that group. In its review of dose reconstructions for individual veterans, the committee encountered a substantial number of cases that clearly involved unusual, or even unique, exposure conditions.

Another challenge in dose reconstructions for atomic veterans is that some of the models and supporting data used to estimate dose, especially in cases of intakes of radionuclides by inhalation or ingestion, represent standard assumptions that were developed for purposes of radiation protection. In radiation protection, which is concerned with controlling exposures and evaluating compliance with dose limits for workers or the public, standard models and databases used to estimate internal dose are assumed to apply to everyone, without uncertainty. However, dose reconstructions for atomic veterans focus on estimating actual doses to specific persons. It is important that uncertainties in standard models used in radiation protection be acknowledged and properly taken into account in dose reconstructions for specific veterans in the context of a compensation program.

I.C.3.2 *Importance of Benefit of the Doubt*

Regulations governing the NTPR program specify that the veteran will be given the benefit of the doubt in estimating dose. Specifically, as stated in 38 *CFR* 3.102:

> When, after careful consideration of all procurable and assembled data, a reasonable doubt arises regarding service origin, the degree of disability, or any other point, such doubt will be resolved in favor of the claimant. By reasonable doubt is meant one which exists because of an approximate balance of positive and negative evidence which does not satisfactorily prove or disprove a claim. It is a substantial doubt and one within the range of probability as distinguished from pure speculation or remote possibility. It is not a means of reconciling actual conflict or a contradiction in the evidence. Mere suspicion or doubt as to the truth of any statements submitted, as distinguished from impeachment or contradiction by evidence or known facts, is not justifiable basis for denying the application of the reasonable doubt doctrine if the entire, complete record otherwise warrants invoking this doctrine. The reasonable doubt doctrine is also applicable even in the absence of official records, particularly if the basic incident allegedly arose under combat, or similarly strenuous conditions, and is consistent with the probable results of such known hardships.

Thus, if there are matters of dispute or a lack of important information about conditions of exposure, plausible assumptions that give the highest estimates of dose should be used in dose reconstructions for atomic veterans. It is not intended that doses be estimated on the basis of assumptions that are beyond reason. Rather, as stated above, assumptions should be "within the range of probability as distinguished from pure speculation or remote possibility." Nonetheless, the need to give the veteran the benefit of the doubt can place considerable demands on the definition of exposure scenarios, the selection of parameter values used in models to estimate dose, and the methods used to account for uncertainty in estimates of dose. The demands are greater if a veteran engaged in unusual activities.

The policy on benefit of the doubt in the use of dose reconstructions to evaluate claims for compensation encourages an approach to estimating dose that focuses on obtaining credible upper bounds, rather than best estimates, of possible doses to atomic veterans. However, in attempting to provide credible upper-bound estimates of dose, uncertainties in methods and assumptions used in the analysis must be considered and evaluated, and proper justification that the estimated doses are credible upper bounds must be provided.

I.C.3.3 *Conduct of Dose Reconstruction over Time*

An unusual aspect of the NTPR program is that it has been going on for 25 years. This can place special demands on the dose reconstruction process with regard to consistency in the technical approach, nondiscriminatory methods of estimating dose, and implementation of changes in methods of estimating dose based on improvements in science.

Ideally, consistent approaches to developing exposure scenarios and consistent methods of estimating doses would be used in dose reconstructions for all atomic veterans. For example, if a bias toward overestimation of dose was deliberately included in some dose reconstructions (for example, on the basis of benefit of the doubt), similar biases should be applied in all cases, consistent with the available information. Likewise, consistent assumptions about models and parameter values should be used in estimating doses to different persons who have been assumed to be exposed in similar ways. It can be difficult to achieve consistency in methods of dose reconstruction when analyses are performed by different people and over an extended period during which the scientific basis of dose reconstruction has been evolving. Preparation of a detailed manual of procedures for the conduct of dose reconstructions can be an important means of achieving the desired degree of consistency among different analysts and over time.

Dose reconstructions for atomic veterans are performed for different purposes. In many cases, as noted previously, dose reconstruction is used to support a claim for compensation. In other cases, however, it is used to inform the veterans about their doses. In general, methods used to estimate dose should not differ on the basis of the purpose of dose reconstruction. An important goal of a dose reconstruction program is to foster a perception that all individuals whose doses are estimated are treated fairly and consistently.

During the 25-year existence of the NTPR program, there have been substantial improvements in the scientific foundations of dose reconstruction and in the tools used to estimate doses and evaluate uncertainties. It is important that these improvements be evaluated and incorporated into the NTPR program as appropriate. Methods of dose reconstruction for atomic veterans do not necessarily need to be changed whenever a new piece of information becomes available, but there must be a deliberate and regular effort to evaluate changes in data and methods of dose reconstruction and to incorporate improvements in the dose reconstruction process as warranted. For example, given the increased capabilities of computers to process large amounts of data rapidly, it is reasonable to expect that the results of dose reconstructions could be presented as probability distributions or confidence intervals of dose, rather than single values, with uncertainties associated with dose reconstructions taken into account. That approach would give veterans a much clearer idea about uncertainties in their estimated doses than the current practice of emphasizing only upper-bound estimates. Peer review of methods of dose reconstruction and the existence of a detailed manual of standard operating procedures, including proper procedures for document control and updating, are important.

The NTPR program should be cognizant of new techniques for measuring or assessing radiation doses that were received by veterans many years ago. The need to rely on uncertain models is reduced to the extent that doses can be estimated on the basis of measurement. An example (see Section VI.D) that has been recognized by the NTPR program is the development of improved bioassay techniques for estimating prior intakes of plutonium. Similarly, new measurement techniques, such as electron paramagnetic resonance, can be used to estimate accumulated radiation dose in tooth enamel (for example, see Romanyukha et al., 2000, 2002). Such developments should be followed, although they may be limited in their sensitivity to the low radiation doses that were received by most atomic veterans and, thus, may provide only limited data of use in evaluating veterans' claims for compensation.

I.D CONCERNS OF VETERANS

Veterans have expressed concerns about the atomic veterans compensation program since its beginning. Their concerns have led to hearings by Congress, independent evaluations of the program, and, as described above, studies by the National Research Council and other organizations. Believing that it is important to understand these concerns and, to the extent possible, to address them in its review of the dose reconstruction program, the committee interacted with veterans and provided an opportunity for them to raise issues for consideration. By listening to the veterans and attempting to address their concerns in this report, the committee hopes that it is providing answers to some of their questions and, more important, helping them to understand the process of dose reconstruction and its role in their compensation program.

Veterans expressed their concerns to the committee through correspondence and by participating in the committee's open meetings. Members of the committee also attended the 2001 annual meeting of the National Association of Atomic Veterans, where a number of questions were asked about the dose reconstruction and claims process.

The following are examples of concerns expressed by the veterans to the committee:

- The perception of an extraordinarily low rate of successful claims under the nonpresumptive regulation.
- The validity of dose reconstruction as a basis for compensation.
- The burden to veterans and their spouses posed by the claims and appeal process.
- The lack of timeliness of claims resolution and responses from DTRA and VA.
- Changes in doses assigned by the NTPR program to individual veterans as they continue to make inquiries or seek help from a legislative official.
- Use of a low-level internal dose screen to eliminate the need for calculation of inhalation dose.
- Failure to account properly for inhalation dose in some scenarios.
- Neglect of possible ingestion doses in dose reconstructions.
- Improper assumptions about scenarios of exposure and failure to consider veterans' own accounts or accounts of companions.

That is not a comprehensive or ranked list, but several of those concerns seem to be overriding and consistent in importance. A few are discussed below, and others are discussed in more detail throughout this report.

The issue that has appeared to be of most concern to the veterans throughout our interaction during the project is the overall effectiveness of the compensation program under 38 *CFR* 3.311, the nonpresumptive regulation. Although that concern seemed to be somewhat peripheral to our scope, it is indirectly related because knowledge about the "accuracy" of the doses, as stated in the committee's charge, could affect decisions about compensation. The veterans have been led to believe that over the course of the compensation program, relatively few claims have been awarded under the nonpresumptive regulation even though more than 4,000 dose reconstructions have been performed. As a consequence, there is an intense distrust and skepticism by the veterans about how dose reconstructions are being performed, and about whether accurate dose reconstruction is even possible, given the lack of historical data and the period of time since exposures occurred. The committee hopes that some of the information provided in this report will be helpful in addressing that issue.

Veterans are also concerned about some of the elements of the dose reconstruction process, including assumptions about scenarios of exposure, improper accounting of internal doses, and the use of a "low-level internal dose screen" that they believe improperly eliminates the need to estimate inhalation doses to some veterans. The concept and use of an internal dose screen is a good example of an issue to which the committee devoted considerable attention in an effort to provide clarification to the veterans. Although DTRA has consistently stated that it does not use an internal dose screen to eliminate the need to estimate inhalation doses, the issue continues to cause considerable confusion among the atomic veterans, which the committee hopes to alleviate.

Some of the concerns expressed by the atomic veterans are beyond the scope of this study, but the committee has endeavored to answer as many of them as possible in the report and appreciates the veterans' willingness to express their concerns to us throughout the project.

II The Committee's Process

To undertake this study, the National Research Council established a committee to review the dose reconstruction program of the Defense Threat Reduction Agency (DTRA). To fulfill its charge, the committee reviewed 99 randomly selected dose reconstructions in some detail and supporting material in the files. The committee supplemented that set with 12 randomly selected dose assessments for occupation forces in Japan after the bombings of Hiroshima and Nagasaki. The committee also interacted with various atomic veterans' representatives, federal agency representatives, and other interested parties.

II.A DISCUSSION OF THE COMMITTEE'S CHARGE

The scope of work of this review, as stated in Section I.A, has from the beginning been somewhat troubling to the committee. The charge required the committee to select a random sample of reconstructed doses and to determine whether they were "accurate" and were "accurately reported." The committee was also to determine whether the historical source data upon which doses are based are, in themselves, "accurate."

Dose reconstruction, because of its historical nature and the large gaps that often exist in available data, cannot be an "accurate" science. The best one can hope for in dose reconstruction is to specify a range of doses that is likely to encompass the true dose. In most cases, when all uncertainties have been taken into account, the dose can be represented by a distribution of possible values, which may spread out over a wide range. Therefore, in responding to its charge, the committee had to work within its understanding of the limits inherent in any

dose reconstruction process. It is important for readers of this report to understand that dose reconstruction generally is an exercise in applying subjective
scientific judgment. In much the same way, the committee's conclusions on the
adequacy of the program of dose reconstruction for atomic veterans rely to a
significant extent on judgment, which is founded on the nature and extent of
information on dose reconstructions that was available to the committee.

Moreover, in the process of adjudicating claims, the Department of Veterans
Affairs (VA), in recognition of the unavoidable uncertainty that attends historical
dose reconstructions (and in the spirit of regulations that specify that the veteran
should be given the benefit of the doubt), relies on specified upper bounds of
doses rather than best or central estimates of dose. The committee was more
concerned with whether uncertainties in estimated doses had been appropriately
addressed than with exactly where on the distribution of possible values a particular actual dose lay. That was the committee's approach to interpreting and
addressing the "accuracy" issue with respect to dose reconstructions.

The question about whether doses were "accurately reported" to the veterans
is much more straightforward to address, but the unavoidable uncertainties inherent in dose reconstruction are important here as well. The committee believes that
uncertainties in assigned doses should be carefully explained and reported to VA
for compensation purposes and also to the veterans.

With regard to whether the source data on nuclear-weapons tests used by
DTRA are accurate, we interpreted this question as asking whether the historical
data that have been comprehensively compiled are sufficiently accurate and complete for use in dose reconstruction. Such original source data would include the
instrument-based measurements made at the time of the tests, weapons-debris data,
film-badge records, and historical records of activities and movements of personnel
participating in the tests. The data that dose reconstructions are based on and the
uncertainties in them clearly are important in the estimation of doses and upper
bounds provided by the Nuclear Test Personnel Review (NTPR) program to VA.

The committee recognizes that the intent of the scope of work was to focus
its attention on the technical methods and assumptions being used in dose reconstructions, the historical data on which dose reconstructions are based, and how
the results are being reported to VA and communicated to the veterans. Those
questions encompass a range of issues, including specific scientific methods,
judgments about scenarios of exposure, and effective communication with veterans. Therefore, the committee has taken a broad view of its scope out of necessity
and its desire to do a thorough, defensible, and enduring job. The committee
hopes that the process it has used will help to answer questions that have lingered
for many years regarding dose reconstructions performed for the atomic veterans
compensation program. The committee also hopes that its process will provide
guidance for making improvements in the program as a whole.

In its statement of task, the committee was asked to issue an interim letter
report with recommendations if after a year it concluded that its findings differed

from previously published and congressionally directed studies or if it found that substantial changes were required in the dose reconstruction procedures and methods. Because of the large number of existing dose reconstructions, the committee elected to use a sampling process (described below) for reviewing the dose reconstruction files. In addition, the committee established a protocol to provide consistent review of the files, and this took time to develop after the preliminary examinations of several files. At the one-year point in the study, the committee had identified specific examples of issues and concerns on the basis of reviews of a limited number of selected cases. However, the committee recognized that it needed to examine many more cases before it could determine whether the specific examples were symptomatic of pervasive problems in dose reconstruction procedures. For those reasons, the committee was not prepared to recommend changes in the program at that time.

II.B SAMPLING PROCEDURES

The first issue in sampling of dose reconstruction files was what to use as the sampling frame. Ideally, one might consider all veterans who had applied for compensation under any of the laws that cover service-connected disability related to radiation exposure in the nuclear-weapons testing program. However, the relevant records are not accessible at VA through any unified database that can be readily queried, as would have been required to construct a list to serve as a comprehensive sampling frame. A database available through the research contractor, currently JAYCOR, contains records for all veterans who requested information about their participation in the program and all veterans for whom documentation was requested by VA in connection with a compensation claim. However, most of the veterans in the JAYCOR database have not filed a claim for compensation, so it was not feasible to use that database as our sampling frame. We elected instead to use the database at Science Applications International Corporation (SAIC), which is the contractor responsible for individualized dose reconstructions and upper-bound dose determinations, as the most appropriate and feasible sampling frame. Most veterans with a record in the SAIC database were there because they had filed a claim for service-connected disability based on radiation exposure in the testing program, although the SAIC list also includes additional assessments for veterans who had not developed a radiogenic disease but simply wanted information about their radiation exposure.

Appendix B describes characteristics of each selected veteran's case file including: the branch of service; the weapons test series (or presence in Japan); the assigned external, internal, and organ doses; the assigned upper bound for the total dose; the date of the most recent dose reconstruction; and whether the veteran filed a claim for compensation, and if so whether it was under the presumptive or the nonpresumptive regulation (see Sections III.B and III.C).

Only 72 of the 99 sampled cases involved claims for alleged radiogenic service-connected diseases.

It should be noted that some veterans who filed claims for compensation would not have been entered into SAIC's database. There are two reasons for this. First, if a claim was filed for a disease that the VA adjudicator considered not to be radiogenic, it might be denied as having no merit, without consideration of dose. Second, if the unit-based dose of record at the time the claim was considered was determined to be very low and there was no indication in the record that the veteran had engaged in any unusual activities that might have increased the potential for radiation exposure, VA might have denied the claim on the basis of a low unit-based dose without requesting an individualized dose reconstruction.

At the time of sampling, in October 2001, the database that was provided to the committee by SAIC contained 3,725 veterans and their assigned doses. To ensure that we had adequate numbers of veterans with a high potential for significant radiation exposure, we carried out a stratified random sampling, sampling at random 66 veterans from the subset of those with an assigned dose of at least 1 rem and 33 from the larger group with a lower assigned dose. The committee thus oversampled veterans whose dose reconstructions may have required a relatively complex approach, and this offered a diverse set of examples for the committee to learn about how scenarios with potential for significant exposure were handled by the dose reconstruction analysts. The weighting also provided greater numbers of veterans who would have had a relatively high potential for radiation exposure and whose exposure may therefore have been high enough for errors in dose assignment to have influenced the compensation-adjudication process. Within each of the two dose-assignment-based strata, selection was based on computer generation of random numbers.

II.C INTERACTION WITH ATOMIC VETERANS

Throughout the course of its work, the committee interacted with DTRA, VA, and atomic veterans. That aspect of our work was important in seeking to answer questions raised and to understand the issues involved. Although interactions with DTRA and VA were essential to obtaining the information needed to fulfill its charge, it was also important to understand the concerns of the veterans, as summarized in Section I.D, and to seek information from them regarding such matters as communication related to the dose reconstruction program and the overall disposition of claims. We found the veterans eager to assist us. In particular, they seemed to be supportive of our study, interested in learning more about the dose reconstruction process, and curious about a number of questions they hoped the committee would be able to answer.

Veterans were invited on several occasions to speak to the committee at its meetings. Such interaction provided a formal, yet open exchange of ideas, questions, and responses with veterans, and in some cases spouses of deceased veterans, and proved useful to us. Early on in the project, we took advantage of an opportunity to expand the interaction.

In 2001, several members of the committee were invited to attend the annual meeting of the National Association of Atomic Veterans (NAAV) in Las Vegas, Nevada. The committee made a brief presentation to the veterans regarding the scope and status of the project, after which the veterans asked a number of questions. We carefully explained our approach in selecting records to review; selection was required under the terms of our charge to be a random sample. Nevertheless, many veterans at the meeting volunteered their records for the committee's use and provided specific concerns that they wished to see addressed. Several members of the committee also briefly visited the NAAV headquarters in Albuquerque, New Mexico, for additional discussions with NAAV officers and to determine which of the various records maintained by NAAV might be of use for the committee's study. NAAV also provided information about our study to veterans through its newsletter and informed veterans about how they could submit their files for the committee's use or express a concern. The committee agreed to accept records that the veterans wished to send to the National Research Council, but only with written permission to use them. If a veteran was deceased, access to records was allowed after verification of death. Furthermore, the committee emphasized that it would continue to use its randomly selected records, and information provided by veterans would be used to supplement its work. Records sent to the committee by the veterans have been helpful in illustrating some key points, such as communication of dose reconstruction results and scenario development, that might not have been available in records maintained by the agencies and their contractors.

Appendix C lists the names of invited speakers and veterans (or spouses) who have interacted with the committee through correspondence, by providing information at meetings, or by providing their files for our use. All together, about 50 veterans and spouses of deceased veterans have interacted with the committee. The type of interaction—for example, e-mail, letter, and attendance at meetings—is also noted. We appreciate and are impressed by the efforts of NAAV and other veterans to work with us during the project, and it has been important to our efforts.

II.D INFORMATION GATHERING

The committee was initially charged with the tasks described in Section I.A and earlier in this chapter. The committee met five times in 2001 and six times in 2002. Between formal meetings, individual committee members were sent copies of files (with names and other identifiers blacked out to protect privacy) to permit detailed review of the 99 sampled dose reconstruction cases.

The committee review of the 99 selected dose reconstructions sometimes led it to records outside a veteran's file, including supporting documentation in the SAIC files. We also examined a number of other dose reconstructions provided by veterans. Committee members reviewed some of the source data and many of the reports on which reported doses were based, including many published reports describing the generic unit dose reconstructions or methods used in dose reconstructions and many unpublished SAIC internal memoranda. Additional information was obtained in formal DTRA and VA replies to written questions from the committee (see Appendix D) and in presentations by DTRA, SAIC, JAYCOR, and VA staff at open committee meetings.

The committee had access to all the information it required during the course of the project. Most information related to the reconstruction of doses to atomic veterans is not classified, but some data on atmospheric nuclear tests require a security clearance for review. The committee had members with appropriate clearances to review the classified records, but access to classified information was not necessary, and we were able to complete our study by using the unclassified records available.

To be consistent with the policies of the National Academies, the committee conducted fact-finding activities involving outside parties in public information-gathering meetings and met in closed session only to develop committee procedures, review documents, and consider findings and recommendations. The information-gathering meetings were structured to solicit information from technical experts, the study sponsor and contractors, and veterans on topics related to the dose reconstruction program. Atomic veterans were also invited to make oral or written statements or to provide written comments to the committee.

Four committee meetings included data-gathering sessions not open to the public, and five committee meetings included information-gathering sessions open to the public. At open meetings, members of the public and veterans were given the opportunity to provide statements, ask questions, or make comments about the dose reconstruction program. A description of meetings at which the committee gathered information is included below.

On the first day of its first meeting, in open session, the committee and observers were briefed by a representative of DTRA. He introduced the charge to the committee and gave background on the NTPR dose reconstruction program. A spokesperson from JAYCOR and another from SAIC were present to address committee questions related to DTRA-contractor support of dose reconstruction. On the second day, the committee visited JAYCOR and SAIC sites to examine sample dose reconstruction records and to identify which aspects of the records would be our focus for auditing the dose reconstruction process.

The second meeting included a full-day open session. The committee heard from a congressional staff person, a representative of the General Accounting Office, the medical doctor responsible for providing medical advice related to claims adjudication for VA, the chief of judicial and advisory review of the VA Compensation and Pension Service, the widow of an atomic veteran who also was an official of NAAV, and an atomic veteran who represented NAAV.

The third meeting included a half-day data-gathering session not open to the public. The committee visited the SAIC site to examine sample dose reconstruction records.

The fourth meeting included a full-day data-gathering session not open to the public. The committee visited the SAIC site to examine the 99 randomly selected dose reconstruction records.

The sixth meeting included a half-day information-gathering session. The committee heard presentations by a retired Navy captain (former manager of the JAYCOR-NTPR effort and chief of the Navy NTPR team) and representatives of VA and DTRA.

The ninth meeting included a full-day data-gathering session not open to the public. Three committee members visited the SAIC site to examine sample dose reconstruction records. The meeting also included a half-day information-gathering session. The committee heard presentations related to the possible use of tooth enamel for radiation dosimetry (see Section I.C.3.3). Representatives of DTRA and VA were present to answer questions related to dose reconstruction and the claims process.

All the information gathered at open meetings is part of the National Research Council's public-access file and is available, on request, to anyone interested.

III The Process of Submitting and Deciding Claims

III.A INTRODUCTION

As discussed in Section I.B.4, various public laws implemented by Title 38, *Code of Federal Regulations*, Part 3 (38 *CFR* Part 3) authorize the Department of Veterans Affairs (VA) to provide medical care and pay compensation benefits to confirmed participants or their survivors for disabilities or death related to exposure to ionizing radiation from US atmospheric nuclear testing, the occupation of Hiroshima or Nagasaki, Japan, or prisoner of war (POW) status in the vicinity of Hiroshima or Nagasaki. Veterans or their survivors can file claims covered by VA regulations (38 *CFR* 3.309 and 3.311) by contacting one of the 58 VA regional offices (VAROs) serving their geographic area. They can also request participation and dose information from the Defense Threat Reduction Agency (DTRA), which is the Department of Defense executive agent for the Nuclear Test Personnel Review (NTPR) program. VA regulations do not, however, require veterans or their survivors to obtain confirmation of participation or dose information from DTRA before initiating a VA claim. Veterans' requests for medical care under VA regulations also do not require the prior filing of claims.

Veterans who believe that they have an illness related to their military service may file a claim under one of the following compensation programs:

- VA nonpresumptive program: Several public laws, as implemented in 38 *CFR* 3.311, provide for VA determination of service connection and benefits for the listed diseases and others that can be documented as radiogenic. VA obtains participation and associated radiation-dose information from DTRA.

• VA presumptive program: Several public laws, as implemented in 38 *CFR* 3.309, authorize VA to pay compensation for 21 types of cancer to participants in US atmospheric nuclear testing or in occupation of Hiroshima or Nagasaki. Filing a VA claim under this regulation does not require dose information from DTRA. Veterans who cannot be confirmed as participants are not eligible for VA compensation under this program.

Veterans must submit competent medical evidence that they have a medical condition covered under the VA regulations. On the basis of the information provided, VA decides under which law to file the claim. The claimant is not required to be familiar with the applicable laws and regulations.

III.B CLAIMS FILED UNDER NONPRESUMPTIVE REGULATION

Figure III.B.1 identifies the organizations that handle a VA claim initiated through a VARO when a claim is filed under the nonpresumptive regulation (38 *CFR* 3.311). The chart depicts the actions and steps through which a claim inquiry passes after the VARO requests information from DTRA. DTRA has a veterans-support effort through a teamed contract with JAYCOR and Science Applications International Corporation (SAIC) to conduct historical research on veterans' participation activities and to determine radiation doses related to those activities. DTRA reviews and approves contractor-prepared work products before submitting them to the VARO. The VARO specifies the above-cited regulation in its inquiry to DTRA for processing a claim.

To start the process, a person must file a claim for disability compensation with the local VARO. As noted above, a VA claim under this regulation does not require the claimant to obtain prior confirmation of participation or information on radiation dose from DTRA. The claimant must furnish medical evidence of a radiogenic disease listed in 38 *CFR* 3.311 or of any other disease *plus* a medical opinion that the unlisted other disease can be caused by radiation exposure. The claimed activity must involve participation in the US atmospheric nuclear testing program, occupation of Hiroshima or Nagasaki, Japan, or POW status in the vicinity of Hiroshima or Nagasaki.

The VARO asks DTRA to confirm the veteran's participation in one of the three activities listed above. If the disease for which the claim is filed is considered a radiogenic disease, the VARO also requests an assessment of the veteran's radiation exposure. The VA letter to DTRA should identify the veteran, describe his claimed participation, name the radiogenic disease, cite the specific controlling regulation, and include available claimant statements and supporting evidence if any.

DTRA through its contractors reviews the information provided by the VARO and conducts historical research, including interviews with the claimant for clarification if necessary, to determine the veteran's participation status. If

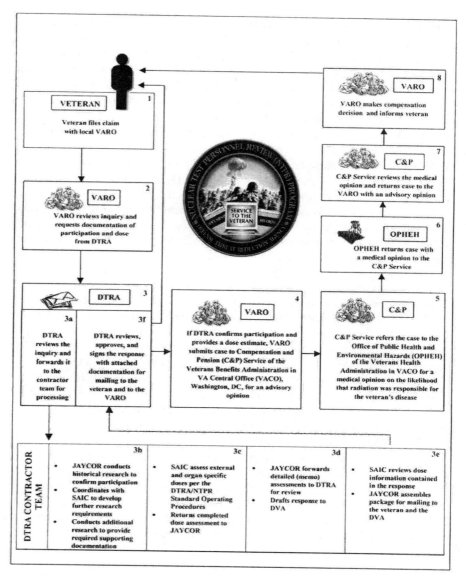

FIGURE III.B.1   How VA and DTRA process a radiation-related claim from a veteran for a nonpresumptive disease.

participation is confirmed, DTRA provides the VARO with a report of the dose assessment. The report includes an external photon dose with upper bound and an external neutron dose if applicable, internal doses to the target organs corresponding to the radiogenic diseases identified in the VA claim, and skin dose and eye dose if applicable. Those doses account for all emissions of alpha, beta, gamma, and neutron radiation from all sources to which the veteran was exposed. A copy of the DTRA response is provided to the claimant.

The VARO sends the claim and DTRA dose assessment to the Compensation and Pension (C and P) Service of the Veterans Benefits Administration in Washington, DC, for an advisory opinion. The C and P Service asks the Office of Public Health and Environmental Hazards (OPHEH) of the Veterans Health Administration for a medical opinion of the likelihood that the veteran's radiation exposure caused his radiogenic disease(s). For each disease, OPHEH considers the upper bound dose to the target organ, age or approximate age at exposure and at diagnosis, family and employment history, and exposure to other toxicants and carcinogens before and after military service.

Thereafter, using DTRA-reported upper-bound doses, OPHEH applies probability of causation methods, such as those based on radioepidemiological tables issued by the National Institutes of Health (NIH) (see Section III.E). Those methods are supplemented by information from other scientific or medical sources and consideration of other factors to evaluate whether it is at least as likely as not that the claimed disease(s) resulted from the reported radiation doses. OPHEH returns the case to the C and P Service with its medical opinion on the claim.

The C and P Service reviews the medical opinion from OPHEH and all the evidence of record and issues an advisory opinion to the VARO. The C and P Service notes whether, according to the regulations, the veteran's disability from the specific diseases is the result of his military service activities.

The VARO informs the claimant about the outcome of the claim. If the claim is granted, the VARO pays compensation based on the current degree of disability resulting from the covered diseases.

Whenever a claim is denied by VA, and the veteran or other claimant believes that VA did not make a good decision because it did not review all the evidence or did not apply the law correctly, the claimant has the right to appeal a decision made by a VARO or medical center. Claimants may appeal a complete or partial denial of a claim or the amount of benefit granted.

III.C CLAIMS FILED UNDER PRESUMPTIVE REGULATION

To initiate a claim under the presumptive regulation (38 *CFR* 3.309), a person must file a claim for disability compensation with the VARO serving his geographic area. A VA claim under this regulation does not require the claimant to obtain prior confirmation of participation from DTRA. The claimant must furnish medical evidence of a radiogenic disease listed in 38 *CFR* 3.309(d). The claimed activity must involve participation in the US atmospheric nuclear testing program, occupation of Hiroshima or Nagasaki, Japan, or POW status in the vicinity of Hiroshima or Nagasaki.

Figure III.C.1 shows that DTRA has a veterans-support effort to conduct historical research on veterans' participation activities. As shown in the figure,

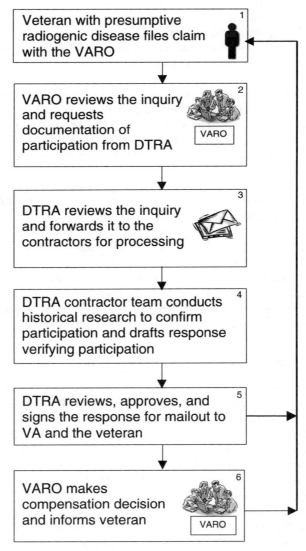

FIGURE III.C.1   How VA and DTRA process a radiation-related claim from a veteran for a presumptive disease.

only verification of participation is required for claims filed under the presumptive regulation; dose information is not required. DTRA reviews and approves contractor-prepared work products before submitting them to the VARO. The requirement for providing only participation verification to support a veteran's claim originates in Public Law (PL) 100-321, as amended. The VARO specifies the above-cited regulation in its inquiry to DTRA for processing a claim in accordance with Figure III.C.1.

If the NTPR program verifies the veteran's status as a participant and he has one or more of the defined presumptive diseases, his claim is granted. If the NTPR program cannot verify his participation, the claim may be refiled under the nonpresumptive regulation.

On verifying that the medical evidence meets the requirements of 38 *CFR* 3.309, the VARO asks DTRA to confirm the veteran's participation in one of the three activities listed above. The VA letter to DTRA should identify the veteran, describe his claimed participation, name the radiogenic disease, cite the specific controlling regulation, and include available claimant statements and supporting evidence if any.

DTRA reviews the information provided by the VARO and conducts historical research to determine the veteran's participation status. DTRA provides a response to the VARO and sends a copy to the claimant.

The VARO reviews the information provided by DTRA and makes a compensation decision. The VARO informs the claimant about the outcome of the claim. If the claim is granted, the VARO pays compensation based on the current degree of medical disability resulting from the covered disease.

If the veteran's disease was diagnosed before the disease was designated as presumptive, compensation is awarded only from the time the law was enacted or modified to add the disease.

III.D COMMUNICATIONS WITH VETERANS

A veteran or his family may either communicate directly with DTRA and request a dose reconstruction or file a claim with VA. If the veteran requests a dose from DTRA and then files a claim with VA, the dose calculation may be revised in light of more specific information gathered for that veteran's case.

In the case of a claim for a presumptive disease, the VARO requests only documentation of participation from DTRA, which then responds by letter to the VARO and sends a copy to the veteran. The VARO then informs the veteran of its compensation decision.

For a nonpresumptive disease, the veteran files a claim with the VARO, which then determines whether the disease can reasonably be considered radiogenic; if so, it requests from DTRA documentation of the veteran's participation in the weapons testing program and an estimate of the veteran's dose. After its research, DTRA responds to the VARO and sends a copy to the veteran. If participation is confirmed and a dose estimate greater than zero has been provided by DTRA, the VARO submits the case to the C and P Service, which seeks an advisory opinion from OPHEH. In light of that medical opinion, the C and P Service sends its advisory opinion back to the VARO, which makes the compensation decision and informs the veteran of it.

Atomic veterans also receive other kinds of information in communications from the NTPR program. For example, veterans who file a claim for compensation or request a dose reconstruction usually receive general information, in the form of "Fact Sheets," on the atomic testing program, the program of dose reconstruction and compensation for atomic veterans, and the magnitude and significance of doses received by the veterans. The NTPR program also communicates indirectly with atomic veterans through press releases and by submitting information for publication in the National Association of Atomic Veterans (NAAV) newsletter, especially when there are important changes in laws and regulations. Further discussion of these communications, including the committee's comments, is given in Section VI.B.

III.E MEDICAL OPINIONS AND PROBABILITY OF CAUSATION

With the passage in 1984 of PL 98-542, compensation was made available to veterans who had radiogenic diseases, primarily cancers. The term "radiogenic" does not mean that the disease was necessarily caused by radiation, but only that credible research has established a link between exposure to radiation and an increased risk of the disease in humans. In any occurrence of cancer, there is no way for medical doctors to determine whether it was caused by radiation exposure. For a veteran to receive compensation for a nonpresumptive disease, a medical opinion is required that considers whether the radiation dose received by the veteran is at least as likely as not to have been the cause of the disease.

To assist in the determination in the case of cancer, PL 97-414 required the Public Health Service to develop radioepidemiological tables that set forth the relationships between probability of causation (PC) and radiation dose for various cancers. The purpose of the tables has been to assist in the medical determination needed for the decision about whether to award compensation. They are not the sole determining factor in the compensation decision. For example, if a radiation-exposed person has lung cancer and has been a long-time cigarette-smoker, there will be a high probability that smoking was the cause of the cancer. The probability is not 100%, and it will depend on the person's age, how long the person has smoked, and the type of cigarettes and number smoked per day. Other factors, such as chemical exposures in the person's occupation, can also affect the probability.

For a given exposure to radiation, scientists have developed methods for estimating the increased chance (probability) that a person will contract a particular type of cancer within a given time. Those methods have been based on extensive studies of the Japanese atomic-bomb survivors and medical patients who received radiation therapy. The methods assume the linear nonthreshold model for radiation effects; that is, it is assumed that there is some chance that a cancer will occur at any dose and that the probability of a cancer is proportional to the dose. In the context of a compensation decision, an important question is: Given that a person has received a particular dose of radiation and later develops a particular type of cancer, what is the likelihood that the radiation caused the cancer? The likelihood is referred to as the probability of causation, or sometimes as the attributable risk or assigned share. Although different names are used, they refer to the same calculation. The PC of a specific cancer is defined as the ratio of the estimated risk of cancer of the particular organ or tissue of concern that is due to radiation exposure to the total risk of cancer of that organ or tissue from all causes, including radiation exposure.

NIH developed the first tables for estimating PC. The tables allow one to look up, for a particular cancer diagnosis at a given age and for the radiation dose received, the estimated probability that the cancer was caused by the radiation exposure. A critical value that is often used for determining responsibility is a PC of 50%. That is, was the radiation exposure at least as likely as not the cause of a cancer? The only other risk factor that is considered in the radiation PC tables is cigarette smoking, and it is applied only to the calculations for lung cancer.

The determination of PC from radiation exposure involves a large amount of uncertainty. The uncertainty has to do with the limited epidemiological data that are available for developing the tables. In recognition of that, the Committee on Interagency Radiation Research and Policy Coordination (CIRRPC) produced additional tables (CIRRPC, 1988) based on the NIH tables. These tables give the estimated radiation dose that would be required for a given cancer type to produce the critical 50% PC value. The 50% PC values are presented in Table III.E.1. The dose associated with a PC of 50% is given for several ages at the time of radiation exposure. Because the risk of leukemia decreases after a number of years post-exposure, CIRRPC calculated different PC values for leukemia occurring within 20 years of an exposure and after 20 years. Because there is uncertainty in the dose associated with this 50% PC estimate, owing to uncertainties in the data on risks in humans, CIRRPC also gives the 5th percentile value of the corresponding uncertainty distribution. At this dose, one can be 95% confident that the dose associated with a 50% PC is not lower. The doses associated with the 95% credibility limit of a 50% PC are given in Table III.E.2. Because there is a requirement of being 95% sure that the PC value is not underestimated, the radiation dose in this case is considerably lower than in the case of estimating the dose associated with a 50% PC in Table III.E.1. CIRRPC also produced doses associated with a 99% credibility level, and these are shown in Table III.E.3.

To give the benefit of the doubt to the veteran, VA has chosen to use the doses associated with 99% credibility limits of PC rather than the 50% PC values. For example, the dose associated with a 50% PC for colon cancer in Table III.E.1

TABLE III.E.1   Organ or Tissue Doses (rad) Corresponding to a PC of 50% Based on NIH Radioepidemiological Tables (CIRRPC, 1988)[a]

Type of Cancer	Age at Exposure, years		
	<20	30	>40
Chronic granulocytic leukemia[b]			
within 20 years of exposure	11.5	16.0	17.6
20 or more years post-exposure	30.8	35.7	59.4
Acute leukemia[b]			
within 20 years of exposure	14.7	22.4	44.5
20 or more years post-exposure	38.7	44.5	55.6
Leukemia (excluding chronic lymphatic)			
within 20 years of exposure	14.4	21.4	37.1
20 or more years post-exposure	37.1	42.4	56.6
Colon cancer	209.4	331.8	497.4
Esophageal cancer	183.8	331.8	458.6
Female breast cancer	92.3	157.0	287.3
Kidney and bladder cancer	258.1	368.3	483.5
Liver cancer	28.0	72.6	138.0
Lung cancer			
known smokers[c]	258.1	409.0	546.8
others[d]	73.2	128.0	178.6
Pancreatic cancer	112.4	202.2	297.9
Stomach cancer	95.6	157.0	225.9
Thyroid cancer	28.9	56.6	63.9

[a]Doses at ages between 20 and 30 years or between 30 and 40 years should be obtained by linear interpolation.

[b]Dose to active bone marrow.

[c]Known to have been a regular smoker (10 or more cigarettes per day) within 5 years of diagnosis. Doses are calculated on the basis of an assumption that the claimant is a member of the average US population that includes smokers and nonsmokers.

[d]Claimant's smoking habits are unknown, or claimant is known to have stopped smoking 5 years or more before diagnosis, or claimant is known to be a nonsmoker. Doses are calculated on the basis of an assumption that the claimant is a nonsmoker.

is 209.4 rad, but VA has used the lower dose of 17 rad associated with the 99% credibility limit of a PC of 50% in Table III.E.3 in evaluating whether it is at least as likely as not that radiation exposure caused a veteran's colon cancer. To give the veterans an additional benefit of the doubt, VA also uses the upper-bound dose reported by the NTPR program as the dose to be used in determining PC. That is, the estimated upper-bound dose to a veteran is compared with the dose from the CIRRPC table giving a 99% credibility limit; if the upper bound is above the table dose, it is presumed that a PC of at least 50% is credible. To the extent that the upper bound is a reasonable representation of the uncertainty in the estimated dose, this procedure generally results in an estimate of PC that even exceeds the 99th percentile credibility limit.

TABLE III.E.2   Doses (rad) to the Affected Organ or Tissue Based on 95% Credibility Limit of PC of 50% (CIRRPC, 1988)[a]

Type of Cancer	Age at Exposure, years		
	<20	30	>40
Chronic granulocytic leukemia[b]			
within 20 years of exposure	1.4	2.0	2.2
20 or more years post-exposure	4.2	5.0	9.3
Acute leukemia[b]			
within 20 years of exposure	1.8	2.9	6.5
20 or more years post-exposure	5.5	6.5	8.5
Leukemia (excluding chronic lymphatic)			
within 20 years of exposure	1.8	2.8	5.2
20 or more years post-exposure	5.2	6.1	8.5
Colon cancer	25.9	48.6	82.7
Esophageal cancer	9.1	22.2	35.8
Female breast cancer	26.7	50.9	104.3
Kidney and bladder cancer	21.1	35.2	51.7
Liver cancer	1.6	5.4	12.9
Lung cancer			
known smokers[c]	37.8	69.6	100.6
others[d]	6.8	14.4	22.8
Pancreatic cancer	10.3	23.4	40.0
Stomach cancer	10.8	21.2	34.8
Thyroid cancer	4.9	10.7	12.7

[a]Doses at ages between 20 and 30 years or between 30 and 40 years should be obtained by linear interpolation. A claimant with a dose less than the dose shown would have less than 5% chance of having a true PC exceeding 50%.

[b]Dose to active bone marrow.

[c]Known to have been a regular smoker (10 or more cigarettes per day) within 5 years of diagnosis. Doses are calculated on the basis of an assumption that the claimant is a member of the average US population that includes smokers and nonsmokers.

[d]Claimant's smoking habits are unknown, or claimant is known to have stopped smoking 5 years or more before diagnosis, or claimant is known to be a nonsmoker. Doses are calculated on the basis of an assumption that the claimant is a nonsmoker.

More epidemiological data on radiogenic cancers have become available in recent years, and the original NIH tables have recently been revised by the National Cancer Institute (NCI) (NCI-CDC, 2002). The National Institute for Occupational Safety and Health (NIOSH) has adopted the NCI tables and has added a few cancers to the original cancers in the NCI update. The method of calculating confidence limits of PC can be found on the NIOSH Web site at http://198.144.166.5/irep_niosh/. On the NIOSH Web site, a person enters sex, year of birth, year in which radiation exposure occurred, year in which cancer was diagnosed, type of cancer, radiation dose, and type of radiation. The Interactive RadioEpidemiological Program (IREP) method also incorporates more sophisticated ways to account for uncertainty in the process of estimating the PC value

TABLE III.E.3   Doses (rad) to the Affected Organ or Tissue Based On 99% Credibility Limit of PC of 50% (CIRRPC, 1988)[a]

Type of Cancer	Age at Exposure, years		
	<20	30	>40
Chronic granulocytic leukemia[b]			
within 20 years of exposure	0.9	1.3	1.4
20 or more years post-exposure	2.7	3.2	5.9
Acute leukemia[b]			
within 20 years of exposure	1.1	1.8	4.1
20 or more years post-exposure	3.5	4.1	5.5
Leukemia (excluding chronic lymphatic)			
within 20 years of exposure	1.1	1.7	3.3
20 or more years post-exposure	3.3	3.9	5.5
Colon cancer	17.0	33.1	58.1
Esophageal cancer	3.9	9.9	16.7
Female breast cancer	18.8	37.0	78.6
Kidney and bladder cancer	13.4	23.1	34.7
Liver cancer	1.0	3.3	8.2
Lung cancer			
known smokers[c]	25.5	48.8	72.1
others[d]	4.3	9.3	15.0
Pancreatic cancer	5.8	13.7	24.3
Stomach cancer	6.9	13.8	23.2
Thyroid cancer	3.3	7.4	8.8

[a]Doses at ages 20 and 30 years or between 30 and 40 years should be obtained by linear interpolation. A claimant with a dose less than the dose shown would have less than 1% percent chance of having a true PC exceeding 50%.

[b]Dose to active bone marrow.

[c]Known to have been a regular smoker (10 or more cigarettes per day) within 5 years of diagnosis. Doses are calculated on the basis of an assumption that the claimant is a member of the average US population that includes smokers and nonsmokers.

[d]Claimant's smoking habits are unknown, or claimant is known to have stopped smoking 5 years or more before diagnosis, or claimant is known to be a nonsmoker. Doses are calculated on the basis of an assumption that the claimant is a nonsmoker.

for a given radiation dose. The computer code allows consideration of uncertainty in an estimated dose and uncertainties in all other parameters that enter into a calculation of PC. The code then calculates not only a central estimate of the PC value for a specified dose and its uncertainty but also the upper 97.5% and 99% credibility limits of PC, taking into account all uncertainties.

The committee has used the current NIOSH computer code (IREP) to estimate doses that can be compared with doses that were used by VA based on the CIRRPC tables as reproduced here in Tables III.E.1 through III.E.3. Table III.E.4 gives the comparison between the doses associated with the 50% PC values in Table III.E.1 and the doses associated with the 99% credibility limits of the 50% PC values in Table III.E.3. The values given in Table III.E.4 are only for a person

TABLE III.E.4 Comparison of CIRRPC Screening Doses (rem) with Values Based on IREP Methodology[a]

Type of Cancer	PC of 50%		99% Credibility Limit	
	CIRRPC[b]	IREP[c]	CIRRPC[b]	IREP[c]
Colon	209.4	72	17	48
Esophagus	183.8	100	3.9	43
Breast (Female)	92.3		18.8	55
Kidney and bladder	258.1		13.4	57[d]
Bladder			NA[e]	66
Liver	28	28	1	14
Lung, smoker	258.1	213	25.5	150
Lung, smoking unknown	73.2		4.3	NA
Lung, never smoked	NA	88	NA	50
Pancreas	112.4		5.8	125
Prostate	NA	479	NA	33
Stomach	95.6	143	6.9	35
Thyroid	28.9	30	3.3	10

[a]Exposure at age of 20 years, and diagnosis at age of 60 years.
[b]Committee on Interagency Radiation Research and Policy Coordination (1988).
[c]Interactive RadioEpidemiological Program (IREP, 2000) developed by NCI and NIOSH.
[d]Kidney only.
[e]Not available.
NOTE: Values from IREP assume chronic exposure to photons of energy > 250 keV.

who was exposed at the age of 20 years and was diagnosed with cancer at the age of 60. The values will be very close to those for a person diagnosed with the cancer at the ages of 50 or 70, except for leukemia. What one observes is that the doses based on the 99% credibility limits of a PC of 50% are higher in the IREP calculations. The reason is that with more knowledge, the uncertainty in estimating PC has decreased, so the dose associated with a PC of 50% has increased. VA is considering using the new and improved NIOSH-NCI tables.

Leukemia is different from solid tumors, such as lung and liver cancer. The difference is that radiation-caused leukemia can occur 1–2 years after exposure, whereas it usually takes at least 10 years for most radiation-caused solid tumors to develop. Furthermore, the radiation risk of leukemia peaks at about 5–10 years after exposure, and there is no longer an observable excess risk after about 25 years; in contrast, the radiation risk of solid tumors often remains proportionally higher than the natural cancer rate throughout one's life after the 10-year latency period. Table III.E.5 shows the doses associated with a PC of 50% and the 99% credibility limit for a PC of 50% for exposure at age 20 and various ages at diagnosis of leukemia. Although leukemia has possibly the greatest cancer risk associated with radiation exposure relative to background cancer rates, the calculations show that after a number of years, the excess risk above background is essentially minimal.

TABLE III.E.5   Doses Corresponding to Different Credibility Limits of PC for Leukemia (rem)[a]

Age at Diagnosis, year	PC of 50%	99% Credibility Limit
30	10	5
40	38	20
50	140	58
60	500	135

[a]Exposure at age of 20 years.

Skin cancer is now also considered to be radiogenic. There are basically three major types of skin cancer. One is melanoma, which can be quite lethal, and the other two are more common nonmelanoma skin cancers, which many people develop primarily from exposure to the sun's ultraviolet radiation. The non-melanoma skin cancers are of two main types: squamous cell carcinoma and basal cell carcinoma. Of those three types, melanoma and basal cell carcinoma can result from exposure to ionizing radiation; squamous cell carcinoma has not been found to be radiogenic. Beta radiation can penetrate only a very short distance in human tissue. Beta radiation, however, can at times be emitted at relatively high levels by radionuclides in nuclear fallout. External beta radiation affects primarily the skin, and the NIOSH IREP program allows consideration of beta radiation in its calculations. The committee has reproduced in Table III.E.6 values of beta doses for skin cancers corresponding to different credibility limits of PC for both black and white men. Background skin cancer risks are considerably smaller in blacks, and this cancer site is the only one that requires racial specification.

Data in Tables III.E.1 through III.E.6 should provide a veteran with an idea of the radiation dose that would be required to achieve a PC of 50% at different credibility limits, including the 99% credibility limit that the 50% PC is not associated with a lower dose, as used by the VA. For a specific case, one can go

TABLE III.E.6   Doses Corresponding to Different Credibility Limits of PC for Skin Cancer (rem)[a]

Skin Cancer Type	Blacks		Whites	
	PC of 50%	99% Credibility Limit	PC of 50%	99% Credibility Limit
Melanoma	39	8	70	10
Basal cell carcinoma	24	4	70	10
Squamous cell carcinoma	NA[b]	190	NA	475

[a]Chronic exposure to beta particles of energy > 15 keV.
[b]No value is calculated, because squamous cell carcinoma of the skin is not considered to be radiogenic.

to the NIOSH Web site and use the IREP program to determine what the 99% credibility limit of PC is for the given upper-bound radiation dose provided in a dose reconstruction and whether the 99% credibility limit of PC is 50% or higher. The values in the tables given here provide some information for the medical decision process. It must be understood that the PC values constitute only one tool used in the medical decision process, and PC is not the only factor used in deciding whether an atomic veteran with a nonpresumptive radiogenic disease receives compensation for service-connected disability.

IV Process of Dose Reconstruction in NTPR Program

This chapter discusses methods of dose reconstruction for atomic veterans that have been used in the NTPR program. Consistent with discussions on the principles and process of dose reconstruction in Section I.C, this chapter is organized as follows: Section IV.A discusses development of exposure scenarios; Section IV.B discusses methods used to estimate external dose from exposure to photons, neutrons, and beta particles; Section IV.C discusses methods used to estimate internal dose from intakes of radionuclides; Section IV.D discusses dose reconstructions for occupation forces in Japan; Section IV.E discusses methods used to account for uncertainties in estimates of external and internal dose; Section IV.F discusses the approach to estimating total dose from external and internal exposure, taking into account all radiation types and exposure pathways of concern; and Section IV.G discusses documentation of dose reconstructions and quality assurance. The committee's evaluations of these aspects of the NTPR dose reconstruction program are presented in Chapters V and VI.

As discussed in the standard operating procedures (DTRA, 1997) and in 32 *CFR* Part 218, the goal of the NTPR program is to obtain upper-bound estimates of dose to atomic veterans, consistent with the policy of giving the veterans the benefit of the doubt in estimating their doses (see Section I.C.3.2). More specifically, the goal is to obtain at least a 95th percentile upper bound of possible doses, taking into account uncertainties in estimating dose. That is, the NTPR program intends that a reported dose should exceed the true dose in at least 95% of all cases and that there should be no more than a 5% chance that the true dose to an individual is higher than the reported value. As discussed in more detail in Sections IV.B and IV.C, it is the current policy of the NTPR program to report a central ("best") estimate of external dose from exposure to photons along with an estimated 95th percentile upper bound, and the same approach has been taken in estimating external dose from exposure to neutrons in dose reconstructions for participant groups. However, only a single estimate of dose is reported when a beta dose to the skin or lens of the eye or an internal dose from intakes of radionuclides is calculated, and this estimate is intended to be at least a 95th percentile upper bound.[1]

IV.A EXPOSURE SCENARIOS

Development of exposure scenarios for participants in the nuclear-weapons testing program involves consideration of assumptions about the locations of the participants of concern, their activities at those locations, and the time spent at each location and assumptions about the radiation environment at the assumed locations of the participants during the time spent at those locations (see Section I.C.2.1). Approaches to development of exposure scenarios used in the NTPR program are described in the standard operating procedures (DTRA, 1997) and in 32 *CFR* Part 218.

The dose reconstruction process requires that the analyst first determine whether a veteran's records support his qualifying as a "participant" according to the definition in applicable laws and regulations. In this initial stage, military records are used to confirm that the veteran was present at the Nevada Test Site (NTS) or the Pacific test sites during designated intervals before and after tests of nuclear devices, was present in Hiroshima or Nagasaki during the occupation of Japan, or was a prisoner of war near Hiroshima or Nagasaki at the time of the atomic bombings. The burden of proof in establishing a veteran's status as a participant is stricter if a veteran's claim is filed for a presumptive disease under 38 *CFR* 3.309 than for a nonpresumptive disease under 38 *CFR* 3.311 (see Section I.B.4).

Once a veteran's participation status has been confirmed and a dose reconstruction has been requested (usually by the Department of Veterans Affairs [VA] in response to a claim filed for alleged radiogenic health conditions), a government contractor (currently JAYCOR) undertakes extensive historical research based on archival records to reconstruct the movements and activities of the veteran during the period of participation.

---

[1]An additional concept used in this report to represent the full range of uncertainty in a quantity is the confidence interval. For example, a 90% confidence interval of an uncertain quantity gives the range of values within which the true value should lie in 90% of all cases, and the lower and upper bounds of this confidence interval are the 5th and 95th percentiles, respectively.

IV.A.1 Unit-Based Dose Reconstructions

If nothing in the historical research suggests that the veteran was involved in unusual activities (that is, activities different from those of the other members of his unit), a "unit-based" dose reconstruction may be carried out. An example is participants in units who observed detonations at the NTS from trenches close to ground zero (see, for example, Figure IV.A.1). This one-size-fits-all strategy assigns the same dose to everyone in a given unit, with an upper bound assigned to allow for uncertainty in the estimated dose.

FIGURE IV.A.1   Observers in the trench from which they observed a nuclear detonation.

If no badge records are available, a unit-based dose reconstruction can be based on radiation-monitoring data that were obtained at the time of an operation as part of the test itself. With computer models, the measurements are interpolated and smoothed across space and time (allowing for the physics of radioactive decay) and then combined with historical summaries of the activities of the unit, including the likely path of the unit through the radiation environment, to reconstruct the dose for the unit. If the exact times spent in various locations are not known, assumptions are sometimes applied on the basis of the presumption that radiation-safety policies in force at the time of the test were followed.

IV.A.2 Individualized Dose Reconstructions

If a participant was involved in unusual activities, an individualized dose reconstruction is required. In such instances, there may have been complete or nearly complete badging during the entire time of participation.[2] If so, and if the issue and turn-in dates for the badges of record are complete and cover the veteran's entire time at the site, the badge readings are simply summed, and their variances are combined with a method called quadrature, in which the variance (error) of the summed dose is taken to be the sum of the variances of the individual readings, assuming independence of errors. The per-badge biases and variances are based on modifications of methods proposed in a previous National Research Council report on film badge dosimetry in atmospheric nuclear tests (NRC, 1989).

One issue that often arises in the dose reconstruction process is related to the fact that participants often had a "permanent" badge, which was supposed to be worn throughout their entire time in an operation, plus occasional "mission" badges, which were issued on particular occasions when radiation safety personnel determined that a participant was likely to encounter an unusual potential for exposure. If the "permanent" badge was not worn on such occasions, the proper way to combine the two types of badge readings would be to sum them. If, instead, the two badges were worn contemporaneously, the mission badges can be ignored because any additional dose experienced on a particular mission presumably was already captured by the permanent badge. A dose reconstruction policy requiring the benefit of the doubt to be given to the veteran would require summing the two, and this was sometimes done.

Because badging often was not complete or uncertainties remained (for example, because the issue or turn-in dates were missing—a common problem), an individualized dose reconstruction is required for some intervals of the veteran's time of participation. The analyst must reconstruct the particular activities and locations of activities that the veteran would have undertaken in the

---

[2]Complete badging of participants happened infrequently, most often for short-term participants during and after 1956 Operation REDWING in the Pacific.

assumed radiation environment and apply modeled radiation levels to those activities and locations. Difficulty often arises in the reconstruction of the veteran's experiences.

In some cases, "cohort" film badging is used to assign a unit-based dose. When only a few members of the unit wore a badge during an operation, the mean of those few badge readings can be assigned to all members of the same cohort. The uncertainty in such dose assignments is computed from the variability in the badge measurements.

Dose assignments based on film-badge data are discussed in more detail in Section IV.B.1 and IV.E.2.

IV.A.3 Individualized Reconstruction of Scenarios

The sources of historical information that can be used to reconstruct exposure scenarios include, in addition to such official documents as morning reports and ship logs, narratives written at the time, such as reports of unexpected changes in wind or fallout that complicated the management of radiation exposure for participants at specific tests. Individual information about a veteran's job type or specific mission responsibilities is sometimes available. Other documents can contribute information on a person's exposure scenario, such as questionnaires filled out by the veteran (especially early in the NTPR Program) or statements the veteran provided in support of his claim. Occasionally, other people are consulted to clarify uncertainties in what was experienced by a particular veteran, such as his commanding officer or others in the same unit. Surviving widows or children are sometimes contacted when the veteran is deceased, although they usually do not provide much detail. Some of the veterans had been sworn to secrecy for national-security reasons and never described their experiences even to their spouses.

IV.B ESTIMATION OF EXTERNAL DOSE

All estimates of external dose to participants are based on film-badge readings or surveys with field instruments. If the participant wore a film badge and the data could be located, the external gamma dose of record is generally based on those data. If no acceptable film-badge data are available or if the film-badge data do not cover all potential exposures, the external dose for these exposures is based on a "scientific" dose reconstruction that relies on survey data. For most participants, the reconstructed gamma and neutron dose from external exposure is based on a generic dose reconstruction performed for their particular units' activities during a given test series, modified as appropriate to conform to a participant's duties and period of exposure.

It is important to note that the method used in the NTPR program to estimate external doses changed over time as shown below (Schaeffer, 2001a).

Year   **Change in Methodology**
1978   Individual services (Army, Navy, Air Force, and Marines Corps) report external doses based on film-badge dosimetry
1980   Neutron dose reconstruction added for unit dose reconstructions
1984   Statistical application of military-unit film-badge readings used when a veteran's film-badge readings are missing
1988   Dose reconstruction applied to periods of incomplete film-badge coverage
1989   Upper-bound doses for individual film-badge data applied
1990   Doses from damaged film badges superseded by reconstructed doses
1992   Total upper-bound doses coupling the estimated uncertainty from film-badge data and reconstructed doses are calculated
1996   Upper-bound doses included in all reports in response to VA claims
1998   Skin-dose calculations added for all skin-cancer claims in response to a VA request

When changes in policy or method are adopted, there is no systematic way to review earlier dose reconstructions or to apply the changes retroactively; this is discussed in more detail in Section VI.E. The method currently used in the NTPR program to estimate the most likely external gamma dose based on film-badge dosimetry, most likely gamma dose based on a scientific dose reconstruction, most likely neutron dose, and upper-bound beta dose to skin or lens of the eye are discussed in the following paragraphs. The method used to estimate upper bounds of gamma and neutron doses is discussed in Section IV.E.2.

IV.B.1 External Dose Estimation from Film-Badge Data

It is the policy of the NTPR program that if a film badge was issued and worn and valid film-badge data can be located, the film-badge reading is to be considered the dose of record for the period when the badge was worn (see Brady and Nelson, 1985). Thus, the policy for reconstructing a dose to a test participant is first to search for film-badge data. However, during the earlier test series, only a small fraction of test participants were badged. Attempts were made beginning in the 1956 Operation REDWING in the Pacific and the 1957 Operation PLUMB-BOB in Nevada to badge all participants. During the earlier test series, mission badges were issued to participants to be worn when some radiation exposure was expected to occur because of the particular duties to be performed, such as maintenance on contaminated aircraft or recovery of contaminated equipment from displays. Civilian and military participants at Operation CROSSROADS numbered about 43,000, but only about 7,000 film badges were issued to the persons thought most likely to receive radiation exposure. Because multiple badges were issued to many of those personnel, only a small percentage of CROSSROADS participants had film-badge records. Often, one or more members of a unit performing similar duties would be issued a "cohort" film badge. For example, only one or two members of each platoon participating in maneuvers during tests at the NTS were badged (for example, see Frank et al., 1982). The data from this cohort badge would provide an estimate of the external dose to the entire group.

Even if permanent or mission badges were issued (see Figure IV.B.1), often the badges or the data from them can no longer be located. For example, although many film badges were issued during Operation UPSHOT-KNOTHOLE in 1953, most of the data from them were lost, and only summary data can be located (for example, see Edwards et al., 1985).

When film-badge data are reported but the data are considered suspect, the NTPR program requests a re-examination of the film by the Department of Energy. If the film is still on file, it is re-examined by a health physicist, and a determination is made as to whether the reading is questionable or highly suspect. Often, film was damaged by heat, water, or humidity, particularly in the Pacific during the REDWING and DOMINIC test series. That was particularly the case if a badge was worn for more than a few weeks (NRC, 1989). Problems with calibration errors also caused film data to be suspect (NRC, 1989). According to current policy of the NTPR program (Schaeffer, 1995; 2002b; 2002e), suspect film-badge data are discarded in favor of a scientific dose reconstruction.

In a 1989 report, a committee of the National Research Council reviewed the method used in the NTPR program to analyze film-badge data (NRC, 1989). The

FIGURE IV.B.1   Example of film badges worn by participants at atomic tests.

review examined possible calibration errors and reported heat and water damage, and it recommended that bias and uncertainty factors be applied for each test series. It also recommended how the film-badge data should be reported and how estimates of uncertainty in individual readings and sums of multiple readings should be treated. The NTPR program claims in its letters to VA and the veterans that its reported film-badge data conform to the recommendations of the National Research Council report. However, the NTPR program does not follow the recommendations exactly but has modified them somewhat. In particular, the report recommended that a reported film-badge reading be divided by a bias factor to convert to a whole-body equivalent dose in rem. Most of the bias factor is intended to convert a film-badge measurement of exposure in air in roentgens (R) to a whole-body equivalent dose in rem. However, the NTPR program assumes that the film-badge exposure in R is a direct estimate of the shielded whole-body dose in rem (Klemm, 1989; Flor, 1992).[3] Furthermore, all badge readings, rather than the bias-corrected doses inferred from them, are summed to get the total dose. The NTPR program asserts that that is done to preserve a one-to-one correlation with the film-badge record so that a veteran can see evidence that original records are being used in the dose reconstruction and to avoid the perception that the program is lowering recorded doses (Flor, 1992; Schaeffer, 2002b). Estimates of film-badge doses used in dose reconstructions thus are higher, by a factor of about 1.3 or more, than if recommendations by the National Research Council committee had been followed precisely.

In assigning doses based on film-badge data, the NTPR program usually assumes that if a participant was issued both a mission badge and a permanent badge encompassing the same interval, the badges were worn concurrently, as required. Thus, doses recorded by mission badges were generally assumed to be included in the permanent-badge reading. However, if the sum of all mission-badge doses is greater than the permanent-badge reading, the higher value is to be used (Flor, 1992). Similarly, when mission badges were issued but no permanent badge was issued, the mission-badge data generally were adjusted by subtracting the reconstructed dose from routine exposure to fallout during the period that the mission badge was supposed to have been worn. This procedure is used because that dose presumably would have been included in the mission-badge reading.

IV.B.2 External Gamma-Dose Estimation Based on Dose Reconstruction

Because only a fraction of participants were issued film badges during the earlier test series and the time when badges were worn often covered only a portion of the time of potential exposure, methods have been developed to recon-

---

[3]The bias correction from exposure in air in roentgens to a whole-body equivalent dose in rem, which is approximately a factor of 0.7, was applied to film-badge data in 1990 as recommended by the 1989 National Research Council report, but was rescinded by the NTPR program in 1992.

struct the external gamma dose to unbadged participants on the basis of available monitoring data and physical models.

Generic average external doses have been estimated for all major units participating in each test series. The results of the generic assessments and the methods used to estimate external dose have been published in a series of reports issued mainly in the 1980s. Examples of such unit dose reconstructions include calculations for observers at NTS tests (Barrett et al., 1987), maneuver troops at NTS tests (Edwards et al., 1983; 1985), garrisons on the headquarters islands at the Pacific test sites on Enewetak and Bikini atolls (Thomas et al., 1982; 1983a; 1983b; 1984), sailors on support ships (Weitz et al., 1982; Thomas, 1983a; 1983b), boarding parties on target vessels (Weitz et al., 1982), and occupation troops at Hiroshima and Nagasaki (McRaney and McGahan, 1980). The published unit-dose reports referred to above are supplemented by internal memoranda in which daily doses are estimated for specific ships and islands and for smaller units at NTS tests (Frank, 1982; Ortlieb, 1991; 1995; Phillips, 1983; Thomas, 1985; Weitz, 1995a; 1995b; 1997). For example, the Ortlieb (1995) memorandum gives daily dose tables for seven support units at Operation UP-SHOT-KNOTHOLE: the 505th Signal Services, the 412th Engineer Construction Battalion, the 3623rd Ordinance Company, the 77th Army Band, the 93rd Army Band, the 371st Evacuation Hospital, and the 163rd Quartermaster Laundry Company.

On the basis of those unit dose reconstructions, the NTPR program assigned a generic dose to all participants in the units. Unless a formal dose reconstruction is requested as a consequence of a VA claim or a specific participant request to the Defense Threat Reduction Agency, the participant's dose of record is generally based on either a film-badge measurement or the assigned average dose for the participant's unit. The applicable unit dose reconstructions are usually the starting point for a scientific dose reconstruction for a specific person.

IV.B.2.1 *Unit Dose Reconstructions at the NTS*

External doses to military units participating as observers or in maneuvers at NTS tests were based on estimates of the location of troops versus time, as obtained from documented unit activity histories, mission plans, and rehearsals (Goetz et al., 1979; 1980; 1981). Shielding provided by vehicles, trenches, and so on, was estimated from radiation transport calculations (Edwards et al., 1983). Computerized interpolation schemes were used to estimate dose rates at various grid locations and times from the available exposure-rate data (Edwards et al., 1985). Separate dose estimates have been made for direct exposure to prompt gamma and neutron radiation in trenches, from exposure to fallout and from overhead debris clouds during observation of a shot from trenches or during post-shot maneuvers, from exposure to fallout-contaminated fields from previous shots during pre-shot rehearsals, and from observation of contaminated displays set up to study blast and radiation effects. For example, Table IV.B.1 summarizes calculated unit doses and upper and lower confidence limits of a 90% confidence interval for maneuver troops participating in tests during the UPSHOT-KNOTHOLE series at the NTS (Edwards et al., 1985). As indicated in the table, external doses from prompt radiation during observation of the tests from trenches were generally smaller than doses received from residual gamma radiation during post-shot maneuvers and touring of contaminated display areas (see Figures IV.B.2 and IV.B.3). The estimated confidence intervals are discussed further in Section IV.E.2.

Shot	BCT	Neutron Dose (rem)	Initial Gamma Dose (rem)[1]	Residual Gamma Dose (rem)[1]	Total Gamma Dose (rem)[1]
ANNIE	A	$0.018^{+0.017}_{-0.011}$	$0.012^{(2)}$	$0.80^{+0.31}_{-0.16}$	$0.81^{+0.31}_{-0.16}$
	B	$0.018^{+0.017}_{-0.011}$	0.012	$0.69^{+0.27}_{-0.14}$	$0.70^{+0.27}_{-0.14}$
NANCY	A	<0.001	0.004	$1.3^{+0.5}_{-0.2}$	$1.3^{+0.5}_{-0.2}$
	B(3)	<0.001	0.004	$2.4^{+1.5}_{-0.7}$	$2.4^{+1.5}_{-0.7}$
	B(4)	<0.001	0.004	$1.1^{+0.5}_{-0.3}$	$1.1^{+0.5}_{-0.3}$
SIMON	A	0.003	0.012	$3.1^{+0.3}_{-0.2}$	$3.1^{+0.3}_{-0.2}$
	B	0.003	0.012	$2.2^{+0.8}_{-0.4}$	$2.2^{+0.8}_{-0.4}$
ENCORE	A/B	--	--	$0.06^{+0.02}_{-0.01}$	$0.06^{+0.02}_{-0.01}$
	Helilifted Platoons	--	--	$0.13^{+0.05}_{-0.03}$	$0.13^{+0.05}_{-0.03}$
GRABLE	A	<0.001	<0.001	0.04	0.04
	B	<0.001	<0.001	$0.54^{+0.54}_{-0.31}$	$0.54^{+0.54}_{-0.31}$
	B(5)	<0.001	<0.001	$2.2^{+2.2}_{-1.4}$	$2.2^{+2.2}_{-1.4}$

(1)Reconstructed film badge dose.
(2)Uncertainties less than 0.01 rem not displayed.
(3)Forward elements.
(4)Rear elements.
(5)Elements that visited the 500-yard display following Shot Grable.

TABLE IV.B.1   Summary of unit dose reconstruction for maneuver troops at Operation UPSHOT-KNOTHOLE (Edwards et al., 1985); BCT = Battalion Combat Team. Estimated lower and upper confidence limits shown are doses to add or subtract from tabulated dose to obtain 5th and 95th percentiles of distribution, respectively.

FIGURE IV.B.2   Troops leaving a trench shortly after a detonation.

FIGURE IV.B.3   Army personnel examining equipment damaged during a nuclear detonation.

IV.B.2.2 *Unit Dose Reconstructions for Pacific Test Sites*

Average generic doses to military units at the Pacific test sites are based on estimates of radiation exposure rate versus time. Those estimates are often based on spare data from post-shot ship and island surveys and from radiation monitors accompanying maneuver troops (see Figure IV.B.4). The total time-integrated dose is estimated from the initial exposure rates and estimated or measured decay rates. Tables of daily doses received by sailors were constructed for each ship participating in tests and for participants on various islands. Doses to ship crews were estimated from exposure to fallout on deck (topside), from direct exposure to contaminated seawater topside, from radiation below decks due to contamination of piping and engineering spaces, and from direct exposure during approach toward contaminated target ships (Weitz et al., 1982; Thomas et al., 1984). Doses to engineering crews were generally estimated to be higher than doses to average crew members because of greater proximity to pipes and condensers contaminated by radionuclides in the seawater below decks. Thus, radiation exposure to some naval participants, such as in Operation CROSSROADS, was actually higher to crew members when below decks than when on decks.

Models were developed to describe doses due to contamination of piping and seawater and to proximity to contaminated vessels (Weitz et al., 1982). Generic estimates of time spent indoors vs outdoors or above versus below decks on ships and generic estimates of shielding provided by housing on islands or ship structures are used to modify the free-in-air exposure-rate estimates for sailors on ships or for island-based personnel. Generally, the reconstructions assume that sailors are topside for an average of 40% of the time during three periods daily and below decks the remainder of the time. It is assumed that exposure below decks is minimal because of shielding by steel decks and structures that reduces the exposure rate to an average of about 10% of the topside rate (Thomas et al., 1982; 1984). It is generally assumed that personnel billeted on islands spent 60% of the time outdoors and 40% indoors and that shielding provided by a building structure reduced indoor doses by an average of 50% compared with the outdoor dose (Thomas et al., 1982; 1984).

Daily dose rates from gamma rays for each ship or island were calculated by decaying the reported mean of initial survey data for a particular ship or island for each fallout episode and applying the generic outdoor and indoor fractions and shielding factors discussed above. A generic decay factor of $t^{-1.2}$ (where t is the time, in hours, after the time of the blast) was generally assumed unless actual data were available. If there were multiple fallout episodes, the residual fallout level from the previous episode was subtracted from the mean of the new survey data to calculate later dose rates in the days after the new event. Thus, the doses from each fallout event are calculated separately for each day and then added. The total dose for any person is calculated either by summing the daily doses during documented periods of potential exposure based on military records (if

FIGURE IV.B.4   Participants conducting radiation survey on deck of a ship.

from multiple events) or by integrating over the period of exposure by using the nominal $t^{-1.2}$ decay rate or the actual decay rate if it is measured. The tabulated daily dose rates for later times are considered "high-sided," in that any decontamination or weathering after the radiation survey usually was not considered in making the estimates.

The 1946 CROSSROADS test series was one of the largest with respect to the number of participants and potential exposures. More than 36,000 persons on

FIGURE IV.B.5   Photograph of underwater Shot BAKER in Bikini Lagoon.

154 support ships participated in this exercise (Weitz et al., 1982). Most of the radiation dose was from Shot BAKER, an underwater test conducted in Bikini Lagoon (see Figure IV.B.5). During that test, a fleet of 88 unmanned target ships was anchored in the lagoon and was heavily contaminated by the wave of contaminated water generated by the blast and by heavy fallout. Personnel of support ships entering the lagoon after the shot were exposed to contaminated seawater, which also contaminated ship internal piping, as discussed above. Other personnel were exposed while serving on boarding parties sent to inspect damage on the target ships or to decontaminate and re-man some of the ships. The NTPR program has constructed curves of daily doses for each support and target ship (Weitz et al., 1982, Volumes 2 and 3) on the basis of monitoring data and decay extrapolations. An estimated decay rate of $t^{-1.3}$ was adopted for the dose rates on target ships; this rate was based on measurements on various target ships and is higher than the nominal $t^{-1.2}$ decay rate normally assumed. The plotted daily doses are means for topside, below decks, and amidships, and often they are based on very sparse data. An example of the average daily dose rate on one such

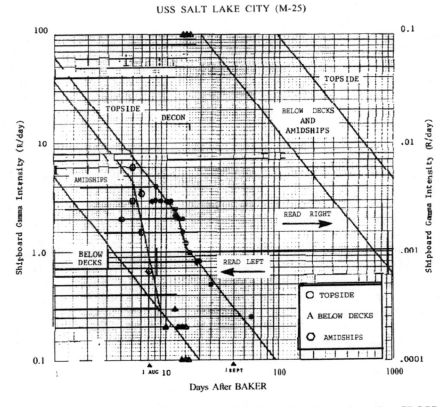

FIGURE IV.B.6   Average daily doses on USS *Salt Lake City* at Operation CROSS-ROADS, Shot BAKER (Weitz et al., 1982, Vol. 2).

target ship, the USS *Salt Lake City*, is shown in Figure IV.B.6. The daily doses to most nonbadged people serving on a particular support ship or boarding a particular target ship were calculated from these curves by using the onboard times that are documented in the support-ship logs. Some daily-dose estimates for particular people were based on cohort film-badge data, if available.

IV.B.3 Estimation of Neutron Doses

Estimates of dose due to external exposure to neutrons produced in a detonation (mainly for some close-in observers at NTS shots and crews of cloud-sampling aircraft) are based on calculations with radiation-transport models. The models calculate the free-in-air exposure in air as a function of distance from ground zero and estimate the shielding provided by trenches and aircraft to calculate a dose in tissue in rad (Goetz et al., 1985). The whole-body equivalent dose in rem is then calculated by applying an estimated spectrum-weighted quality factor (QF) of 13 (Goetz et al., 1985); QF accounts for the biological effectiveness of neutrons relative to that of gamma rays in inducing stochastic radiation effects, such as cancer. However, a QF of 8.5 was used for participants at the PLUMBBOB test series (Goetz et al., 1979). The dose to any organ or tissue from external exposure to neutrons is assumed to be the same as the equivalent dose to the whole body.

IV.B.4 Estimation of Beta Dose to Skin and Lens of the Eye

Energetic electrons are emitted by most fission and activation products and were an intrinsic component of the radiation to which atomic veterans were exposed. Most of the electrons are beta particles, which are electrons emitted as a direct result of a radioactive decay process in which a neutron transmutes to a proton and an electron is discharged. As beta particles from sources outside the body enter tissue, the dose falls off rapidly with depth, and tissues and organs lying deeper than 10 mm in the body are unaffected. Thus, beta particles are appropriately ignored in considering external dose to most tissues and organs, which lie deeper than 10 mm, and for them the appropriate quantities are gamma dose and neutron dose. The two exceptions are the skin, with its sensitive component (basal cells) at a depth of 0.07 mm, and the eye, with its sensitive component (lens) at a depth of 3 mm. The potential contribution from beta particles should be considered whenever the dose to skin or the lens of the eye is assessed.

The current method used in the NTPR program to assess beta-particle doses from sources outside the body is described by Barss (2000). Before 1998, skin doses were not estimated in the NTPR program except on a case-by-case basis (Schaeffer, 2002c). Before the 2000 publication of the Barss report, beta doses were computed by using information from references cited in the publication (Schaeffer, 2002c), principally the user's manual for the CEPXS radiation transport code (Lorence et al., 1989) and a report that specified the beta and gamma energy spectra as functions of time after a detonation (Finn et al., 1979). Those two references were often cited in the individual beta-dose reconstructions before 2000. Although the data used in calculations have changed, the general method has remained substantially the same since routine assessment of skin dose began in 1998. Accordingly, the Barss report can be considered to generally document the method used in 1998 and 1999 and to present the method used since January 2000.

In the Barss (2000) report, external beta dose from standing on a contaminated surface, from being in contaminated air and water, and from contaminated skin is considered in some detail. The report describes models for calculation and, for the case of external beta dose from standing on contaminated ground, makes numerous comparisons with other methods of calculation and with measurement data. Pertinent information from the report is summarized below.

IV.B.4.1 *External Beta Dose from Standing on Contaminated Surface*

External beta doses from standing on contaminated ground or other surfaces are calculated by applying a beta-to-gamma dose ratio to an estimated upper-bound gamma dose, which is determined from film-badge data or dose reconstruction. Doses to the skin or lens of the eye are then calculated by adding the beta and gamma doses. The fundamentals of the method for estimating external dose to the skin and eye from standing on contaminated ground are summarized as follows:

• Beta dose to the skin or lens of the eye from external sources is accrued with the gamma dose from radioactive fallout, contamination, or neutron-induced radionuclides. As a result, the beta dose is proportional to the gamma dose, and its relative magnitude can be expressed by a beta-to-gamma dose ratio.

• The beta-to-gamma dose ratio depends on radionuclide decay and distribution and on geometric relationships between the exposed individual and the radiation source. Gamma and beta energy spectra are interdependent functions of time. Consequently, the beta-to-gamma dose ratio depends on time since detonation.

• Because of the attenuation characteristics of electrons in matter, beta dose assessments depend more critically than gamma dose assessments on geometry and the shielding material between the radioactive source and the exposed individual. Consequently, the nature of specific job- or task-related activities and their associated protective measures entails special attention and evaluation in determining a beta dose component.

• Beta doses to skin are evaluated at the anatomic location where a skin cancer has been diagnosed. The depth for evaluation is 0.07 mm, the conventional depth of the basal-cell layer of the skin, which is assumed to be the tissue at risk for skin cancer. Beta doses to the lens of the eye are assessed at a depth of 3 mm below the front surface of the eye, where the tissue at risk for posterior subcapsular cataract development is assumed to be.

• A beta energy greater than 0.07 MeV is required to penetrate the dead epidermal layer, so beta particles with energies less than that are not included in dose assessments.

• Skin and eye doses are assessed as the sum of the applicable "high-sided" beta and "high-sided" gamma doses (neutron doses presumably are included if they are significant).

• Fallout deposited on a surface is considered to be a semi-infinite plane isotropic source, and decontamination activities are considered in evaluating beta doses. Assessments of skin doses are simplified by ignoring attenuation of electrons by large fallout particles that contain volume-distributed activity, particle and photon scattering due to surface roughness, particle and photon attenuation due to penetration into a radioactive surface, and radioactive-source depletion due to weathering, chemical dissociation, or environmental transport (concentration or dispersion). Those simplifications have the effect of making the calculated doses overestimates ("high-siding").

Barss (2000) provides methods and tables useful for assessing beta dose. They include separate tables of beta-to-gamma dose ratios from exposure to fission products on the ground as a function of time since detonation for NTS and Pacific tests (in the Pacific, one table applies to Operation CASTLE, Shot BRAVO fallout, and a second table applies to all other shots). A separate table of beta-to-gamma dose ratios is provided for activation products in soil. The beta-to-gamma dose ratios are calculated from published beta-particle and gamma-ray spectra. For illustration, the beta-to-gamma dose ratios from the table for Pacific tests are plotted in Figure IV.B.7. The figure illustrates the substantial variation of beta-to-gamma dose ratios with time after detonation and height above ground. The beta-to-gamma dose ratios for the lens of the eye are much smaller because the greater depth of the sensitive tissue (3 mm for lens vs 0.07 mm for skin) leads to much more attenuation.

Activation products are distributed with depth in soil because they originate primarily by interactions with neutrons that penetrate into the soil rather than in deposition from the atmosphere. Thus, beta-to-gamma dose ratios of activation products are small because most of the activation products are deeper in the soil than the range of the emitted beta particles.

Dose to skin is reduced by clothing, and Barss (2000) provides a method for determining the reduction as a function of clothing thickness. Many other factors affect beta dose, such as sizes and shapes of the source materials, position of the body relative to the source (for example, standing, sitting, or kneeling), whether the person is above or below deck on a ship, and whether the person is in an aircraft. Barss provides separate tables to guide adjustment of beta dose when skin is covered by light clothing or denser materials (see Tables 13 and 14 in that report).

The simplest calculation is for a short-term exposure to beta and gamma radiation. The total beta plus gamma dose to the skin or lens of the eye is estimated as

$$\text{Dose}_{\text{skin/lens}} = [D(t)_{\gamma/\text{ub/fall}} \times R_{\beta/\gamma}(x,t) \times M(x,t)] + D(t)_{\gamma/\text{ub/total}}$$

where

$D(t)_{\gamma/\text{ub/fall}}$ = upper-bound gamma dose due to external exposure to fallout or other beta radiation field,

$R_{\beta/\gamma}(x,t)$ = β/γ ratio = bare-skin or lens beta-to-gamma dose ratio at distance x and time t,

$M(x,t)$ = combined modifying factor that accounts for differences from the simple case of standing on contaminated ground with bare skin (for example, attenuation by clothing, position of the body, and location above or below decks on

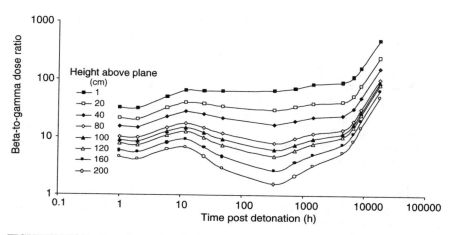

FIGURE IV.B.7   Beta-to-gamma dose ratios for contaminated surfaces used at Pacific tests [see Barss (2000), Table 5].

a ship or in an aircraft) [see Barss (2000), Appendix C, Table 12], and

$D(t)_{\gamma/ub/total}$ = upper-bound gamma dose from all sources.

If an exposure extends over a period sufficient for any term in the equation to change significantly, the individual beta doses must be summed or integrated to obtain a total beta dose. Alternatively, if $M(x,t)$ is constant, an average value of $R_{\beta/\gamma}(x,t)$ may be calculated and then multiplied by $D(t)_{\gamma/ub/fall}$ and $M(x,t)$ to obtain total beta dose. In some cases, $R_{\beta/\gamma}(x,t)$ may be zero, such as when gamma dose is accrued in uncontaminated interior spaces of a ship.

In an effort to "high-side" the total skin or eye dose, the beta-to-gamma dose ratio is applied to the upper-bound gamma dose from exposure to fallout or other beta-radiation field to calculate beta dose, and the upper-bound gamma dose from all sources is then added to the beta dose.

IV.B.4.2 *External Beta Dose from Immersion in Contaminated Air or Water*

Immersion in a descending fallout-debris cloud was a less frequent circumstance than exposure to fallout after deposition on the ground or other surface. However, several potential applications for this exposure condition are identified by Barss (2000), such as ground troops advancing toward ground zero at the NTS immediately after a detonation, naval personnel outside on ship weather decks when fallout began, or contaminated air penetrating into the interior of an aircraft during cloud sampling or tracking activities. Some participants on residence islands in the Pacific also were immersed in descending fallout.

Beta-to-gamma dose ratios, such as those used to estimate beta dose from exposure to a contaminated ground surface, are not used in cases of exposure to descending fallout. Instead, beta dose is estimated by using calculated dose coefficients, which give equivalent dose rates from electrons per unit concentration of radionuclides in air,[4] combined with the duration of exposure and composite beta-spectrum radiation energies associated with a reconstructed gamma exposure or film-badge reading. The calculated beta dose is added to the upper-bound gamma dose for the corresponding period. The approach is to be tailored case by case for any person exposed to descending fallout. The dose coefficients assume uniform suspension of radioactive material in an infinite air source, and the current method is said to overestimate the corresponding beta dose because it includes no provision for shielding from electrons or for self-attenuation by fallout particles.

Composite energy spectra from mixed fission-product beta particles were calculated as a function of time after a detonation, and the resulting beta energies were binned and combined with calculated beta-particle dose coefficients for air immersion (Kocher and Eckerman, 1981). The resulting composite dose coefficients for the assumed mixtures of radionuclides were calculated as a function of time after the detonation in units of rem y^{-1} per μCi cm^{-3} of air. The dose coefficients were tabulated by Barss (2000) for specific periods after detonation, because descending fallout typically lasted anywhere from less than a few hours to 1 day. The composite dose coefficients for assumed mixtures of fission products in descending fallout are illustrated graphically in Figure IV.B.8. Multiplication of the composite beta dose coefficient by the corresponding total activity concentration results in an initial beta dose rate that can be integrated over the time of exposure to give the total beta dose. The resulting beta dose is summed with the upper-bound gamma dose over the same period of time to yield the skin dose due to exposure in air. Clothing modification factors can be applied as appropriate. Barss also describes special methods for the case of being in contaminated air in an aircraft performing cloud sampling or tracking activities and provides an approach for the case of immersion in water. For aircraft, it is appropriate to account for beta-particle backscatter from the interior of the fuselage. The approach for submersion in water is similar to that for air, but it accounts for the different densities of water and air.

IV.B.4.3 *Skin Contamination*

Barss (2000) correctly indicates that for skin contamination (for example, from fallout or resuspended radioactive soil), the film-badge gamma dose is a

---

[4]In the NTPR program, external dose rates per unit concentration of radionuclides in a source region are referred to as dose conversion factors. In this report, however, they are referred to as dose coefficients to conform to the terminology currently used in internal dosimetry by the International Commission on Radiological Protection (ICRP, 1991a).

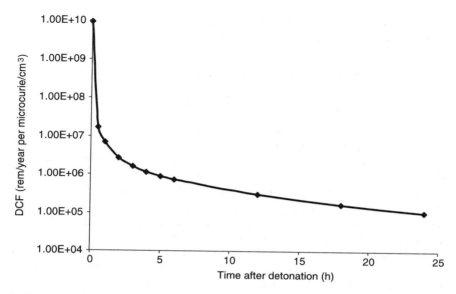

FIGURE IV.B.8   Composite beta dose coefficient for immersion in fallout-contaminated air based on data in Table 16 of Barss (2000).

highly inaccurate indicator of skin dose, so beta-to-gamma ratios are not appropriate for such applications. As in the method used for surface deposition and exposure in air, the beta energy spectrum due to radioactive material on the surface of the skin can be determined as a function of time after detonation, and this allows beta doses to be directly calculated by using dose coefficients from Kocher and Eckerman (1987). The dose coefficients are based on radionuclides deposited on or near the skin surface, and they are nearly constant for average beta energies greater than 0.1 MeV. Hence, it is possible to estimate an average dose coefficient for skin of about 9 rem h^{-1} (beta plus gamma) per µCi cm^{-2} of skin, which includes a 30-35% overestimate due to the potential presence of an external backscatter surface (for example, a contaminated surface or tool) and a gamma contribution of about 5%. For contaminated gloves, a dose-reduction factor of 0.5 is assumed.

Barss (2000) recommends that levels of skin contamination be based on measurements when they are available and provides information to guide estimates of skin contamination based on measurements expressed in terms of dose or exposure rate. The report also indicates that the VARSKIN code (Durham, 1992) can be used for additional calculations of skin dose, presumably for source geometries where an assumption of uniform large-area contamination is inappropriate. Barss does not discuss contamination estimates that might be made for troops that are potentially contaminated by resuspended fallout while marching or performing other work, but there are some early articles on the subject (Schwendiman, 1958; Black, 1962).

IV.C ESTIMATION OF INTERNAL DOSE

IV.C.1 Introduction

This section discusses methods that have been used in the NTPR program to estimate internal doses to atomic veterans. In general, participants at the NTS, in the Pacific, or in Japan could have received internal doses as a result of intakes of radionuclides by inhalation, ingestion, or absorption through the skin or open wounds (see Section I.C.2.2.2). Intakes by inhalation are expected to be the most important for most participants, and the most important exposure scenarios usually involve inhalation of descending fallout or fallout that was deposited on the ground or other surfaces and then resuspended into the air.

Estimation of internal doses to atomic veterans is inherently more difficult than estimation of external doses. External doses usually can be estimated directly on the basis of measurements of external exposure with film badges worn by participants or measurements of external exposure rates at various locations and times in a participant's exposure environment with field instruments (see Sections IV.B.1 and IV.B.2). In general, however, internal dose cannot be measured directly but must be estimated based on models and other assumptions.

Ideally, estimates of internal dose should be based on relevant monitoring data that were obtained at the time of exposure or shortly thereafter. For example, intakes by inhalation can be monitored on the basis of measured concentrations of radionuclides in air during times of exposure, and intakes by any pathway can be monitored on the basis of measured activities of radionuclides excreted in urine or feces at known times after exposure. Given such data, internal doses can be estimated by using mathematical models that describe the behavior of radionuclides in the human body over time after an intake and the doses delivered to various organs and tissues.

In practice, however, suitable monitoring data to estimate intakes of radionuclides by atomic veterans generally were not obtained. The inhalation hazard posed by plutonium was a concern during the period of atomic testing, especially at the NTS (for example, see Dick and Baker, 1961, and Luna et al., 1969). There also were efforts at some tests to monitor intakes with urinanalysis (NRC, 1985b), and urinanalysis was used to estimate intakes of ^{131}I by participants who received very high doses from exposure to fallout on Rongerik Atoll in the Pacific (Goetz et al., 1987). However, in all dose reconstructions reviewed by the committee, no information was provided on airborne concentrations of radionuclides or contemporaneous measurements of radionuclides in urine or feces that could be used to estimate internal doses. Therefore, indirect methods based on other data are required to estimate intakes of radionuclides.

Section IV.C.2 describes methods and databases that have been used to estimate inhalation doses to atomic veterans. Since the middle 1980s, methods documented by Barrett et al. (1986) and incorporated into the FIIDOS computer code (Egbert et al., 1985) have been used; the committee did not review methods used earlier. The approach to addressing uncertainty in estimates of inhalation dose is described in Section IV.E.4. Section IV.C.3 discusses methods used to estimate ingestion doses to atomic veterans in rare cases in which ingestion is considered, and Section IV.C.4 discusses the possibility of internal doses due to absorption through the skin or open wounds. The committee's evaluation of methods used in the NTPR program to estimate internal doses to atomic veterans and to account for uncertainty in the estimates is presented in Section V.C. The so-called low-level internal dose screen, which was developed to assess the potential importance of inhalation exposures of participant groups (Barrett et al., 1986), and a bioassay program, which was recently undertaken in an effort to assess internal exposures to plutonium on the basis of urinanalysis, are discussed in Section VI.C and VI.D, respectively.

IV.C.2 Methods of Estimating Inhalation Dose

Inhalation is the only pathway of radionuclide intake normally considered in dose reconstructions for atomic veterans. Four basic scenarios of inhalation exposure have been defined and used in dose reconstructions (Barrett et al., 1986):

- Inhalation of radioactive material in fallout that was deposited on the ground or other surfaces and then resuspended into the air by mechanical or natural disturbances;
- Inhalation of neutron-induced radioactive material in soil that was lofted (suspended) into the air by mechanical or natural disturbances;
- Inhalation of radioactive material in descending fallout;
- Inhalation of radioactive material in an atmospheric cloud.

In any scenario, the dose from inhalation of a particular radionuclide is estimated by using the following equation in the notation of Barrett et al. (1986):

$$D = AA \times BR \times T \times DF, \qquad (IV.C-1)$$

where

D = equivalent dose to organ or tissue of concern (rem),[5]
AA = activity concentration of radionuclide in air (Ci m^{-3}),
BR = breathing rate (m^3 h^{-1}),

---

[5]In the NTPR program, the dosimetric quantity calculated in dose reconstructions is called "dose equivalent," but the term "equivalent dose" is used in this report (see Note on Units and Section I.C.2.3).

T = duration of exposure (h),
DF = equivalent dose to organ or tissue of concern per unit activity
         intake of the radionuclide inhaled (rem Ci^{-1}).[6]

Equation IV.C-1 gives the dose from inhalation of a single radionuclide. However, any scenario for inhalation exposure of atomic veterans involves intakes of mixtures of radionuclides, and the dose to an organ or tissue of concern is the sum of the doses from intakes of all of them. Thus, composite dose coefficients that apply to assumed mixtures (activity concentrations) of radionuclides are used in all dose reconstructions. Concentrations of radionuclides in air and the associated composite dose coefficients generally are time-dependent because of radioactive decay.

When a veteran does not file a claim for compensation for a specific disease (such as a cancer in a particular organ) but a dose reconstruction is requested, the quantity often calculated and reported to the veteran to represent inhalation dose is the effective dose equivalent, rather than the equivalent dose to a specific organ or tissue. The effective dose equivalent is a weighted average of equivalent doses to several organs and tissues that was developed for use in radiation protection (ICRP, 1977).

A breathing rate of 1.2 m^3 h^{-1} normally is assumed in estimating inhalation doses (Egbert et al., 1985). That value was recommended for use in assessing inhalation doses to average adult workers engaged in light activity (ICRP, 1975), and it represents the breathing rate during walking on a flat surface at about 3 mph (4.8 km h^{-1}) (TGLD, 1966). A higher breathing rate of 1.5 or 2 m^3 h^{-1} may be assumed if veterans engaged in moderate or heavy activity, respectively (Egbert et al., 1985).[7]

The following sections discuss, first, methods that have been used to estimate concentrations of radionuclides in air in the different scenarios of inhalation exposure listed above and, second, equivalent doses to specific organs or tissues per unit activity intake of inhaled radionuclides (inhalation dose coefficients).

IV.C.2.1 *Methods of Estimating Concentrations of Radionuclides in Air*

Concentrations of radionuclides in air generally were not measured at locations and times of exposure of atomic veterans. Therefore, indirect methods of estimating them must be used in dose reconstructions. Methods and data used in the NTPR program to estimate airborne concentrations of radionuclides in each

---

[6]In the NTPR program, the quantity *DF* is referred to as a "dose conversion factor." In the present report, however, this quantity is referred to as a "dose coefficient" to conform to the terminology currently used by the International Commission on Radiological Protection (ICRP, 1991a).

[7]On the basis of the committee's review of dose reconstructions, a breathing rate higher than 1.2 m^3 h^{-1} apparently is used only rarely.

of the four scenarios of inhalation exposure listed above are described in the following sections.

IV.C.2.1.1 Scenario involving inhalation of resuspended fallout

A four-step procedure is used to estimate concentrations of radionuclides in air due to resuspension of fallout particles that were deposited on the ground or other surfaces after a particular nuclear detonation.

First, the relative activities of radionuclides produced in a detonation are estimated on the basis of radiochemical data obtained from sampling of the atmospheric cloud soon after the detonation and calculations of the activities of different fission and activation products that result from an assumed number of fissions in the weapon type of concern. Radionuclides considered in the analysis are fission products, radionuclides produced by neutron activation of plutonium or uranium and other components of the weapon and its support systems, and plutonium or uranium that did not undergo fission or activation. Given the estimated relative activities of radionuclides at the time of detonation, the relative activities at later times, including activities of any radioactive decay products, are calculated on the basis of the half-lives of the radionuclides produced in the detonation and their decay products.

Second, an assumption is made about the relative activity concentrations of radionuclides in fallout deposited on the ground or other surfaces compared with the estimated relative activities in the atmospheric cloud. In all shots, chemical and physical separation of different radionuclides, a process referred to as fractionation and discussed below, is assumed not to occur before deposition, except that noble-gas radionuclides (isotopes of krypton and xenon) in the cloud are removed from fallout. Thus, except for the effects of radioactive decay and the exclusion of noble gases, the relative activity concentrations of radionuclides in fallout are assumed to be the same as the estimated relative activities in the atmospheric cloud.[8]

Third, the estimated relative activity concentrations of radionuclides in fallout deposited on the ground or other surface are renormalized (scaled) to obtain an estimate of the absolute activity concentrations (Ci m^{-2}). Renormalization is based either on measurements of total photon exposure in air above the surface in roentgens (R) with film badges worn by participants who were present in the fallout field at known times and locations or on measurements of photon exposure rates (R h^{-1}) at various locations and times with field instruments, combined with calculations of the exposure rate in air per unit activity concentration of each

---

[8]The FIIDOS computer code (Egbert et al., 1985) includes an option to account for fractionation in estimating concentrations of radionuclides in deposited fallout on the basis of data provided by the analyst. However, the committee found no evidence that an assumption of fractionation in fallout, other than removal of noble gases, has been used in any dose reconstructions.

radionuclide on the surface (R h^{-1} per Ci m^{-2}). That is, estimated relative activity concentrations of radionuclides in deposited fallout obtained in the first two steps are adjusted to give a calculated total exposure or exposure rate above ground that matches measurements obtained by using film badges or field instruments, taking into account radioactive decay and the known times of measurement and exposure. Mathematically, the activity concentration on the surface is $(SA/I) \times I$, where I is the exposure rate and SA/I is the reciprocal of the calculated exposure rate per unit concentration on the surface. Exposure rates per unit concentration of radionuclides on the surface are calculated on the basis of known energies and intensities of photons emitted by the radionuclides assumed to be present in fallout, an assumption that radionuclides are distributed uniformly on a surface of infinite extent, and theoretical considerations of photon transport from the source region on the surface to the assumed height of a film badge or field instrument above the surface and the resulting exposure in air, taking into account the shielding effect of ground roughness.[9] An important condition in applying this method is that exposures measured with film badges or field instruments must have been due primarily to deposited fallout.

Fourth, the activity concentrations of radionuclides in air resulting from resuspension of deposited fallout are estimated by using a simple resuspension-factor model given by

$$AA = SA \times K, \tag{IV.C-2}$$

where AA is the activity concentration of a radionuclide in air (Ci m^{-3}); SA is the activity concentration on the ground or other surface (Ci m^{-2}), as estimated in the first three steps described above; and K is the resuspension factor (m^{-1}). The resuspension factor is assumed to be the same for all radionuclides on a surface, and it is estimated on the basis of relevant experimental studies. Resuspension factors that are used in different exposure scenarios are described later.

Thus, the equivalent dose to an organ or tissue from inhalation of resuspended fallout, D, is represented by:

$$D = I \times (SA/I) \times K \times BR \times T \times DF, \tag{IV.C-3}$$

where

I = measured photon exposure rate in fallout field (R h^{-1}), adjusted for radioactive decay between time of measurement and time of exposure;

---

[9]To account for ground roughness, which typically reduces exposure rates from fallout deposited on the ground by a factor of 0.7 compared with an unshielded plane source, exposure rates are calculated by assuming that the source region consists of two plane sources containing equal concentrations of radionuclides, one plane source located at a depth of 0.25 cm in soil and the other at a depth of 0.75 cm (Egbert et al., 1985).

SA/I = reciprocal of calculated exposure rate per unit activity
concentration of radionuclides on the ground or other surface (Ci
m^{-2} per R h^{-1});

K = resuspension factor (m^{-1});

BR = breathing rate (m^3 h^{-1});

T = duration of exposure (h); and

DF = inhalation dose coefficient (rem Ci^{-1}).

As in Equation IV.C-1, calculation of inhalation dose is more complicated than indicated by the representation in Equation IV.C-3 because the quantities SA/I and DF are estimated on the basis of an assumed mixture of radionuclides in fallout and I, SA/I, and DF are time-dependent. The dependence of the photon exposure rate (I) on time usually is assumed to be $t^{-1.2}$, where t is the time after detonation in hours.

In some dose reconstructions for participants at a particular shot, estimates of inhalation dose due to resuspension of deposited fallout also take into account the presence of fallout that was deposited in the same area after previous shots. Inhalation doses due to resuspension of previously deposited fallout are estimated with the same methods described above, with radioactive decay since the times of the previous shots taken into account. Dose reconstructions for participants on residence islands in the Pacific consider resuspension of all fallout that was present during the time of residence, but only during times when the measured exposure rate exceeded background. At the NTS, exposures of participants in areas where fallout from previous shots occurred are often a concern. Shots at the NTS at which inhalation of resuspended fallout from previous shots has been taken into account in dose reconstructions (Barrett et al., 1986) are listed in Table IV.C.1. In most of these cases, the previous shot occurred relatively recently in the same test series. Fallout that occurred after shots in previous test series usually is not taken into account, because it is assumed that the resuspension factor would have diminished by at least a factor of 10 through weathering, including penetration into surface soil and attachment to large soil particles (Barrett et al., 1986).

To recapitulate, airborne concentrations of radionuclides due to resuspension of previously deposited fallout are estimated in the NTPR program in the following way. Estimates of relative activities of radionuclides produced in a nuclear detonation combined with an assumption of no fractionation in fallout, except for removal of noble gases, are used to estimate relative activities of radionuclides in fallout deposited on the ground or other surfaces. Activity concentrations of radionuclides in deposited fallout (Ci m^{-2}) are then estimated by using measured photon exposures or exposure rates at the locations of fallout and calculations of the exposure rate per unit activity concentration of each radionuclide on the surface, which is assumed to be uniformly contaminated and infinite in extent. Finally, activity concentrations of radionuclides in air (Ci m^{-3}) due to resuspension are estimated by using the estimated concentrations in deposited fallout

TABLE IV.C.1   Shots at Nevada Test Site Where Inhalation of Resuspended Fallout from Previous Shots Is Taken into Account in Dose Reconstructions[a]

Operation	Shot of concern in dose reconstruction (date)	Previous shot accounted for in dose reconstruction (date)
BUSTER-JANGLE	UNCLE (11/29/51)	SUGAR (11/19/51)
TUMBLER-SNAPPER	FOX (05/25/52)	EASY (05/07/52)
UPSHOT-KNOTHOLE	RAY (04/11/53)	NANCY (03/24/53)
	BADGER (04/18/53)	NANCY (03/24/53)
	HARRY (05/19/53)	BADGER (04/18/53), SIMON (04/25/53)
	CLIMAX (06/04/53)	BADGER (04/18/53)
TEAPOT	HORNET (03/12/55)	MOTH (02/22/55)
	APPLE I (03/29/55)	TURK (03/07/55)
	POST (04/09/55)	TESLA (03/01/55)
PLUMBBOB	LASSEN (06/05/57)	BOLTZMANN (05/28/57)
	WILSON (06/18/57)	BOLTZMANN (05/28/57)
	OWENS (07/25/57)	BOLTZMANN (05/28/57)
	SMOKY (08/31/57)	BOLTZMANN (05/28/57), DIABLO (07/15/57), SHASTA (08/18/57)
	WHEELER (09/06/57)	BOLTZMANN (05/28/57), SMOKY (08/31/57)
	CHARLESTON (09/28/57)	BOLTZMANN (05/28/57), SMOKY (08/31/57)
	MORGAN (10/07/57)	SMOKY (08/31/57)[b]

[a]See Section 2.2 of Barrett et al. (1986). In addition, impact of fallout from TEAPOT Shot TESLA at location of PLUMBBOB Shots LASSEN, WILSON, HOOD, OWENS, WHEELER, CHARLESTON, and MORGAN is taken into account, except resuspension factor applied to TESLA fallout 2 years later is factor of 10 lower than value applied in other cases.
[b]Impact of fallout from PLUMBBOB Shot BOLTZMANN on location of Shot MORGAN—which was at same ground zero as earlier PLUMBBOB Shots LASSEN, WILSON, OWENS, WHEELER, and CHARLESTON—apparently was considered to be insignificant.

and an assumed resuspension factor. Radioactive decay is taken into account from the time of detonation to the time exposure is assumed to occur. That method is applied in all cases of inhalation of resuspended fallout that was deposited on the ground surface, on exterior or interior surfaces of ships, or on surfaces of airplanes used in cloud sampling.

Further discussion that focuses on the intention of the NTPR program that the method should result in overestimates of airborne concentrations of radionuclides in resuspended fallout is presented in Section IV.C.2.1.7.

IV.C.2.1.2 *Fractionation of radionuclides*

The term fractionation refers to the chemical and physical separation of radionuclides produced in a detonation. As noted in the previous section, an

assumption about fractionation is needed to estimate concentrations of radionuclides in fallout deposited on the ground or other surfaces on the basis of estimated relative activities of radionuclides in the atmosphere after a detonation.
The following description of fractionation and its effect on the radionuclide
composition of fallout is based mainly on a discussion by Hicks (1982).

Chemical separation of radionuclides in the atmosphere occurs in the first
few minutes after a detonation as a result of differences in the boiling points of
the various elements that make up the radioactive materials. Initially, all materials in the fireball are vaporized, because of the extreme temperatures. As the
fireball cools to about 3000°C, iron oxides and soil materials that make up most
of the mass in the fireball form liquid droplets in which so-called refractory
elements are dissolved. Refractory elements have relatively high boiling points,
comparable to or greater than the melting points of materials that form the liquid
droplets; examples are barium, cerium, uranium, and plutonium. As the liquid
droplets continue to cool, they solidify at about 1500°C, and over the next few
minutes, as the solid particles cool to the ambient temperature of about 50°C,
volatile elements that have boiling points lower than the melting points of the
particles condense on particle surfaces. Examples of volatile elements are strontium, iodine, and cesium. As a result of the chemical separation of radionuclides
between those which are dispersed mainly throughout the volume of solid particles (refractory elements) and those which mainly condense on the surface
(volatile elements), and taking into account that the surface area-to-volume ratio
of a particle is about $1/r$ (where r is the radius), smaller particles contain a
relatively high proportion of volatile radionuclides, compared with the initial
composition of the atmospheric cloud, and larger particles contain a relatively
high proportion of refractory radionuclides.

Physical separation of radionuclides then occurs as a result of differences in
the fall velocities of small and large particles in the atmosphere. Larger particles
fall to Earth more rapidly than small particles, and this results in a physical
separation by particle size along the path of travel of the atmospheric cloud.
Thus, fallout close to ground zero consists mostly of relatively large particles that
contain a higher proportion of refractory radionuclides, whereas fallout at more
distant locations consists mostly of relatively small particles that contain a higher
proportion of volatile radionuclides.

Another factor that contributes to fractionation of radionuclides in fallout is
the presence of the noble gases krypton and xenon among the fission products.
Separation of noble gases from other materials begins as soon as liquid droplets
form and continues as the droplets solidify and begin to fall to Earth. As noted
previously, this type of fractionation is taken into account by the NTPR program
in estimating concentrations of radionuclides in fallout deposited on the ground
or other surfaces at the NTS and in the Pacific. In addition, differences in half-
lives of noble-gas radionuclides can result in fractionation of chemically similar
elements. A case in point involves the important fission products ^{90}Sr and ^{137}Cs,

both of which are volatile. Those radionuclides are produced mainly by decay of shorter-lived fission products in the mass 90 and 137 decay chains, which include noble-gas radionuclides with substantially different half-lives (^{90}Kr with a half-life of 32 s, and ^{137}Xe with a half-life of 3.84 min). As a result, condensation of ^{90}Sr and ^{137}Cs on surfaces of solid particles tends to occur at different times, and fractionation occurs as particles containing ^{90}Sr are physically separated from ^{137}Xe before ^{137}Cs is produced in the atmospheric cloud. That effect often results in a severe depletion of ^{137}Cs, compared with the amounts of ^{90}Sr and other volatile radionuclides, in fallout close to ground zero. Fractionation of ^{90}Sr and ^{137}Cs also can be important in detonations just below the ground surface.[10] In that type of shot, longer-lived ^{137}Xe can vent to the atmosphere to a much greater extent than ^{90}Kr, and this results in increased ^{137}Cs in fallout near ground zero compared with ^{90}Sr, in contrast to the depletion of ^{137}Cs relative to ^{90}Sr that often occurs in aboveground shots.

As noted previously, in estimating inhalation doses to atomic veterans, the NTPR program generally assumes no fractionation of radionuclides in fallout except for removal of noble gases. On the basis of the foregoing discussions, that assumption normally should result in overestimates of the concentrations of volatile radionuclides deposited on the ground or other surfaces but underestimates of the concentrations of refractory radionuclides, because participants at the NTS and in the Pacific usually were exposed at locations where fallout was dominated by larger particles that contained a high proportion of refractory radionuclides. The method also does not account for fractionation of the volatile radionuclides ^{90}Sr and ^{137}Cs, which is due to the difference in half-lives of their noble-gas precursors.

IV.C.2.1.3 *Resuspension of radionuclides in deposited fallout*

The resuspension factor that should be applied to fallout deposited on the ground or other surfaces is recognized to be a highly uncertain parameter that depends on the nature of the activity causing the resuspension (Egbert et al., 1985; Barrett et al., 1986). On the basis of a review of available data and the use of subjective judgment, different resuspension factors are used in the NTPR program to estimate airborne concentrations of radionuclides in various exposure scenarios. Resuspension factors commonly used in dose reconstructions at the NTS (Barrett et al., 1986) and on ships in the Pacific (Phillips et al., 1985) are summarized in Tables IV.C.2 and IV.C.3, respectively. A resuspension factor of 10^{-5} m^{-1} often is used in estimating inhalation exposures on residence islands in the Pacific, with the additional assumption that inhalation of resuspended fallout occurs only during the fraction of the time spent outdoors but not during time

---

[10]An example of such a near-surface detonation is Operation BUSTER-JANGLE, Shot SUGAR.

spent indoors. In cases of exposure in an airplane whose interior was contaminated during cloud sampling, a resuspension factor of 10^{-4} m^{-1} is used.

Resuspension factors different from those summarized in Tables IV.C.2 and IV.C.3 and described above have been used in some cases. At Operation UPSHOT-KNOTHOLE, Shot GRABLE, for example, a higher resuspension factor of 10^{-2} m^{-1} was assumed in dose reconstructions for participant groups that encountered a severe dust storm in an area where neutron activation of soil occurred (Goetz et al., 1981; Edwards et al., 1985). But the resuspension factor often is reduced by a factor of 10, compared with the standard assumption of 10^{-5} m^{-1} used in many dose reconstructions, or to zero in cases of exposure to aged fallout that was deposited well before the time of exposure; this is especially the case on residence islands in the Pacific.

TABLE IV.C.2 Resuspension Factors Normally Assumed for Various Activities of Atomic Veterans in Contaminated Areas at NTS[a]

Category	Activity	Resuspension factor (m^{-1})
Observers and maneuver troops	Touring display area	10^{-5}
Maneuver troops	Maneuvers involving helicopter landings and takeoffs	10^{-3}
	Assaults or marches behind armored vehicles	10^{-3}
	Crawling through open terrain	10^{-4}
	Digging foxholes, and such	10^{-4}
	Ground assaults (no vehicle)	10^{-5}
	Trucking	10^{-5}
Project troops	Digging out buried instrumentation and equipment	10^{-4}
	Equipment and data recovery	$10^{-4} - 10^{-5}$
	Decontamination projects (bulldozing, and such)	10^{-4}
	Visit project area (on foot or in vehicle)	10^{-5}
Support troops		
– Engineers and ordinance	Digging trenches, installing and dismantling displays	10^{-4}
– Communications	Laying wire (communication network)	10^{-4}
– Decontamination	Equipment and personnel decontamination	10^{-4}
– Transportation	Trucking or bussing	10^{-5}
– MPs	Traffic control, security sweeps	10^{-5}
– Radiation safety	Surveying area on foot or from vehicle	10^{-5}

[a]See Table 5 of Barrett et al. (1986).

TABLE IV.C.3   Resuspension Factors Normally Assumed for Various
Activities of Participants on Contaminated Ships in Pacific[a]

Location	Activity	Resuspension factor (m^{-1})
Bikini	Inspection and repair, below decks	10^{-5}
	Inspection and repair, topside	10^{-6}
	Decontamination, below decks	10^{-6}
	Decontamination, topside	10^{-6}
Kwajalein	Ammunition unloading, below decks	10^{-4}
	Ammunition loading, topside	10^{-5}
Kwajalein	Maintenance and security, below decks	10^{-5}
	Maintenance and security, topside	10^{-6}
Naval shipyards	Inspection and maintenance, below decks	10^{-5}
	Inspection and maintenance, topside	10^{-6}

[a]See Table 2 of Phillips et al. (1985).

Thus, in general, judgment has been applied by the NTPR program in estimating resuspension factors used in dose reconstructions for particular exposure scenarios that involve previously deposited fallout. As discussed further in Section IV.C.2.1.7, the intent of the NTPR program has been to select resuspension factors that result in overestimates of airborne concentrations of radionuclides in resuspended fallout to which participants at the NTS and in the Pacific were exposed.

IV.C.2.1.4 *Scenario involving inhalation of suspended neutron activation products in soil*

The method of estimating airborne concentrations of radionuclides due to suspension of radioactive material that was produced by neutron activation in soil, which occurred only at the NTS, is essentially the same as the method used in the scenario involving resuspension of deposited fallout described above (Barrett et al., 1986). The method is summarized as follows.

Relative activities of activation products in soil are estimated on the basis of field measurements of known activation products, known elemental compositions of soil, calculations of neutron transport in air and soil, and calculations of neutron capture in nuclei of stable elements in soil. Activities of radionuclides per unit volume of surface soil (Ci m^{-3}) then are estimated by renormalization (scaling) of the estimated relative activities to give a calculated photon exposure in air above ground or exposure rate at a particular time that matches available measurements with film badges or field instruments, taking radioactive decay into account. Finally, a resuspension factor is applied to the estimated concentrations of radionuclides in surface soil to obtain an estimate of the concentrations in air. Radionuclide activities per unit volume of surface soil are converted to equivalent surface concentrations (Ci m^{-2}) by assuming that the top 1 cm of soil can be suspended in the air. As in the scenario involving resuspension of deposited fallout, an important condition in applying this method is that the film-badge or instrument readings must be due primarily to activation products in soil.

Resuspension factors that are applied to activation products in surface soil at the NTS are the same as those that are applied to deposited fallout (see Table IV.C.2). That is, a resuspension factor of 10^{-5} m^{-1} often is assumed, and higher values are assumed when suspension is caused by an unusually vigorous disturbance of surface soil. Again, the intent of the NTPR program has been to select resuspension factors that result in overestimates of airborne concentrations of activation products in soil to which participants at the NTS were exposed.

IV.C.2.1.5 *Scenario involving inhalation of descending fallout*

Airborne concentrations of radionuclides in descending fallout are estimated on the basis of estimates of concentrations on the ground or other surfaces that resulted from deposition of the fallout. The basic concept underlying the method used in the NTPR program is that activities of radionuclides per unit area on a surface resulted from deposition of the activities per unit volume in air averaged over the height of the atmospheric cloud produced by the detonation.

With reference to Equations IV.C-1 and IV.C-3, the concentration of a radionuclide in air (AA) can be represented as $I \times (SA/I) \times (AA/SA)$, where I is the maximum photon exposure rate in air above the ground or other surface due to the deposited fallout at the time after detonation, t (in hours), when fallout is complete, and SA/I is the reciprocal of the calculated exposure rate per unit activity concentration on the surface (SA) at time t. Again, SA/I is based on an assumption that the surface is uniformly contaminated and infinite in extent, and the calculation accounts for the shielding effect of ground roughness. Because the airborne concentration (AA) persists for a time T during which fallout is assumed to descend with a velocity V at the location of concern, the activity concentration on the surface at the time fallout ends can be expressed as $SA = AA \times V \times T$.[11] Thus, if exposure is assumed to occur throughout the duration of descending fallout, Equation IV.C-1 becomes

$$D = I \times (SA/I) \times (1/V) \times BR \times DF, \qquad \text{(IV.C-4)}$$

where the duration of exposure, T, which corresponds to the duration of descending fallout at the location of concern, is eliminated.

---

[11]AA and V are time-dependent, so the deposition rate, $AA \times V$, also is time-dependent and the total deposition generally is given by the time integral of the deposition rate over the period of descent. However, this simplified expression for surface activity gives the correct total deposition if AA and V represent average values over the entire period of descent, T.

The deposition velocity, V, in Equation IV.C-4 is based on an assumption that fallout particles descended from a height of around 10 km (10^4 m). Thus, $V \sim (10^4 \text{ m})/t$, where t again is the time (in hours) after a detonation when deposition of fallout is complete. The dose due to inhalation of descending fallout then is estimated as

$$D = I \times (SA/I) \times (10^{-4} \text{ m}^{-1}) \times t \times BR \times DF. \qquad \text{(IV.C-5)}$$

Equation IV.C-5 is similar to Equation IV.C-3, which is used to estimate the dose from inhalation of resuspended fallout. The differences are that the term (10^{-4} m^{-1}) is an "effective" resuspension factor, rather than an actual value, and t is the time (in hours) after a detonation when fallout deposition is complete and the maximum exposure rate occurs, rather than the duration of exposure. Again, the quantities I, SA/I, and DF are time-dependent, and SA/I and DF are calculated based on an assumed mixture of radionuclides in fallout.

Thus, activities of radionuclides per unit volume of air in descending fallout are estimated on the basis of estimated concentrations on the ground or other surface that resulted from fallout, given by $I \times (SA/I)$, and an effective resuspension factor given by the reciprocal of the assumed height of the atmospheric cloud. In essence, resuspension is deposition reversed in time, and if all deposited material is assumed to be resuspended, activity concentrations in air are given by the concentrations on the surface divided by the assumed height of the resuspended plume. This roundabout method is necessitated by the lack of data on concentrations of radionuclides in descending fallout.

In this scenario, concentrations of radionuclides deposited on the ground generally are based on measurements of external photon exposure rates at particular locations and times with field instruments rather than measurements of total external exposure with film badges worn by participants who were exposed to descending fallout (Barrett et al., 1986). The use of film-badge readings is inappropriate in this scenario because exposures generally continued for considerable periods of time after deposition of the fallout ceased, and the readings would lead to unreasonable overestimates of airborne concentrations of radionuclides in descending fallout. An important condition in applying the method is that exposure rates measured with field instruments must have been due primarily to deposited, rather than descending, fallout. Measurements of maximum exposure rates at times that descent of fallout is completed are the most appropriate.

In many dose reconstructions, it is assumed that intakes of fully inhalable fallout particles occurred only at times, t, later than 10 h after a detonation. At such times, calculations based on assumed mixtures of radionuclides in fallout indicate that the quantity SA/I is roughly constant, with a value of 0.16 Ci m^{-2} per R h^{-1}. If the usual breathing rate of 1.2 m^3 h^{-1} is assumed, the dose due to inhalation of descending fallout is calculated in these cases as

$$D = (2.0 \times 10^{-5}) \times I \times t \times DF. \qquad \text{(IV.C-6)}$$

In contrast to the methods of estimating airborne concentrations of radionuclides in resuspended fallout or suspended activation products in soil described previously, the assumption of an effective resuspension factor of 10^{-4} m^{-1} in this scenario is not intended to result in overestimates of airborne concentrations of radionuclides in descending fallout. Rather, the NTPR program assumes that the presumption of exposure throughout the entire period of deposition and use of the maximum photon exposure rate at the time when fallout is complete will result in overestimates of airborne concentrations (Barrett et al., 1986).

IV.C.2.1.6 *Scenario involving inhalation in an atmospheric cloud*

The scenario involving inhalation of radionuclides in an atmospheric cloud is applied only for participants who flew through a cloud in an airplane or helicopter. In this scenario, the general expression for inhalation dose given in Equation IV.C-1 is written as

$$D = I \times (AA/I) \times BR \times T \times DF, \qquad \text{(IV.C-7)}$$

where AA/I is the reciprocal of the calculated exposure rate per unit activity concentration of a radionuclide in air. Values of AA/I generally are different from calculated values of SA/I used to estimate inhalation doses from resuspended or descending fallout, and they are calculated based on an assumption that the atmospheric cloud is uniformly contaminated and infinite in extent. An effective resuspension factor is not used in this scenario, because the quantity I is a measured photon exposure rate due to airborne radionuclides rather than radionuclides deposited on a surface. Again, the quantities I, AA/I, and DF are time-dependent, and AA/I and DF are based on an assumed mixture of radionuclides in air. The exposure rate, I, generally can be based on readings of film badges.

Calculations based on assumed mixtures of radionuclides in an atmospheric cloud indicate that the quantity AA/I is only weakly time-dependent and is inversely proportional to the average energy of emitted photons, E. Thus, in all dose reconstructions, AA/I is assumed to be given by $(5 \times 10^{-4})/E$, where E is in MeV per disintegration. If a breathing rate of 1.2 m^3 h^{-1} is again assumed, the dose due to inhalation in an atmospheric cloud is calculated in all cases as

$$D = I \times (6 \times 10^{-4}) \times T \times DF/E. \qquad \text{(IV.C-8)}$$

When exposure of a person to an atmospheric cloud while the person was in an aircraft was inadvertent, inhalation doses generally are estimated based on concentrations of radionuclides in air obtained as described above and an assumption that no respiratory protection was used. However, in cases of planned exposure, such as occurred during cloud sampling, it generally is assumed that respiratory protection was used and furthermore that the equipment provided complete protection. Thus, the inhalation dose in such cases is assumed to be zero.

The method of estimating concentrations of radionuclides in an atmospheric cloud described above does not involve assumptions that intentionally result in overestimates of inhalation exposure. However, the NTPR program believes that use of film-badge readings in the method tends to result in some overestimation of inhalation exposure; the reason is that aircraft presumably would have avoided regions of highest activity concentrations in air, but these regions nonetheless contribute to external exposures measured in regions of lower activity concentrations where inhalation exposure occurred (Barrett et al., 1986).

IV.C.2.1.7 *Summary of methods of estimating concentrations of radionuclides in air*

In the various scenarios for inhalation exposure of atomic veterans, estimates of concentrations of radionuclides in air are based on the following data and assumptions:

• Data on relative activities of different radionuclides produced in a particular nuclear detonation (all scenarios).
• An assumption of no fractionation of radionuclides, except for removal of noble gases in fallout (all scenarios).
• Data on external photon exposures or exposure rates due to radionuclides on the ground or other surfaces or in the air, as obtained from film-badge readings or field instruments, and calculations of exposure rates per unit concentration of radionuclides on a surface or in the air (all scenarios).
• Assumptions about the amounts of radionuclides in fallout deposited on the ground or other surfaces or amounts of activation products in soil that were resuspended or suspended in the air, with resuspension also used to mimic deposition in the scenario involving inhalation of descending fallout (all scenarios except inhalation in an atmospheric cloud).

Thus, the data that underlie all calculations of inhalation dose are estimates of the amounts of different radionuclides in an atmospheric cloud or in soil that were produced in a detonation and measured external photon exposures or exposure rates due to radionuclides deposited on the ground or other surfaces (fallout), distributed over a depth of surface soil (neutron-activation products), or distributed in the cloud. Especially in scenarios involving inhalation of resuspended or descending fallout, a key assumption in the methods used to calculate inhalation dose is that there is no fractionation of radionuclides other than removal of noble gases; these two scenarios are the most important in inhalation exposures of many atomic veterans.

As emphasized at the beginning of Chapter IV, the goal of the NTPR program is to obtain estimates of dose to atomic veterans that are upper bounds (at least 95% confidence limits) of possible doses; that is, the estimated doses are intended to be "high-sided." In attempting to meet that goal when estimating inhalation doses, the NTPR program relies to a large extent on an assumption that resuspension factors used in some scenarios result in substantial overestimates of airborne concentrations of radionuclides to which most participants were exposed. That is, data on relative activities of radionuclides produced in a particular detonation, the assumption of no fractionation of radionuclides except for removal of noble gases, data on external exposures or exposure rates obtained with film badges or field instruments, calculations of exposure rates per unit concentration of radionuclides on a surface or in the air, and assumed breathing rates used in dose reconstructions usually are not intentionally biased in an effort to yield overestimates of inhalation exposure. But assumptions about resuspension often are intended to be "high-sided" to an extent sufficient to compensate for possible errors and uncertainties in all other factors that enter into a calculation of inhalation dose, including the inhalation dose coefficients discussed in the following section.

IV.C.2.2 *Dose Coefficients for Inhalation of Radionuclides*

Inhalation dose coefficients (equivalent doses to specific organs or tissues per unit activity of radionuclides inhaled) used in all dose reconstructions are 50-year committed doses. That is, they represent the total dose received over a period of 50 years after an acute intake. The use of committed doses takes into account that an acute intake of a radionuclide may result in a dose that is received over many years after intake, and the 50-year period for calculating committed doses represents an average life expectancy of a young adult. For some radionuclides, the entire 50-year committed dose is received within a short time after an intake. For example, the committed dose due to intakes of ^{131}I, with a half-life of 8 days, is received within a few months. However, in cases of intakes of long-lived radionuclides that are tenaciously retained in the body, such as plutonium, the dose to an organ or tissue of concern can be protracted throughout the rest of life, with only a small fraction of the committed dose received in each year. In all dose reconstructions, 50-year committed equivalent doses are assigned to the year in which exposure occurred regardless of the dependence of the total dose received on time after an intake (the time dependence of the dose rate).

All inhalation dose coefficients used in dose reconstructions are calculated with two kinds of models:

• Biokinetic models that describe the deposition, retention, translocation, and absorption of inhaled radionuclides in the respiratory tract or ingested radionuclides in the gastrointestinal (GI) tract and the transfer, deposition, and retention of absorbed radionuclides in different organs and tissues of the body.
• Dosimetric models that are used to calculate equivalent doses to different organs or tissues (the target organs) per decay of a radionuclide in each source organ (site of deposition or transit through the body).

Inhalation dose coefficients used in all dose reconstructions were calculated based essentially on dosimetric and biokinetic models described in ICRP *Publication 30* (ICRP, 1979a; 1979b; 1980a; 1981a; 1981b; 1982a; 1982b). In some dose reconstructions, inhalation dose coefficients are those given in the ICRP *Publication 30* reports; these dose coefficients also are given by Eckerman et al. (1988). In many dose reconstructions, however, inhalation dose coefficients were obtained from Oak Ridge National Laboratory (ORNL) reports (Killough et al., 1978a; Dunning et al., 1979). The models used to calculate inhalation dose coefficients in the ORNL reports (Killough et al., 1978b) are in many respects the same as the models described in ICRP *Publication 30*. In particular, the model of the behavior of inhaled radionuclides in the respiratory tract is the same.

However, there are differences between the two models. For example, the behavior of radioactive decay products in the body is treated differently. In the ORNL models (Killough et al., 1978b), radioactive decay products that are produced in the body after intakes of a parent radionuclide are assumed to behave according to the biokinetic model for the decay product, whereas in the ICRP models (ICRP, 1979a), decay products that are produced in the body are assumed to behave according to the biokinetic model for the parent. Also, for some radionuclides, there are differences in parameter values used in the biokinetic models. For example, the fraction of ingested plutonium absorbed in the GI tract used in the ORNL calculations (Dunning et al., 1979) is a factor of 3 higher than the value assumed by ICRP (1979a). The same difference occurs in dose coefficients for organs or tissues other than those in the GI tract when plutonium is directly ingested or inhaled plutonium is swallowed.

The committee is not aware of any written policy in the NTPR program to assist an analyst in choosing between ICRP and ORNL dose coefficients for use in dose reconstructions for atomic veterans; both sets of data are included in the FIIDOS computer code (Egbert et al., 1985) and are readily available. Nor did the committee find explanations for choosing one data set over the other in dose reconstructions for individuals or participant groups.

In selecting inhalation dose coefficients from the ORNL or ICRP reports, several assumptions about conditions of exposure and other factors are required, including the size of inhaled particles, chemical form of inhaled radionuclides, and biological effectiveness of alpha particles relative to that of photons and electrons. Assumptions used in dose reconstructions for atomic veterans are described below.

IV.C.2.2.1 *Size of inhaled particles*

Inhalation dose coefficients used in dose reconstructions for atomic veterans assume that inhaled radionuclides are attached to particles. The size of inhaled particles is specified by a quantity called the activity median aerodynamic diameter (AMAD). In any inhaled materials, there is a distribution of particle sizes (usually assumed to be lognormal), and AMAD is defined as the diameter in an aerodynamic particle-size distribution for which the total activities attached to particles above and below this size are equal (NCRP, 1997). The particle size (AMAD) used in calculating inhalation dose is taken to be the diameter of a sphere of unit density that has the same terminal settling velocity in air as that of the particle whose activity is the median for the distribution of particle sizes (ICRP, 1979a; NCRP, 1997).

In many dose reconstructions, a particle size (AMAD) of 1 μm is assumed, on the basis of a standard assumption recommended for use in radiation protection in ICRP *Publication 30* (ICRP, 1979a). In some dose reconstructions, however, inhalation dose coefficients for large, essentially nonrespirable particles based on an assumed AMAD of 20 μm are used, especially when this choice results in higher doses to the organ or tissue in which a cancer in an atomic veteran occurred. Particle size affects inhalation dose coefficients in the following way.

The pattern of deposition of inhaled particles in different regions of the respiratory tract depends on particle size. For an AMAD of 1 μm, the lung model developed by ICRP (1979a) assumes that 30% of the inhaled activity is deposited in the nose and throat, 8% in the trachea and bronchial tree, and 25% in the pulmonary lung; the remainder is assumed to be exhaled without deposition in the respiratory tract. For an AMAD of 20 μm, however, nearly all the inhaled activity is assumed to be deposited in the nose and throat, with little activity deposited in other regions of the respiratory tract or exhaled (ICRP, 1979a). Furthermore, when radionuclides are deposited in the nose and throat, rates of absorption into blood or transfer to the GI tract by swallowing are assumed to be substantially higher than when deposition in the pulmonary region occurs (ICRP, 1979a). Thus, for example, the dose to the thyroid from inhalation of ^{131}I attached to 20-μm particles is higher than when an AMAD of 1 μm is assumed because, first, a larger fraction of the inhaled material is assumed to be deposited in the respiratory tract and then absorbed into blood from the nose and throat or GI tract and, second, a substantial fraction of the smaller particles is deposited in the pulmonary lung, where the activity is reduced by decay before absorption into blood occurs; the effect of the larger deposition fraction is the more important. Similarly, the dose to organs of the GI tract is higher when an AMAD of 20 μm is assumed, because a larger fraction of the inhaled activity is assumed to be transferred to the GI tract by swallowing (ICRP, 1979a). When radionuclides attached to large particles are moderately or highly insoluble, for example, nearly all the activity is assumed to be transferred from the nose and throat to the GI tract, but the fraction of such transfers is substantially lower when an AMAD of 1 μm is assumed.

On the basis of considerations of the dependence of inhalation doses on particle size, an AMAD of 20 μm is assumed in some dose reconstructions, especially when the organ or tissue of concern is the thyroid—in which case the dose is due mainly to intakes of ^{131}I—or one of the organs of the GI tract. That assumption may be used in estimating doses due to inhalation of descending or resuspended fallout, based on the consideration that fallout to which participants were exposed consisted mainly of large particles. However, inhalation dose coefficients for large particles are not used when the lung is the organ of concern, even when inhalation of large particles may be more important, because an assumed AMAD of 1 µm always results in a higher estimate of dose to the lung, due to the greater deposition of smaller particles in this part of the respiratory tract. In addition, an AMAD of 1 µm normally is used in estimating inhalation doses due to suspension of activation products in soil, regardless of the organ or tissue of concern, on the basis of an assumption that resuspended soil particles were relatively small. The lower particle size is sometimes used in cases of resuspension of deposited fallout, especially when fallout was deposited well before the time of exposure, on the basis of an assumption that large particles in fallout were reduced in size by weathering.

Thus, with the exceptions noted above, the standard practice in the NTPR program is to choose the particle size that gives the higher estimate of inhalation dose to the organ or tissue of concern (Schaeffer, 2002b), and dose coefficients for particle sizes (AMAD) of 1 and 20 µm are included in the FIIDOS code (Egbert et al., 1985) for this purpose.[12]

IV.C.2.2.2 *Chemical form of inhaled radionuclides*

Once inhaled radionuclides are deposited in different regions of the respiratory tract, rates of transfer to blood or to the GI tract by swallowing depend on the assumed chemical form of the radionuclides. Different chemical forms have different solubilities, and materials that have a higher solubility are transferred more rapidly than materials that have a lower solubility. In the lung model used in the NTPR program (ICRP, 1979a), three solubility (lung clearance) classes of inhaled radionuclides in particulate form are defined: highly soluble materials that are assumed to be cleared from the respiratory tract in a matter of days (Class D), moderately insoluble materials that are cleared in a matter of weeks (Class W), and insoluble materials that are cleared in a matter of years (Class Y).[13]

---

[12]The committee notes, however, that the standard practice has been not followed in all dose reconstructions. For example, the committee found cases of dose reconstructions performed since 1991 in which a particle size of 1 µm was assumed in estimating inhalation dose to the colon or a particle size of 20 µm was assumed in estimating dose to the prostate, with the bladder wall used as a surrogate for the prostate. In either case, selection of the other particle size would have increased the estimated inhalation dose to the organ of concern.

[13]In the lung model described in ICRP *Publication 30* (ICRP, 1979a), assumed rates of clearance of radionuclides from different regions of the respiratory tract take into account mechanical clearance and absorption into blood.

Inhalation of soluble Class D materials usually results in relatively low doses to tissues of the respiratory tract and relatively high doses to other organs and tissues. As the solubility decreases, doses to tissues of the respiratory tract increase and doses to other organs and tissues decrease, although there are some exceptions in organs of the GI tract.

Inhalation dose coefficients that apply to radionuclides in oxide form are used in all dose reconstructions (Egbert et al., 1985), on the basis of the expected insolubility of radioactive materials in fallout and the insolubility of oxide forms of many radionuclides compared with other chemical forms. That assumption usually results in the highest estimates of dose to the lung and organs in the GI tract, but in these cases, it often results in lower estimates of dose to other organs and tissues compared with doses based on an assumption of a more soluble chemical form.[14] However, there are exceptions, including isotopes of strontium for which the oxide form is assumed to be Class D rather than Class Y (ICRP, 1979a). For some elements, including iodine, the same clearance class is assumed for all chemical forms.

IV.C.2.2.3 *Biological effectiveness of alpha particles*

On the basis of numerous studies in animals, alpha particles are known to be more effective than photons and electrons in inducing biological responses (NCRP, 1990). That is, if the same absorbed dose of alpha particles and of photons or electrons is delivered to an organ or tissue, the probability that a cancer or other stochastic radiation effect will result is substantially higher in the case of exposure to alpha particles. The biological effectiveness of alpha particles is an important concern, for example, in estimating equivalent doses due to inhalation of plutonium, which is an alpha-emitter.

Two sets of dose coefficients for inhalation (and ingestion) of alpha-emitting radionuclides were calculated by the ORNL group (Dunning et al., 1979); one set assumed a biological effectiveness of 10 for alpha particles compared with photons and electrons, and the other assumed a biological effectiveness of 20. However, only the dose coefficients calculated by the ORNL group that incorporate the higher biological effectiveness of alpha particles (20), thus giving higher equivalent doses, are included in the FIIDOS code (Egbert et al., 1985) and have been used in dose reconstructions since then. This assumption conforms to the current recommendation by ICRP (1991a), and is incorporated in all dose coefficients used in dose reconstructions that are obtained from ICRP *Publication 30* (ICRP, 1979a).

---

[14]Dose coefficients for organs of the GI tract do not show a predictable dependence on chemical form of a radionuclide, because the comparison also can depend on the half-life of the radionuclide, its decay properties, and sites of deposition after absorption into blood.

IV.C.2.2.4 *Summary of inhalation dose coefficients used in dose reconstructions*

All dose coefficients for inhalation of radionuclides used in dose reconstructions for atomic veterans are based essentially on dosimetric and biokinetic models developed in ICRP *Publication 30* (ICRP, 1979a) and are given in documentation of the FIIDOS computer code (Egbert et al., 1985). The dose coefficients assume that inhaled radionuclides are in oxide form and, in all cases of inhalation of alpha-emitting radionuclides, assume a biological effectiveness of alpha particles of 20 relative to photons and electrons.

In estimating inhalation doses to atomic veterans, an analyst must choose between dose coefficients calculated by the ORNL group (Killough et al., 1978a; Dunning et al., 1979) and values calculated by ICRP (1979a; 1979b; 1980a; 1981a; 1981b; 1982a; 1982b). There are some differences in inhalation dose coefficients in the two data sets, but the committee is not aware of any written policy in the NTPR program to guide the choice. An analyst also must choose inhalation dose coefficients based on an assumed particle size (AMAD) of 1 or 20 μm. With some exceptions that are based on reasoned arguments, the stated policy is that the particle size that gives the higher estimate of dose to the organ or tissue of concern will be used (Schaeffer, 2002b).

Inhalation dose coefficients used by the NTPR program are not intended to overestimate doses that would result from given intakes of radionuclides of known particle size and solubility. In using these dose coefficients, only the assumed particle size could result in substantial overestimates of inhalation dose to particular organs or tissues. Thus, as emphasized in Section IV.C.2.1.7, the NTPR program, in its effort to obtain "high-sided" estimates of inhalation dose, relies primarily on assumptions that are intended to result in overestimates of airborne concentrations of radionuclides to which participants were exposed, especially assumptions about resuspension factors that are applied to deposited fallout or neutron-activation products in surface soil.

IV.C.3 Methods of Estimating Ingestion Dose

When ingestion of radionuclides is assumed to occur, such as by consumption of food or water contaminated by fallout, equivalent dose to an organ or tissue of concern is calculated by using an estimated activity of each radionuclide ingested and the appropriate ingestion dose coefficient. Dose coefficients are obtained either from ORNL reports (Killough et al., 1978a; Dunning et al., 1979) or from ICRP *Publication 30* (ICRP, 1979a; 1979b; 1980a; 1981a; 1981b; 1982a; 1982b). The latter set of dose coefficients also is given by Eckerman et al. (1988).

Ingestion of radioactive materials apparently is considered only rarely in dose reconstructions for atomic veterans. The one case encountered by the committee concerned a small group of participants on Rongerik Atoll in the Marshall Islands who were exposed to very high levels of fallout after Operation CASTLE,

Shot BRAVO (Goetz et al., 1987). In the dose reconstruction for that group, ingestion of radionuclides that had deposited on food eaten during the period of fallout was considered to be the most important intake pathway.[15]

It also should be noted that ingestion dose coefficients are used in estimating doses due to inhalation of large particles (AMAD, 20 µm). When inhalation of large particles is assumed, ingestion dose coefficients are applied to the large fraction of inhaled activity that is assumed to be transferred from the nose and throat to the GI tract by swallowing.

IV.C.4 Absorption Through the Skin or Open Wound

Intakes of radionuclides due to absorption through the skin or an open wound are rarely considered in any dose assessment. The one exception is in cases of exposure to an atmospheric cloud of ^{3}H in the form of tritiated water vapor, when intakes by skin absorption usually are assumed to be half the estimated intakes by inhalation (ICRP, 1979a).

Documentation of the FIIDOS computer code (Egbert et al., 1985) does not include consideration of intakes of radionuclides by absorption through the skin or an open wound. Therefore, the NTPR program presumably has not considered those intake pathways in dose reconstructions for atomic veterans. Absorption through intact skin is unlikely to be important, but absorption through an open wound or cut could have occurred when a veteran had such a condition and radioactive materials were deposited on the affected part of the body surface. That situation could occur, for example, while a person was digging a trench or crawling in a contaminated area.[16]

IV.D DOSE RECONSTRUCTIONS FOR OCCUPATION FORCES IN JAPAN

After Japan surrendered on September 2, 1945, US military forces occupied Hiroshima and Nagasaki and therefore may have been exposed to residual radioactive contamination from the August 6 and 9 atomic bombings. In addition, some prisoners of war (POWs) were exposed before arrival of occupation forces. This section describes methods used in the NTPR program to estimate doses to service personnel in Japan.

The Manhattan Engineering District conducted radiological surveys in Nagasaki from September 20 to October 6 and in Hiroshima from October 3 to 7. The Naval Medical Research Institute conducted surveys in Nagasaki from October 15

---

[15]The committee found one other case of a dose reconstruction for a participant in which the possibility of ingestion exposure was mentioned, but an ingestion dose was not estimated.

[16]In its review of dose reconstructions, the committee encountered very few cases in which the existence of an open wound was mentioned by a veteran.

to 27 and in Hiroshima on November 1 and 2. Their data and later measurements made by the Japanese constitute the basis of dose reconstructions for participants who were in either city after the war ended (McRaney and McGahan, 1980).

The radiation surveys identified two areas of contamination in each city. One was at and near ground zero, and the other was some distance downwind. Measured radiation levels were low; the exposure rate was usually less than 1 mR h^{-1}. The residual radiation near ground zero was most likely due to neutron-induced activity. According to laboratory studies of neutron activation in soil and building materials, the only radionuclides of significance at the time of the surveys were ^{46}Sc and ^{60}Co. The measured exposure rates at ground zero were less than 0.075 mR h^{-1} at Nagasaki and 0.1 mR h^{-1} at Hiroshima.

The second area of contamination was due to the reported downwind "black rain" resulting from deposition of fission products. The Nagasaki surveys showed a maximum exposure rate of 2 mR h^{-1} at the Nishiyama Reservoir. In the downwind area at Hiroshima, the maximum exposure rate in the Kita-Kogo area (west Hiroshima) was 0.045 mR h^{-1} and so was much less than at Nagasaki's Nishiyama Reservoir.

Before occupation of Hiroshima and Nagasaki, a POW recovery team arrived in Nagasaki on September 11. During the period September 11-23, about 10,000 US and allied POWs who were on Kyushu Island were processed through Nagasaki for evacuation to hospital ships. Dates of arrival and departure of major units of the occupation forces and their assigned strength and basic assignments are known. Locations of command posts and billets were believed to be outside the radiologically contaminated areas.

"Worst-case" scenarios were developed for the servicemen (McRaney and McGahan, 1980). Those scenarios assume that a participant was with his unit for the entire duration of the unit's occupation period and that he spent 8 h d^{-1} within the small area defined by the highest radiation intensity. External doses at Hiroshima and Nagasaki were estimated by integrating measured exposure rates over time. Because fallout in Hiroshima was insignificant compared with that in Nagasaki, inhalation and ingestion doses were calculated only in Nagasaki. To estimate inhalation doses, the surface activity was estimated from the measured photon exposure above ground (see Section IV.C.2.1). A "high-sided" resuspension factor of 10^{-4} m^{-1} was used in the vicinity of ground zero to account for the relatively high resuspension caused by mechanical activities, and a resuspension factor of 10^{-5} m^{-1} was used at locations in fallout fields away from ground zero. In the Nagasaki fallout field, the worst-case calculation of ingestion dose assumed direct ingestion of water from the Nishiyama Reservoir. Concentrations of fallout radionuclides in water were calculated by assuming complete mixing in the reservoir. This assumption should overestimate concentrations in water because fallout particles presumably were highly insoluble and tended to deposit in sediment.

A summary of upper-bound estimates of external, inhalation, and ingestion doses to the occupiers of Japan is given in Table IV.D.1 (McRaney and McGahan,

TABLE IV.D.1 Upper-Bound Estimates of Dose to Occupiers of Hiroshima and Nagasaki[a]

| | Hiroshima | | | | Nagasaki | | |
| | Ground Zero | | Fallout Area | | Ground Zero | Nishiyama Area | |
	41st Division	24th Division	41st Division	24th Division	2nd Division	2nd Division (RCT-2)	2nd Division (Artillery Gp)
External dose	0.030 rem	0.030 rem	0.019 rem	0.014 rem	0.081 rem	0.47 rem	0.63 rem
Inhalation dose[b]							
Bone	0.004 rem	0.004 rem			0.018 rem	0.14 rem	0.58 rem
Red bone marrow	c	c			c	0.033 rem	0.13 rem
Whole body	0.003 rem	0.003 rem			0.014 rem	0.017 rem	0.068 rem
Ingestion dose[b]							
Bone					0.09 rem	0.02 rem	0.07 rem
Red bone marrow					0.05 rem	0.01 rem	0.04 rem
Whole body					0.03 rem	0.01 rem	0.02 rem

[a]See Table 4 of McRaney and McGahan (1980).
[b]50-year committed equivalent dose.
[c]Dose coefficient for red bone marrow for 46Sc was not available; dose to red bone marrow should be less than dose to bone but greater than dose to whole body.

1980). The belief that the estimated doses are upper bounds is based on consideration of two assumptions used in the analysis. First, the assumption that a person spent 8 h d^{-1} in the most contaminated area for the entire period of occupation by the person's unit should overestimate actual exposures because the area of maximum contamination was only about 1% of the entire area occupied; this assumption probably results in an overestimate of dose by at least a factor of 2. Second, it is highly unlikely that units were stationed near the Nishiyama Reservoir over the entire period; this assumption probably results in an overestimate of dose by a factor of about 10.

The results in Table IV.D.1 indicate that the maximum estimated dose was received by the 2nd Division Artillery Group in Nagasaki; the estimated external dose is 0.63 rem, and the estimated 50-year committed internal dose to bone is 0.65 rem, with lower internal doses to other organs. Again, those estimates are expected to be upper bounds of actual doses. At Hiroshima, the maximum dose occurred near ground zero. At that location, the estimated upper-bound external dose is 0.03 rem, and estimated internal doses are less than 0.005 rem. Thus, occupation forces in Hiroshima experienced little radiation exposure.

The results of the generic assessment by McRaney and McGahan (1980) described above and summarized in Table IV.D.1 represent upper-bound estimates of dose to participants who were exposed in Japan. However, when a participant files a claim for compensation or requests a dose reconstruction, an individualized dose reconstruction is often performed, based on knowledge of the participant's times of exposure, locations, and activities. Doses assigned in individual dose reconstructions often are substantially lower than upper bounds obtained in the generic assessment.

IV.E METHODS OF ESTIMATING OR ACCOUNTING FOR UNCERTAINTY

The following sections consider the methods used in the NTPR program to estimate or otherwise take into account uncertainty in exposure scenarios and in estimates of external and internal dose. Consideration of uncertainty in all aspects of a dose reconstruction is important when the veteran is to be given the benefit of the doubt and the policy of the NTPR program is to obtain credible upper-bound estimates of dose.

IV.E.1 Uncertainty in Exposure Scenarios

Uncertainty in exposure scenarios is inherently difficult to quantify, especially uncertainty in a veteran's activities and locations of the activities during periods of potential radiation exposure. When uncertainty in a veteran's activities involves two or more specific choices, such as two well-characterized ships that the veteran might have been on during a given interval of time, the veteran could be given the benefit of the doubt by assuming that he was on the more heavily contaminated ship. Sometimes, however, the uncertainty is harder to represent so specifically and has to be resolved in favor of assumptions regarding what, in the judgment of the analyst, the veteran was most likely to have done. In those cases, the attending uncertainties in dose are not quantified, and sometimes there is a comment that the assumptions being applied regarding the scenario are inherently "high-sided." Uncertainty regarding off-duty behavior—for example, souvenir hunting, using small boats to visit contaminated islands, and engaging in dusty sports activities—is not part of any official records and is typically ignored.

Uncertainties in the assumed radiation environment in which a veteran was exposed are assessed separately. Estimated uncertainties in external exposure rates measured with field instruments are imputed to the veteran for his specific path through space and time at the location of a particular test. Such uncertainties are based on what is known about the quality of instruments used at the time to monitor radiation combined with uncertainties in the interpolation methods that are used to smooth the spatial distribution of exposure rates and model their changing patterns over time, reflecting dispersion, fallout, and known rates of radioactive decay.

IV.E.2 Uncertainty in Estimates of External Gamma and Neutron Doses

The current policy of the NTPR program is to report an upper-bound estimate of external dose with the central ("best") estimate. The upper bound is intended to be a 95th percentile, meaning that if one calculates the distribution of possible doses for the participant, the true dose is expected to be lower than the assigned upper bound in 95% of cases with an identical exposure scenario. Although some uncertainties were estimated in early unit reports, particularly if the dose was based primarily on a unit dose reconstruction, the NTPR program did not generally report upper bounds of total external dose to the veteran or the VA before 1992. However, upper bounds were estimated for some reconstructed doses on the basis of unit dose reconstructions beginning as early as 1979. From 1989 to 1992, after the National Research Council (NRC, 1989) film badge report was issued, upper bounds were reported on the individual components of the total dose, that is, film-badge results and reconstructed doses. That approach was taken because until the 1989 film-badge report, there was no established method of reporting uncertainty in film-badge data, and until 1992 no established protocol for combining film-badge and reconstructed uncertainties (Flor, 1992; Schaeffer, 2001a). Beginning in 1992, a combined central estimate of external dose and an upper bound has been reported (Schaeffer, 2002b). According to the NTPR program, by 1996 all external gamma and neutron dose estimates reported to VA have included an upper-bound estimate.

IV.E.2.1 *Film-Badge Upper-Bound Estimates*

Film-badge uncertainty estimates used by the NTPR program are based on a slight modification of the National Research Council recommendations (NRC, 1989). The uncertainty for a single film-badge reading is calculated as recommended by the NRC (1989). The film-badge reading is divided by the recommended bias factor for that test series but modified by assuming that the bias factor relating exposure (in roentgen) to deep equivalent dose (in rem) is 1.0 rather than 1.3. The upper bound for an individual film-badge reading is then calculated by multiplying the bias-corrected reading by the recommended 97.5th percentile uncertainty factor $K = GSD^{1.96}$ (NRC, 1989), where GSD is the geometric standard deviation. The result is then converted to a 95th percentile upper bound to achieve correspondence with the measure of an upper bound adopted by the NTPR program for use in dose reconstructions. Because of the lower net bias applied, the upper bound for a single film-badge reading is about 30% higher than had the NRC (1989) recommendations been followed exactly.

The National Research Council report (NRC, 1989) suggested that a method of estimating the upper bound of a combination of lognormally distributed film-badge data that would overestimate a desired confidence limit (be generous to the veteran) would be to sum the upper bounds of each reading. Alternatively, the report suggested that the variances of the individual bias-corrected readings could be summed, assuming that the data were independent and uncorrelated lognormal variants. The result of summing multiple badge variances generally is much lower than the sum of upper bounds. The NTPR program combines the estimated variances of the individual readings to obtain an upper bound. However, it is assumed that some of the uncertainty estimates are correlated (the uncertainty due to how the badge is worn) or partially correlated for badges from the same test series ("environmental" uncertainty and uncertainty in the "R to rem conversion"). That modification is not documented to explain the rationale for choosing the degree of correlation assumed but was conveyed to the committee via a copy of the computer program used since 1992 to calculate the upper bound for a sum of badge data (Schaeffer, 2002d). The assumption that some of the uncertainty is correlated results in a higher upper bound than assuming no correlations but gives a lower result than summing the upper bounds of the individual readings.

IV.E.2.2 *Upper-Bound Estimates in Unit Dose Reconstructions*

Unit dose reconstructions usually estimate the uncertainty in the average or "best-estimate" dose to a unit or subunit on the basis of estimated uncertainties in measured exposure rates, estimated decay rates, estimated positions of troops versus time, shielding, the fraction of the day spent indoors versus outdoors, and so on. The estimated 95th percentile upper bound is assumed to appropriately reflect the upper-bound dose to each individual in the unit (Goetz et al., 1981).

Most of the estimated uncertainty is generally due to uncertainty in the assumed average exposure rate on a ship or an island or to uncertainty in field measurements of exposure rates for maneuver troops or observers. However, those averages are often based on sparse data.

IV.E.2.2.1 *Upper bounds in unit dose reconstructions at the NTS*

The generic unit dose reconstructions discussed in Section IV.B.2 generally also included estimates of uncertainty. As stated in Barrett et al. (1987), "where errors are determined, they are estimates of uncertainty in mean dose associated with the group activities. No attempt is made to predict the distribution of dose within a unit [and] departures by individuals from the average activity scenario are not considered."

Table IV.B.1 in Section IV.B.2.1 lists the total uncertainty (actually, the doses to be added or subtracted from the indicated dose to obtain the 5th and 95th percentile doses) estimated in a unit dose reconstruction for maneuver troops at Operation UPSHOT-KNOTHOLE at the NTS (Edwards et al., 1985). Maneuver troops observed the tests from trenches several thousand meters from ground zero. Shortly after the blast, they left the trenches and marched toward ground zero to predetermined objectives. After reaching their objectives, the troops proceeded to display areas to observe the effects of the test on various types of equipment. The calculated uncertainty range arises from two basic sources: the uncertainty in the gamma-radiation field and the uncertainty in the space-time scenario of troop movements. Errors in position, time, and gamma intensity are not independent, owing to the radiation-safety (rad-safe) constraint that limited troops to areas with exposure rates less than 2.5 R h^{-1} (except at Shot GRABLE). It was assumed that there was no violation of rad-safe limits, except for Battalion Combat Team (BCT)-B at Shot NANCY, so the assumption that troops kept close to the limits when detouring is expected to result in "high-sided" estimates of dose. Thus, the exposure rate of 2.5 R h^{-1} is assumed to apply without error under these conditions. The uncertainty is all assumed to be in the duration and path length of any detour.

In the display areas, limits of advance were not always reported. An assumption was then made that all the displays within rad-safe limits were inspected; this assumption should tend to overestimate exposures. The timing of the troops' march was based on the reported time of attack, time of arrival at the objective, and arrival at the pickup point. Reasonable march speeds (usually about 70 ± 20 m min^{-1}) and display-area stay times (usually 5 min with a range of 2.5–10 min) were assumed to construct a scenario consistent with the known times. The most important influence of timing on the uncertainty in dose was assumed to be the time spent at the positions of highest exposure.

The various sources of uncertainty were combined approximately because it was asserted that they could not be combined rigorously, owing to the disparity of their associated distributions. For each source of uncertainty, the limits on dose are interpreted in terms of error factors on the best-estimate doses given in Table IV.B.1. For example, for BCT-B (forward unit) at Shot NANCY, the dose to add to or subtract from the total dose of 2.4 rem (−0.7 rem and +1.5 rem) was determined by combining the error factors for the components described below:[17]

- The contribution due to uncertainty in stay times at halt points and displays: −0.5 to +1.0 rem.
  - The contribution due to uncertainty in march speed: −0.3 to +0.5 rem.
  - The contribution due to uncertainty in gamma exposure rate: −0.0 to +0.4 rem.

For both BCTs at Shot SIMON, the total uncertainty was assumed to be dominated by the time spent at the 2.5-R h⁻¹ rad-safe limit. The total uncertainty in this case thus is assumed to be due entirely to the uncertainty in march speed and arrival and stay times. The error factors due to march-time uncertainty were estimated to be −0.06 and +0.12 rem, and the estimated error factors due to stay time were −0.13 and +0.27 rem, with a combined uncertainty range in total dose of $3.1 - 0.2 (= 2.9)$ to $3.1 + 0.3 (= 3.4)$ rem.

The uncertainty analyses for other NTS unit dose reconstructions are similar to that described above, and the central ("best") estimates generally are asserted to be "high-sided." However, the estimated uncertainty ranges for various parameters vary from shot to shot and among units, depending on the available information and specific exposure scenario.

IV.E.2.2.2 *Upper bounds in unit dose reconstructions at Pacific test sites*

The Pacific-test-site unit dose uncertainties are usually based on the estimated coefficient of variation (CV) in the measured post-shot exposure rate at various locations. In some of the dose reconstructions for the later test series, where most participants were badged, the upper bound for reconstructed doses is based on a comparison with the standard deviation of the available film-badge measurements (for example, see Weitz, 1995b; 1997). The upper bounds based on the measured post-shot monitoring data generally assume that multiple exposures to fallout from the same shot are independent, that is, that a participant is exposed at random locations each time he is outdoors or topside on a ship. The upper bound is calculated by assuming that the uncertainty in the dose incurred during each interval of exposure on deck or outdoors to the same fallout (assuming three intervals per day totaling 40% of the day topside for ships and 60% outdoors for islands) is completely independent of the previous exposure, that is, that the participant's location with respect to the distribution of fallout exposure-rate

---

[17]The details of how the error factors were combined and any correlations assumed are not given by Edwards et al. (1985).

measurements is completely random. The uncertainty in the sum of all doses outdoors or on deck is calculated by summing the variances in each individual dose (for example, see Thomas et al., 1982; 1984). The dose incurred below decks is assumed to be very low because of an assumed 90% shielding factor. The consequence of those assumptions is that the more time the participant is assumed to be exposed to the same fallout, the smaller the estimated fractional uncertainty in the total dose received.

A default CV of 50% is usually applied to measured average exposure rates on the basis of survey data from ships on which about 30 or more measurements were made (for example, see Thomas et al., 1984). However, the CV based on the distribution of measurements on some ships or islands for some fallout events often was much greater. For many ships, only an average exposure rate was reported. If data are not available for a particular ship, data from a nearby ship or island are used because it is assumed that the amount of fallout would have been similar. Although an additional systematic uncertainty of a factor of 1.2 is asserted for the effective shielding factor[18] (Thomas et al., 1984), it appears that this was not usually applied in practice. No uncertainty is apparently assumed in the decay rate or in the default CV in calculating these upper bounds.

Upper-bound factors (ratios of upper bounds to central estimates) for participants exposed for an entire test series on a particular ship or island based on the above assumptions are tabulated in the relevant unit dose-reconstruction reports. The values are usually applied and referenced in the individual dose-reconstruction reports. Although the upper-bound factor should be greater than the tabulated values for participants exposed for only a part of the test series, it appears that the same upper-bound factor was used regardless of the time exposed. If it is assumed that all exposures outdoors (and indoors on islands) are random, an estimated average upper bound for participants exposed over an entire test series will be only about 10–20% greater than the central estimate (Thomas et al., 1982; 1984).

IV.E.2.3 *Upper-Bound Estimates of Neutron Dose*

An upper bound of the dose from external exposure to neutrons is also estimated in the relevant unit dose reports and is based on the estimated uncertainty in the transport-calculated exposure and the uncertainty in the shielding correction. Uncertainty in the quality factor for neutrons (an uncertainty in the relative biological effectiveness of fission neutrons versus gamma rays) has not been taken into account in estimating the upper-bound neutron dose.

---

[18]The effective shielding factor is the product of the shielding factor weighted by the fraction of time spent indoors or below decks. For the default shielding factor of 0.5 used for troops billeted on residence islands and assumed to be indoors 40% of the time, the effective shielding factor is (0.5 × 0.4) + 0.6 = 0.8; for sailors on ships and assumed to be below deck 60% of the time, it is (0.1 × 0.6) + 0.4 = 0.46.

IV.E.3 Uncertainty in Estimates of External Beta Dose

The NTPR program does not perform uncertainty analyses for beta-particle dosimetry, relying instead on an assumption that estimates of dose are upper bounds ("high-sided"). Beta doses from contaminated ground or other surfaces, for example, are calculated by multiplying a presumably overestimated beta-to-gamma dose ratio by an upper-bound gamma dose. As noted in Section IV.B.4, the current methodology for assessment of beta-particle dose from sources external to the body is described in Barss (2000). The method has remained substantially the same since routine assessment of skin dose began in 1998, although numerical values of the beta-to-gamma dose ratios have evolved.

IV.E.3.1 *Exposure to Contaminated Ground*

As noted in Section IV.B.4.1, beta dose to the skin or lens of the eye from external sources is accrued simultaneously with gamma dose from radioactive fallout, contamination, or neutron-induced radionuclides. As a result, the beta dose is proportional to the gamma dose, and its magnitude can be mathematically expressed by a beta-to-gamma dose ratio.

Uncertainty in the beta-to-gamma dose ratios is discussed by Barss (2000), and sources of uncertainty are identified, with emphasis on how simplification of the assessment process leads to estimates that are higher than the likely actual doses. It is stated that the uncertainty most difficult to quantify is that in the reduction in beta-particle fluence between the source and receptor locations, which is much more dependent than gamma fluence on shielding material, chemical and physical properties of the radionuclide and the surface, and distance from the source deposited on a surface.

An example offered by Barss (2000) concerns deposition locally on the ground, for which large particles associated with tower shots provided substantial self-shielding. As a result of weathering, environmental transport, and dispersion, fallout particles may also penetrate to such a depth in soil as to substantially reduce beta doses compared with such material being on the surface. The inability to model the magnitude of each of these reduction effects adequately, and their degree of interdependence, are said to limit the usefulness of any attempt to quantitatively model their associated uncertainties.

Another example offered by Barss (2000) concerns initial decontamination techniques used on ships and aircraft surfaces (washdown systems and fire hoses), which tended to remove loosely bound or attached particles but would "fix" residual material to the surface to an extent proportional to the surface porosity or accessible surface area. The implication is that after washdown the remaining contamination would be largely fixed in surficial pores, causing more attenuation of beta particles than of gamma rays and resulting in a lower beta-to-gamma dose ratio than if the material were truly on top of a surface.

The Barss (2000) report indicates that calculated beta-to-gamma dose ratios 1 and 2 years after detonation are substantial overestimates because they ignore the effects mentioned in the two foregoing examples. The report further indicates that an enormous expenditure of resources would be needed to adequately describe and quantify the uncertainties in model parameters, given the high degree of variability in the environmental interaction with residual radionuclides. Additional resources would be required to further propagate the uncertainty associated with each model parameter to obtain an estimate of the overall uncertainty in a calculated beta dose.

The report notes that although there are environmental models that attempt to achieve the objectives (quantify parameters and propagate uncertainty), their usefulness remains inversely proportional to their degree of scientific debate and interpretation. The discussion concludes that the best resolution of the dilemma (quantification of uncertainty), in the absence of a rigorous and scientifically appropriate approach to quantify and apply modification factors for environmental and particle effects, is consideration of field measurements. It is stated that, in some comparisons, the current beta-to-gamma dose ratios are in reasonably good agreement with previous calculations and available measurements and, at worst, overestimate the measurements by a factor of 2-3.

There is no discussion of factors that might cause underestimation of beta doses, such as errors in estimating time since detonation and underestimates of distances from contaminated surfaces or exposure times. It is clear from Barss (2000) that application of the method is expected to result in a "high-sided" dose.

IV.E.3.2 *Immersion in Contaminated Air or Water*

As noted in Section IV.B.4.2, beta doses from immersion in contaminated air or water are calculated by using dose coefficients, durations of exposure, and composite beta-spectrum radiation energies associated with a reconstructed gamma exposure or film-badge reading. The calculated beta dose is added to the upper-bound gamma dose for the corresponding period.

There is no discussion of uncertainty by Barss (2000) related to beta-particle doses from immersion, although it seems clear from examination of Figure IV.B.8 (see Section IV.B.4.2) that a small uncertainty in the time of onset of exposure could lead to a large uncertainty in a composite dose coefficient, particularly during the period shortly after detonation.

IV.E.3.3 *Skin Contamination*

As noted in Section IV.B.4.3, dose coefficients from Kocher and Eckerman (1987) can be used to calculate beta dose from skin contamination, with adjustments for backscatter and for the case in which a glove is contaminated. The

VARSKIN code (Durham, 1992) can be used to calculate skin dose for specific source geometries. Skin-contamination measurements are recommended as the best source of contamination data from which to calculate dose, but methods are also suggested for using dose or exposure-rate measurements to estimate contamination.

Barss (2000) does not discuss uncertainty related to beta-particle doses from skin contamination.

IV.E.4 Uncertainty in Estimates of Internal Dose

Estimates of uncertainty in calculated internal doses are not presented in dose reconstructions for individual atomic veterans or in unit dose reconstructions for participant groups. In all dose reconstructions that include an estimate of internal dose, the calculated dose is presented as a single value without uncertainty. Uncertainties in internal doses are also not evaluated or discussed in any detail in reports documenting the calculation methods (Egbert et al., 1985; Barrett et al., 1986).

Thus, the treatment of uncertainty in estimated internal doses differs from the approach to addressing uncertainty in estimated doses from external exposure to photons. As discussed in Section IV.E.2, dose reconstructions for individual veterans often provide an estimated upper bound of the external photon dose, especially if the veteran filed a claim for compensation. Many generic dose reconstructions for participant groups also include a quantitative analysis of uncertainty in external photon dose. An upper-bound estimate of external photon dose is intended to represent a 95% confidence limit, and the difference between the upper bound and the central estimate indicates the magnitude of uncertainty. Upper-bound estimates of dose are important because, in accordance with the policy that the veteran will be given the benefit of the doubt (see Section I.C.3.2), the NTPR program intends that upper bounds will be used in evaluating claims for compensation.

In the absence of a quantitative analysis of uncertainty in estimated internal doses, this uncertainty is addressed in the NTPR program by using an alternative approach mentioned in Section I.C.2.4. An argument is made that methods used to estimate internal doses incorporate assumptions that should result in overestimates of internal doses to participants. For example, the method of estimating dose from inhalation of resuspended fallout that was previously deposited on the ground or other surfaces (see Section IV.C.2.1) relies mainly on an assumption that resuspension factors that are applied in various exposure scenarios greatly overestimate the actual extent of resuspension of deposited fallout. On the basis of that type of argument, estimates of internal dose obtained in dose reconstructions are assumed to represent suitable upper bounds for use in evaluating claims for compensation; that is, the estimated doses are assumed to be "high-sided." As discussed in Section IV.E.3, essentially the same approach to accounting for uncertainty is used in estimating external beta dose to the skin or lens of the eye.

The committee reiterates that an approach of relying on "high-sided" assumptions to estimate credible upper bounds of possible doses, rather than an approach involving a quantitative analysis of uncertainty in a central estimate, is a reasonable way to address uncertainty. Furthermore, an upper bound so obtained is appropriate for use in evaluating claims for compensation. However, it is a valid approach to addressing uncertainty only if estimated doses are indeed "high-sided." That is, on the basis of available information and scientific judgment, there must be a high degree of confidence that calculated internal doses do not underestimate actual doses to participants.

Thus, an evaluation of methods used in the NTPR program to estimate internal doses to atomic veterans essentially involves an assessment of the extent to which the methods are likely to overestimate doses. The committee's evaluation of the methods of estimating internal dose is presented in Section V.C.

IV.F ESTIMATES OF TOTAL DOSE AND UNCERTAINTY FOR INDIVIDUAL PARTICIPANTS

Although many participants have received a dose assessment from the NTPR program based on film-badge data in their medical records or their unit's generic dose reconstruction, VA may request a formal dose reconstruction from DTRA to evaluate a claim for compensation (see Section III.B). A veteran may also request a detailed dose reconstruction by directly contacting DTRA. An individual dose reconstruction attempts to determine all possible pathways and sources of exposure for the participant on the basis of his military records and personal statement. The analyst reviews the assumed exposure scenario for the participant and modifies or recalculates the unit dose reconstruction according to the time exposed (which may have differed from the time assumed in the generic reconstruction), special duties or missions, available film-badge data, and so on. The analysis and dose estimate are reported in a detailed memorandum that specifies the assumed exposure scenario, exposure rates and decay rates, references to the applicable unit dose reports, and any other analysis methods applied (see Section IV.G.1). If the participant was exposed at various times and places, the memorandum reports the estimated dose from each exposure and sums the individual doses to determine the total for all exposures from all test series that the veteran participated in.[19] The neutron dose in rem is added to the estimated whole-body gamma dose, and the total is reported to VA or the veteran when an external dose is reported.

---

[19]For example, some veterans participated in multiple test series, both in the Pacific and at the NTS; some were exposed at various locations during the same test series, such as on different ships or islands; and some were on leave during parts of the time during a test series.

Reconstructions of external dose that were done in recent years (1998 and later) also often include an assessment of dose to the skin or lens of the eye, particularly for participants claiming compensation for skin cancer or cataract, and these assessments include the contributions from beta exposure. Appendix A contains two examples of dose-reconstruction memoranda from sample cases reviewed by the committee.

Some individual dose reconstructions are unique—that is, not covered by a generic unit dose-reconstruction method—and require a fairly complex dose assessment. For these cases, the dose-reconstruction memorandum details the assumptions made in estimating the dose.

The total dose reported to VA or the veteran consists of the best estimate of the gamma-plus-neutron equivalent dose from all sources of external exposure and an upper-bound (95th percentile) estimate that combines the estimated upper bounds from each source of exposure and from estimates based on film-badge data and reconstructions. The neutron and gamma upper-bound estimates also are combined to estimate an upper bound for the sum. As discussed in Section IV.B, total upper bounds for external gamma-plus-neutron dose have been consistently reported in a formal dose reconstruction since 1996. The upper-bound calculations typically assume that exposures to different shots, or at different locations, are not correlated and can be combined in quadrature by summing the variances (Flor, 1992). In addition, DTRA is often asked by VA to provide upper-bound estimates for generic or film-badge doses that were previously provided to the veteran or VA for which upper bounds had not been estimated. Those upper-bound requests often result in the reporting of a revised central estimate based on a new method or new exposure scenario information.

If a skin or eye dose from beta exposure is calculated, it is reported separately. Estimates of beta dose are already considered to be "high-sided," so no additional upper bound is reported. If a claim involves a disease of a specific organ, and an internal (inhalation) dose has been calculated for that organ, the calculated organ dose is also reported separately. As discussed in Sections IV.C.2.1.7 and IV.E.4, this estimate is also intended to be "high-sided," so no additional upper bound is reported. (Often, even though an actual "high-sided" inhalation dose estimate is provided, the inhalation dose is also reported as less than the screening criterion of 0.15 rem; see discussion of the low-level internal dose screen in Section VI.C). When no specific organ dose is calculated, a committed effective dose equivalent from inhalation is sometimes calculated (see Section IV.C.2). Again, it considered to be a "high-sided" estimate, and no additional upper bound is reported.

Although the NTPR program does not combine external and inhalation dose estimates to estimate the total and upper-bound doses to a specific organ, the VA practice is to sum the reported external-dose upper bound (if an upper bound is provided) with the reported "high-sided" inhalation (internal) dose estimate to obtain an estimate of the upper bound of the organ dose. As discussed in Section III.E, the sum is used in the process of evaluating whether it was at least as likely as not that a veteran's disease was caused by the radiation exposure.

IV.G DOCUMENTATION AND QUALITY ASSURANCE

IV.G.1 Documentation of Dose Reconstructions

The documentation of dose reconstructions for individuals required by the NTPR program is specified in the standard operating procedures (SOPs) (DTRA, 1997). However, the discussion in the SOPs appears to be limited to the documentation that is to be sent to the veteran or his representative, rather than a complete documentation requirement. The SOPs state that: "In order to consistently serve the veteran, the veteran (or his representative) needs disclosure of the information that leads to his dose." The documentation requirements are summarized as follows:

- Documentation pertaining to relevant generic (unit) dose reconstructions.
- All scenario and radiological information pertinent to the dose determination (explicitly or by reference).
- Detailed information or analysis not fully covered in previous documents, which is to be communicated in an individual dose memorandum attached to the case correspondence or in the body of the correspondence.
- Information that is too complex or generic or that otherwise detracts from the presentation of the individual dose memorandum or correspondence, which is to be covered by fact sheets (or other written material) distributed to the correspondent.
- Information on availability of cited formal reports and unpublished documents (subject to Privacy Act-related redactions), which is to be included in the case correspondence.
- Appropriate disclosure of other information, including representations from the time of the operation, such as operational summary data or data entered into individual records, even if such information is not corroborated; explanation of when this information, if it is not the most credible, is not retained in the final analysis; and other types of information that do not necessarily furnish the dose of record, such as dose entries in medical records and information contradicted elsewhere in records.

Beyond the referenced information, an individual dose reconstruction or synopsis should include an explanation of what is specific to the veteran's case, for example:

- The adaptation from a published report of the dose for the veteran's period of participation.

- The principal source of uncertainty that affects the upper-bound dose.
- That the potential for an internal dose has been considered, even if the finding is of no internal dose.
- That a finding obtained from internal dose screening (see Section VI.C) applies to the veteran's target organ.
- That internal dose assessments do not apply to assessments of dose to skin or lens of the eye.
- The reason for a change in estimated dose from previous correspondence if the change is based on new data (a change that results from a procedural change is addressed in the correspondence but not in a dose-reconstruction memorandum).
- Reporting of total doses in a dose summary, which is in tabular form if there are multiple contributions to external gamma or neutron dose.

The specifications for documentation discussed above are related to what should be provided to the veteran. Also of concern to the committee is the detailed internal documentation of the dose reconstructions themselves, which is necessary to make detailed, independent reviews possible. This is discussed in Section VI.A.

IV.G.2 Quality Assurance

The committee did not see details of a formal quality assurance (QA) program in the SOPs (DTRA, 1997), and the files of individual dose assessments reviewed by the committee did not contain the expected indications of a systematic QA process. The committee was informed (Schaeffer, 2001a) that: "There are no additional quality assurance written procedures other than those provided to the committee. The SOP indicates what constitutes a quality dose reconstruction and directs review for conformity with the SOP's procedures, and appropriateness and responsiveness to the correspondence or request received by the NTPR program. The DTRA Program Manager conducts the final review/approval."

The committee notes that the SOPs (DTRA, 1997) constitute more of a program overview than a detailed document that could guide specific day-to-day work, and they have little to say about QA. They do, however, specify the documentation discussed in the previous section that should accompany a dose assessment, to serve the veteran consistently, and that the assessment is to be reviewed (but not by whom).

On further inquiry by the committee, additional information on QA for the dose reconstruction program was given in a letter from DTRA (Schaeffer, 2002e), which is provided in Appendix D. The letter indicates that QA had always been a key element in management and direction of the NTPR program and that the DTRA solicitation for NTPR program support contained a program-management requirement for QA monitoring, which was one of the contract-evaluation factors for award. In response to the solicitation, the contractor submitted a technical proposal that specified QA measures.

The statement of work included in the DTRA solicitation for NTPR program support indeed contains the following requirement for quality assurance: "The contractor shall provide quality assurance monitoring for the NTPR Program in the areas of database management, dose assessment, and veteran assistance." As stated by DTRA (Schaeffer, 2002e): "In response to the solicitation, JAYCOR/ SAIC submitted a technical proposal that specified quality assurance measures in the program task areas of database management, radiation exposure assessment, and veteran assistance." The committee did not have the opportunity to review the technical proposal submitted by JAYCOR and SAIC and consequently did not see the specified QA measures.

V Committee's Findings Related to NTPR Dose Reconstruction Program

The committee's evaluation of the NTPR dose reconstruction program considered not only the validity of central and upper-bound estimates of dose for the assumed exposure scenarios obtained in dose reconstructions, but also the approaches used to determine the veteran's exposure scenario. The committee's findings regarding scenario determinations, estimates of external and internal doses and related uncertainty, and estimates of total organ doses from all pathways are discussed below, with examples taken from the 99 individual dose reconstruction cases sampled and from reconstructions for other veterans who provided written consent for use of their records. In parallel with the discussions in Chapter IV, Section V.A discusses scenario determination, Section V.B the estimation of external dose, Section V.C the methods of estimating internal dose, Section V.D the dose reconstructions for occupation forces in Japan, and Section V.E the estimates of uncertainty and upper-bound doses from all radiations and exposure pathways combined. Section V.F summarizes the committee's findings regarding dose and uncertainty estimates obtained by the NTPR program.

V.A DETERMINATION OF EXPOSURE SCENARIOS

V.A.1 Introduction

As discussed in Section I.C, the most important part of the dose reconstruction process is the determination of a participant's exposure scenario. Because exact histories do not exist for individual veterans, the analyst often has to reconstruct a scenario or a set of possible scenarios on the basis of plausible assumptions. Problems arise because "plausibility" can be subjective. It is often difficult, 50 years after most of the atmospheric tests, to verify even a veteran's participation status with certainty. For example, the original list of veterans provided for the earlier Five Series study (see Section I.B.6) was to have indicated all participants in five test series, but it erroneously omitted more than 20,000 participants and included some 8,000 who were later determined to be nonparticipants.

The committee was generally impressed with the extensive historical research carried out by JAYCOR to document the whereabouts and roles of veterans who took part in the testing program. JAYCOR had to locate and piece together deteriorating, obscure, and often almost-unreadable records (morning reports, ship logs, unit histories, and so on) from diverse archival sources. With such sources, the dates of arrival and departure, where a veteran was quartered, and so on, could usually be documented. In contrast, the veteran's specific duties and the time he spent in various locations (such as on contaminated ships) were typically difficult to document with certainty.

Procedures to be followed by the NTPR program for dose reconstructions, as laid out in 32 *CFR* 218.3, specify that "possible variations in the activities, as well as possible individual deviations from group activities, with respect to both time and location, are considered in the uncertainty analysis of the radiation dose calculations." There is also an expectation that a veteran will be given the benefit of the doubt in determinations used to adjudicate a claim for a nonpresumptive disease under 38 *CFR* 3.311. As stated in 38 *CFR* 3.102, "when, after careful consideration of all procurable and assembled data, a reasonable doubt arises regarding service origin, the degree of disability, or any other point, such doubt will be resolved in favor of the claimant" (see also Section I.C.3.2).

In many of the records examined by the committee, however, the participant did not appear to have been given the benefit of the doubt regarding the assumed exposure scenario or film-badge dose, including the time and place of exposure. In reviewing the 99 cases, which were randomly sampled within strata, the committee found at least 20 in which a veteran's external exposure scenario appeared to be incorrect, incomplete, or suspect (for example, see cases #15, 22, 27, 32, 33, 37, 40, 47, 53, 73, 77, 81, 83, 84, 87, 88, 89, 93, 97, 98, and 99). The inaccuracies were often due to insufficient follow-up by an analyst with the participant or other members of his unit. Examples are discussed below.

One tendency the committee saw in the 99 cases was for the analyst to assume that an activity that allegedly violated radiation safety (rad-safe) or operational guidelines in place at the time did not happen. For example, an analyst often assumed that decontamination crews did not stay longer than the allowed times on contaminated ships, that radiation safety monitors and other personnel did not go beyond the 10 R h^{-1} demarcation line, or that badges that were issued and then returned had, in fact, been worn (not left in a drawer). If the date of issue of a film badge was missing, it was often assumed to have been the recorded date of turn-in of the veteran's previous badge.

Such pragmatic assumptions reflect the analyst's need to complete the calculations and seem also to reflect a tendency to idealize human behavior, particularly military behavior. Such assumptions tend to deny that chaos, confusion, and a perceived need among leaders to ignore rules to complete the task at hand may drive what happens in the field, particularly when a nuclear weapon has just been detonated. The commander of a decontamination crew may have been focused on getting a ship decontaminated and may have considered the rad-safe guidelines to be unnecessarily restrictive and thus not to be taken literally. The rad-safe limit line was not "drawn in the sand," and forward units were sometimes unsure about their exact location relative to that line and to ground zero. Communication of radiation intensity from rad-safe monitoring personnel to commanding officers in the field was sometimes unreliable.

Generic estimates of shielding and time spent indoors versus outdoors used to estimate external dose are questionable for some participants. For example, some participants on ships claimed that because of the heat they slept on deck, where they would not have been shielded at all (see case #28). The assumed 50% shielding factor for participants on Pacific islands may be too high for those who were billeted in tents or thin metal structures that may have had many open windows at night (see Figures V.A.1 and V.A.2). Thus, as discussed later in this chapter, generic dose estimates on ships and islands may not be reasonable estimates of the doses to some unit members.

FIGURE V.A.1   Typical metal buildings used at Enewetak during Operation CASTLE.

FIGURE V.A.2   Tents on Parry Island at Operation CASTLE.

Some sources of information about veterans were not used as well as they might have been. For example, it is not apparent that information in "File A" (see Section I.B.3) for individual veterans was always considered. Additionally, the veteran himself and his buddies were rarely contacted, nor were civilian radiation-safety personnel who often accompanied participant groups during planned activities. That approach might reflect a difference in worldview between a researcher and a claims adjudicator or government contractor, but it is our view that additional and sometimes useful information could have been obtained from the veterans themselves. The questionnaire that was administered in the early days of the NTPR program was very sketchy. It included such questions as "Were you issued a badge?" and "Did you wear it?" When questions came up in the scenario reconstruction about what specific activities a veteran was involved in, the veteran apparently was almost never asked for clarification. The committee's impression is that the contractor assumes that the veteran himself should not be regarded as a reliable source of information. When, on occasion, a veteran came forward with an account of what happened on the sometimes-chaotic day of a weapon test, his story may have been discounted by the analyst and may not even have influenced the calculation of uncertainty, that is, the assigned upper bound of the dose. Examples illustrative of those points are detailed below.

V.A.2 Discussion of Selected Cases Illustrating Scenario Determination Problems

In this section, we discuss some of the 99 sampled cases and additional files submitted by veterans. These cases are listed in Appendix B.

Case #22: The participant claimed that he was present at Operation IVY. However, his service records had been damaged, and his claim that he participated in IVY could not be verified. He was not given the benefit of the doubt in evaluating his claim for a nonpresumptive disease, and no dose was calculated for possible participation in IVY. Nor was the estimated upper bound of his assigned total dose (from his participation in other test series) adjusted to reflect his possible participation in IVY. He was not contacted to investigate his claim further.

Case #53: This case provides a good example of inconsistent application of assumptions used in estimating the external dose and upper bound from boarding target ships at Operation CROSSROADS. The dose memorandum states that the veteran was given the benefit of the doubt by assuming that he participated in two-thirds of the target-ship boardings by his unit. However, the calculations in the case file are based on only one-third of the boardings. In other cases involving target-ship boarding (for example, cases #45 and 49), the veterans were usually given the benefit of the doubt by assuming that they participated in all boardings (see Figure V.A.3).

FIGURE V.A.3   Sailors sweeping deck of ship.

Case #77: This veteran was a member of the 50th Chemical Platoon at Operation TEAPOT, and much of his film-badge information has been lost. From film-badge data summaries that have been found, it is known that several members of the 50th Chemical Platoon, which made up the Desert Rock Radiological Safety (Rad-Safe) Section for TEAPOT, received external doses that greatly exceeded the operational limit of 6 rem, but it is not known who those individuals were. The veteran in question was informed that a reconstructed dose of 3.12 rem was his "most probable dose," but he was given the benefit of the doubt by assigning him the operational limit of 6 rem instead. The fact that no upper bound was provided implies that the dose of 6 rem would be considered as a 95th percentile of this veteran's dose in any adjudication process (the veteran did not file a claim for compensation).

The veteran's personal narrative was provided to the analyst. He stated that he was assigned as rad-safe monitor for two colonels from the Pentagon, who were "dressed in silver suits covering every part of their body, including shoes. They taped all seams with a tape comparable in appearance with duct tape. I watched all this while wearing only a T-shirt and fatigue pants. I was curious as to what they knew, what I didn't know, and what they weren't telling me." He goes on to describe what happened next (apparently, this incident occurred at Shot MET):

> About ten (10) minutes after detonation of the 22 kiloton device, I entered the blast area to find instruments the two Colonels had placed in the area. (We had previously met to acquaint me with the location and critique the recovery.) When I arrived at the site it was very dark, dusty and windy. I can't recall the exact readings, but they were high. I returned to meet the Colonels, who had driven their van onto a road leading into the site. I reported that the recovery area was very hot, and they would have to work very quickly. I led them back to the instrument location. They recovered their instruments and packed them into boxes. In the recovery area fires were still burning. There was a lot of smoke and dust, and the mushroom cloud was still visible. The ground around us was black. The winds were strong. We had passed several mannequins burning, and I learned later this was a test of fireproof clothing. The mannequins burned and the clothes did not. We were several minutes in the area. We then left and I never saw the two Colonels again. I stopped to brush myself off, as I was covered with dust. I made a note of the time spent in the area. I remember thinking that the two Colonels had exceeded 5 Roentgens – more like 6 – and that my double trip into the area would place me even higher. (For example, see Figure V.A.4.)

The veteran goes on to give details about several other tests, one of which again suggests the potential for an inhalation dose:

> One of the major studies undertaken by the 50th Chemical Platoon was to try to correlate a radiation pattern between the ground and the air. In order to be sure these readings from the air were accurate, it was necessary to have men on the

FIGURE V.A.4   US Army observers examining dummies set up about 3,000 yards from ground zero during dry run for Operation TEAPOT Shot MET.

ground to check them. As part of this group, I was assigned to be a ground monitor. The exercise took place at a site where a nuclear detonation had occurred. I am unsure of the exact reading, but our location was radioactive enough to gather data from aircraft flyovers.

After the first series, it was decided that the aircraft probe was not accurate. We stopped for a couple of days while a lead shield was built to protect the probe in the aircraft from every angle except straight down. We then spent a few more days testing this new device. Adjustments were made, and we were in and out of the area several more times. We took readings all the way to ground zero.

This exercise was the dirtiest of my stay. Every day we were covered with dust from our travels through the test site. We had no protection and were inhaling dust constantly. I remember thinking our lungs must have looked like our clothes. I do not remember if we had film badges.

He then describes an operation (apparently at Shot APPLE-II) in which he became disoriented near ground zero:

During the test known as the Survival City Shot, I was assigned to locate a large group of military vehicles. I made several trips through the area prior to the test to orient myself to the location of these vehicles. They consisted primarily of 2½ ton – ¾ ton trucks and jeeps.

I especially remember the layout of Survival City with its city street and completely furnished houses. There were even families of mannequins set in the houses. There was a two story brick building which had been built especially for the test. It was kind of a landmark because it was the tallest structure on the desert except for the bomb towers. Farther from ground zero was a completely equipped mobile home park. A large number of civil defense people were at this test.

I entered the test site shortly after the blast, with a team, seeking the ten Roentgen line [10 R h^{-1}]. I could not find the vehicles. They had been parked less than a mile from ground zero. The ground was black, the two story building was gone, and I became disoriented for a few minutes as I drove around looking for some trace of the vehicles. While I was looking, a call came over the radio that all troops were being pulled from the area due to a wind shift. When I found my way back I had been inside the ten roentgen line. I did not stop my jeep to take a reading. I was alone at this time, and was relieved to find my way back. I believe my exposure was quite high for this event. It was very windy, with dust and smoke. I had no protective clothing or equipment.

The analyst only peripherally considered this narrative in the dose reconstruction. Regarding the first account, about accompanying the two colonels after Shot MET, the analyst writes that the veteran "did not provide sufficient information to identify the specific project that he supported on shot day." Because the veteran commented on seeing burning mannequins, the analyst decided to assign him to Project 40.20, the Clothing Test Project, and accordingly assigned him a dose of 0.20 rem appropriate to that group, apparently discounting the veteran's statement that "I learned later this was a test of fire-proof clothing." Evidently, no inhalation dose was considered.

Regarding the project to assess the correlation between readings on the ground and air-based readings, the analyst comments that although the veteran described this as a "major study," "such a project is not listed, per se, among the Desert Rock projects at operation TEAPOT." The closest documented match that the analyst could find was Project 40.19, CBR Defense Team Training, and the veteran's dose from that activity was accordingly based on a reconstruction that had been done for that group, with the comment that his "dose resulting from this activity was certainly less than 1.7 rem." Again, no inhalation exposure was considered, nor was any allowance made for the possibility that the veteran's account may reflect an activity that was not represented in other surviving records from the time. A note in the file states that because this veteran was a PFC (private first class) at the time, he could not have been involved in CBR team training and, therefore, the dose of 1.7 rem noted above should be subtracted from his dose. However, the 1.7-rem piece of his dose was not replaced with a more accurate estimate.

Regarding the third narrative, related to Survival City in connection with Shot APPLE-II, the analyst found other records that supported the veteran's claim that he was involved. However, some details of the veteran's account were evidently discounted. The analyst's report states that:

> the scenario is questionable since rad-safe monitors did not travel alone in jeeps and there was no reason to send anyone into the shot area to 'search for' the test vehicles since their locations were well known. Moreover, it was not the function of the 50th Chemical Platoon to locate vehicles, but merely to accompany project personnel who were to evaluate damaged vehicles.

The analyst goes on to assign the veteran a dose for this shot on the basis of a reconstruction that had been done for 573rd Ordnance Company personnel and accompanying rad-safe monitors.

In the end, the analyst made an argument that the veteran's overall dose could not have exceeded the operational limit of 6.0 rem. The argument was based on information that seven members of the 50th Chemical Platoon evidently did exceed the limit and were restricted from further radiation-related work, but this veteran evidently was not restricted. The analyst states that "the dose calculation . . . does not consider [the veteran's] allegation that he became disoriented while searching for some test vehicles and spent a few minutes in a high-radiation area. The dose resulting from such an excursion cannot be estimated without more specific information."

This narrative illustrates two points. First, if given the opportunity, veterans sometimes can provide detailed and compelling accounts about their experiences. The men who participated in these atomic tests knew that they were making history at the threshold of the nuclear age. Although memory is not totally reliable, such experiences are not easily forgotten. Second, although it is inherently difficult for an analyst to take scenario uncertainty into account quantitatively, a better effort could be made to acknowledge that such uncertainty exists and to account for it. Although the committee did not try to recompute the veteran's dose, there was consensus that his true external dose could have greatly exceeded the assigned 6 rem, and that there was also the potential for substantial inhalation dose and beta dose to the skin, exposure routes that were not considered.

Contributed case: Another example, not among the 99 sampled cases but a record that was randomly pulled from the Science Applications International Corporation (SAIC) files and then used with the permission of the veteran, concerns an Air Force helicopter technician. In this case, assumptions made throughout the dose reconstruction did not appear to give the veteran the benefit of the doubt. Other personnel involved, whose names and ranks were provided to the analyst by the veteran, could have provided supplemental information, but the record does not indicate that any follow-up contacts were attempted. The case is particularly interesting because it involved highly unusual, or possibly unique, conditions of exposure, which can place considerable demands on the analyst in developing an exposure scenario that fits the particular circumstances.

The veteran had been trained in maintenance of F-84G aircraft that were used for cloud sampling after nuclear detonations in the South Pacific (see Figure V.A.5). He arrived at Kwajalein on September 30, 1952, and was present for both detonations in Operation IVY. After Shot MIKE (November 1, 1952), two F-84G sampler planes had to leave the radioactive cloud because one got into trouble and "went into a spin" and the other followed it. The first one could not return to land and the pilot went down in the sea with his plane. The other plane just made it to Enewetak but had a rough landing, blowing out two of its tires. The veteran was flown to Enewetak to change the wheels and tires, refuel the plane, and use a power source to restart its engine so that it could return to Kwajalein. The downed F-84G must have still been holding its very hot air samplers on its wings and nose. On his return to Kwajalein, the veteran recalled that he required more than 4 h of showering before the Geiger-counter reading on him came down to acceptable levels.

The veteran's initial dose reconstruction, as reported to him in 1983, assigned him a dose of 0.000 rem. He complained right away. In 2000, he filed a claim for service-connected disability. The analysts revisited the calculations at that time, and a revised dose assessment was reported.

The second dose reconstruction began with the fact that 4 days after Shot MIKE, the external exposure rate at 4 in. from the pylon of the F-84G that he had serviced was recorded as 0.10 R h^{-1}. That was extrapolated back in time (on the basis of a decay rate of $t^{-1.2}$) to the time when the veteran would have been on Enewetak changing the tires, but this extrapolation evidently did not take into

FIGURE V.A.5 F-84G cloud-sampling aircraft.

account the presence of cloud samples while the veteran was working on the plane and the likelihood that the plane lost some of its radioactivity in a washdown after its return to Kwajalein. It was assumed that it had taken him 1 h of work close to the hot plane to get both tires changed, refuel it, and restart it. It was assumed that the veteran spent that time near the landing gear at a distance of 1 m from the contaminated fuselage ("his arms being extended"). The landing gear and blown tires were assumed to be uncontaminated because they would have been "tucked inside" the plane.

The committee did not attempt to do a dose reconstruction for the veteran, but the committee took issue with every assumption that was applied and considers the assigned upper bound of 0.8 rem to be much too low to adequately reflect the uncertainties in scenario definition and estimation of dose. The extrapolation of the measured exposure rate backward in time is complicated by the fact that the plane would have had its highly radioactive air samplers removed immediately on its return and the possibility that it cooled off during the 2-h flight back to Kwajalein and was hosed down before day 4 to begin its decontamination.

Elsewhere in the dose reconstruction report, the analyst calculates doses that the veteran might have received in later work where he decontaminated F-84s, mentioning that the planes were routinely decontaminated within a day of their return from flying through the mushroom cloud. The analyst states that:

> During the mornings following both shots (2 November and 17 November) the F-84G aircraft were moved to a decontamination ramp at Kwajalein, where they would be thoroughly scrubbed and washed down. The average radiation intensity upon landing of the F-84G's was 2.5 [R h^{-1}]. As the readings were taken of various aircraft parts, the average was likely indicative of radiation levels at 4 inches from the surfaces of aircraft components that personnel were likely to spend the majority of their time maintaining. Engine/intake area decontamination effectiveness was about 50 percent; smooth surfaces were about 95-98 percent. The highest surface contamination zones on the aircraft were leading edges, air intakes, and engines.

Even if the wings were decontaminated with an effectiveness of only 90%, it follows that the measured reading of 0.1 R h^{-1} on day 4 should have been multiplied by 10 before extrapolating it back to shot-day levels. On that basis, it seems reasonable to suppose that the estimated dose during the tire-changing event was too low by at least a factor of 10. Again, this conclusion does not take into account the presence of air samplers, which would increase the extent of underestimation of the veteran's dose.

Other assumptions made in the scenario reconstruction do not seem to give the veteran the benefit of the doubt. The assumption that the landing gear and tires were not contaminated seems doubtful. Potentially, the well in which the landing gear is housed during flight may serve as a trap for radioactive particles. The metal cover over the wheel well swings down when the landing gear is extended, and the cover presumably was contaminated. Finally, the assumption that it took 1 h to complete the maintenance of this plane (with a half-hour spent away from it) is not well established by the record.

This example shows that despite assertions by the NTPR program, assumptions about exposure scenarios used in reconstructing doses are not necessarily "high-sided" and do not necessarily give the benefit of the doubt to the veteran.

Case #73: In other examples among the cases reviewed by the committee, assumptions applied did not seem to give the veteran the benefit of the doubt. In this case, an assumption was made that because there was no record of badging for some missions and the analyst believed that the policy would have been to badge all participants with potential for exposure, the veteran must not have had the potential for any measurable exposure during his missions. The case involved a participant in the CROSSROADS test series who was stationed aboard the USS *Prinz Eugen*. He was cited for outstanding work in removing ammunition from contaminated vessels under difficult and hazardous conditions. In the reconstructed scenario, the veteran was assumed to have done decontamination work for only 4-h shifts every second day; little basis for the assumption was offered. When there were gaps between days with badging, the veteran evidently was assumed to have zero dose. As discussed in Section IV.B.2, doses on ships are assigned on the basis of the mean exposure rate recorded on known dates on the ship with allowance for decay according to the fitted time course and extrapolation backward in time based on the physics of radioactive decay. No original badge records are included in the file, nor does the file document the radiation levels on the dozen target ships this veteran had worked on. Little information came directly from the veteran, as the file includes only a brief questionnaire with his terse responses. Very little uncertainty was assigned to the estimated dose for the veteran. The estimated dose and upper bound in this case do not adequately reflect plausible conditions of exposure and uncertainties in estimating the veteran's dose.

Case #47: Another veteran participated in Operations CASTLE and IVY. Many film badge records of CASTLE evidently were lost, and the records of many participants do not include dates of issue or turn-in. The records also show confusion over who was wearing a particular badge, and the veteran in this case evidently wore more film badges than could be found. A memorandum in the case file to VA states that "at this point, there is reason to suspect the entire CASTLE database as having the potential for serious errors that could be very embarrassing in litigation." The veteran's main responsibility was driving people from place to place on the islands. His story about his experiences included a statement that he was flown over ground zero 6 days after the blast and a claim that he visited an island about 15 miles from ground zero 3 days after a test, stayed for about 8 h, and was then ordered off the island because of concern about his radiation dose. Those stories were discounted and were not used in the scenario reconstructions after the analyst contacted his two former superior officers, who did not recall the alleged events and considered them unlikely. The dose returned to VA was assigned no upper bound (at that time, in 1989, upper bounds were not being calculated), so the adjudication of his claim presumably did not take dose uncertainties into account but used only the reported central estimate of dose.

The general who was contacted stated that the usual practice had been to leave the "permanent" badge behind on missions for which a mission badge was issued. (This is in contrast to a report on the REDWING test series (Bruce-Henderson et al., 1982), which states that permanent badges were to be worn at all times.) If that was the usual procedure, dose estimation should routinely have added mission badges to permanent badges (in line with giving the veteran the benefit of the doubt) rather than treating them as redundant measures and ignoring them, as was usually done (see Section V.B.1.1).

Case #55: Occasional difficulties in reading original film-badge data were not always resolved in a way that gave the veteran the benefit of the doubt. For example, the veteran in this case has a record for a badge issued July 7, 1956, that seems to indicate a reading of 3.105 rem, but the analyst apparently believed that the handwritten number "3" could also have been a "2." It was originally treated as a "3" but was treated in the final dose assessment as a "2." Nevertheless, uncertainty of less than 1 rem was assigned to the dose estimate.

Case #40: Another example of a scenario reconstruction that did not appear to give the veteran the benefit of the doubt involved a participant who served with the Army as a smoke-generator specialist in Operation UPSHOT-KNOTHOLE at the NTS and was present during a fallout event at Camp Mercury. He contacted the program initially in 1989 to request information on his dose. He was part of a group of 17 men who carried out two experiments in spring 1953 to determine whether smoke screens can protect against thermal radiation (heat). One of the experiments was associated with Shot ENCORE (May 8, 1953), and the other with Shot GRABLE (May 25, 1953); they had the same intended ground zero. His group had to go out to the site during the hours before detonation and set up hundreds of smoke pots and smoke generators along specified lines near the intended ground zero, trigger the smoke-generation system remotely at the time of the blast, and then collect all the contaminated equipment on shot day. The file includes no statement from the veteran about his precise role in the experiments, and no badge data exist for him though others in his unit had measurable doses ranging up to 0.9 rem. On the basis of that information and a presumption that badge records for this test series are essentially complete, a letter from DTRA to the veteran states that according to rad-safe requirements at the time of UPSHOT-KNOTHOLE, he would not have been allowed in an area with radiation intensity above 0.01 R h^{-1} without a badge.

Thus, the rationale in this case is that the veteran's dose must have been low because he apparently was not badged. His total assigned dose of 0.1 rem is based on reconstruction.

Case #37: A photographer for the Army who served in Operations TEAPOT (1955) and PLUMBBOB (1957) might not have been given the benefit of the doubt. One uncertainty concerned his date of arrival at the NTS. On the same day (August 21, 2000), the same person at JAYCOR evidently wrote two memoranda to the same analyst at SAIC, one stamped "Received" citing the veteran's date of arrival at Camp Desert Rock as March 23, 1955, and one not stamped "Received" giving his date of arrival as April 18, 1955. The analyst evidently treated April 18 as the correct date, thereby excluding the possibility that the veteran participated in the several shots in TEAPOT that fell between those dates. Additional uncertainty attends his dates of participation in 1957. The file contains no direct statement from the veteran, but there is a note that he claimed that he was present by special orders (of which he had a numbered record, issued April 19, 1955) in a tank at ground zero within 1 h of a detonation. Evidently, some records related to these special orders have survived, but they are not in the SAIC file. The mission would probably have occurred at Shot APPLE-II, which took place on May 5, 1955. No film badge record remains for estimating the veteran's dose. There was apparently also a question regarding the veteran's unit. An initial dose assessment gave him the benefit of the doubt by assuming that he was a member of the most highly exposed unit, but a later assessment reduced his dose by assigning him a weighted average of the doses to the various units participating in the exercise.

Case #87: A number of veterans had a clear potential for skin contamination. One Army veteran operated earth-moving equipment during Operation UPSHOT-KNOTHOLE and later developed skin cancer (see Figure V.A.6). Earth-moving was required in building roadways, setting up target areas, clearing sites after shots, and digging trenches in preparation for new tests. This kind of work was presumably very dusty in the Nevada desert, and there were regular opportunities for both skin dose (through being dirty all day) and inhalation of radioactive dust produced by resuspension of radionuclides in previously deposited fallout. The veteran also was an observer in the trenches during shots. In the dose reconstruction, the veteran seems to have been assigned a generic dose on the basis of averaging the daily person-time that engineering units would have spent in clearing operations and estimating the probability that each member participated. In short, an average dose for the unit was calculated and assigned to the veteran. However, there may have been considerable variation in dose among the participants in this work, and the unavoidable uncertainty about what this particular veteran was assigned to do remains unaccounted for, given that the assigned upper-bound dose is within a factor of 2 of the central estimate.

FIGURE V.A.6   Photograph of a bulldozer clearing a path through a contaminated area.

Case #99: The uncertainty in scenario reconstruction sometimes goes beyond the specific tasks and conditions that a veteran experienced and goes to some basic questions: Was he even there? Which series was he present for? For example, one veteran who was a major in the Army filed a claim in 1980, alleging participation in Operation UPSHOT-KNOTHOLE. He was evidently an ordnance officer with the Special Weapons Command. His presence at the NTS initially could not be verified, although he claimed to have witnessed 21 shots. No film-badge records were found, so the dose was based entirely on reconstruction. The analyst made some educated guesses about what the veteran's responsibilities might have been and credited him with being present at 11 shots, describing this approach as "high-sided" despite the veteran's claim that he was present at 21 tests. (The analyst clearly was frustrated by this case and offered in a memorandum that the veteran also may have been present at another series of tests, perhaps TUM-BLER-SNAPPER.) Because the veteran's case was analyzed in 1983, he was assigned a dose but not an upper bound.

Case #84: Another case in which some basic facts are unclear involves an Army sergeant who participated in Operation UPSHOT-KNOTHOLE. The veteran had no film-badge data, so his estimated dose was based entirely on a generic (unit) dose reconstruction and consideration of his personal circumstances. The file contains little information to support the unit-based assignment of dose, but the veteran evidently was part of Battalion Combat Team B (BCT-B), which was a forward unit at Shot NANCY. The reconstruction appears to ignore published information about what happened at Shot NANCY, which indicates that troops accidentally ventured into a radiation field well above the stated limit (14 R h^{-1} compared with the 2.5 R h^{-1} limit). The excursion was attributed to the fact that monitors had not kept their commanders informed of the radiation environment and then experienced difficulty in withdrawing the troops from the high-radiation area. It was asserted that the effect on the dose estimate was small because little time was spent in the area (Edwards et al., 1985). That might not have been the case, however. After the shot, events apparently were chaotic and confused. The unit found itself too far forward; without knowing it, the unit was in an area with radiation intensities in excess of 14 R h^{-1}. There was inefficient communication between the rad-safe monitoring personnel and the commanders, and once the commanders recognized that they were in trouble, they had difficulty in moving the troops out quickly. In the view of the committee, the upper bound that was assigned to this veteran's dose does not adequately reflect the uncertainty in the estimate due to uncertainty in the exposure scenario.

Case #93: A personal account that was discredited concerned a veteran who had worked as a laundryman for the Army at Operation BUSTER-JANGLE (1951). He stated that he wore a badge for both shots that he witnessed, but no records of his film-badge readings remain. The scenario reconstruction discredited his account of being about 2 miles from ground zero and instead assumed that he participated only as an observer in Shots UNCLE and SUGAR, which would have put him many miles from ground zero and beyond the range where measurable exposure would have occurred. The analyst documented that the veteran was on a 20-day emergency leave and accordingly could have been present only at Shots SUGAR and UNCLE. However, an error may have been made in the assumptions. His original emergency leave was effective October 17, 1951. A 10-day extension was granted on October 27 but seems to have been modified in a later morning report to add 10 days of leave but to begin it effective November 6 rather than October 27. With those assumptions, the modified account of the dates would fit better with the veteran's own account because he could have been present as a witness at Shots DOG (November 1) and EASY (November 5). The veteran's actual dose therefore could have been much higher than the dose that he was assigned.

V.A.3 Conclusions on Adequacy of Scenario Determinations

The preceding discussion clearly illustrates that dose reconstructions performed in the NTPR program often fail to adequately establish the exposure scenarios that are the basis of the veterans' dose estimates. The committee found clear examples in which a veteran's location and duration of exposure were not unambiguously determined, a veteran was not given the benefit of the doubt with respect to his exposure scenario, adequate follow-up with a veteran or members of his unit was not carried out to define the scenario, or some potential exposure pathways were not considered. The following sections, which discuss the committee's findings with respect to the methodology used in the NTPR program to reconstruct external and internal doses and related uncertainties, provide further examples where uncertainty in the exposure scenario impacts estimated doses and upper bounds.

The committee recognizes that development of exposure scenarios can be challenging, given the lack of information on a veteran's activities and exposure environment in many cases. However, in accordance with applicable regulations, a veteran must be given the benefit of the doubt in the development of an exposure scenario. In the committee's view, that means that an analyst must consider plausible conditions of exposure that are consistent with available information, including statements by a veteran and other people with knowledge of the veteran's activities, and then select a plausible exposure scenario that results in the highest estimate of dose to the veteran. Selection of exposure scenarios should not be constrained by rad-safe guidelines or plans of operation when there is evidence that they were not followed. The committee's evaluation of individual cases discussed in this section suggests that selection of plausible exposure scenarios based on giving the veteran the benefit of the doubt is not an unreasonably burdensome task. If that approach is not followed, it is unlikely that credible upper bounds of doses will be obtained in many dose reconstructions, as intended by the NTPR program.

V.B EXTERNAL DOSE ESTIMATION

V.B.1 Introduction

Reconstruction of external doses by the NTPR program includes gamma doses estimated from film-badge data and scientific dose reconstructions in cases in which film-badge data are not available. Neutron dose is generally considered separately because film badges were relatively insensitive to neutrons. Beta skin and eye doses are also considered separately. Thus, the committee's findings with respect to both the central and the upper-bound estimates will be discussed separately for dose reconstructions based primarily on gamma exposure measured with film badges and for gamma doses estimated from unit dose reconstructions. Neutron and beta dose reconstructions in the NTPR program are also discussed separately. External doses based on film-badge data are discussed in Section V.B.2, reconstructed external gamma doses in Section V.B.3, and a summary of the committee's findings regarding external gamma dose estimates in Section

V.B.4. Neutron doses are discussed in Section V.B.5, and beta skin and eye doses in Section V.B.6.

As discussed in Section II.B, the committee, in responding to its charge, reviewed 99 randomly selected dose reconstructions in detail, including the supporting documentation in the SAIC files. A number of other dose reconstructions submitted by veterans were also examined. Committee members reviewed many of the data and reports on which the estimated doses for those cases were based, including published reports describing the generic unit dose reconstructions and unpublished internal memoranda. Additional information was obtained from formal replies by DTRA to written questions from the committee (see Appendix D) and presentations by NTPR program and VA staff at open committee meetings.

In 29 of the 99 cases examined by the committee, the veteran's reported external gamma dose was based primarily on film-badge data; 21 of these were for participants at Pacific tests. Upper bounds were reported for all but four of the 29. In 51 of the 99 cases, the veteran's external gamma dose was based primarily on his unit's generic dose reconstruction (22 associated with NTS testing and 29 with Pacific testing). In 14 of the 99, all Pacific-test cases, the veteran's dose reconstruction was based on a mixture of film-badge data and generic unit doses; the generic doses were modified as necessary to reflect the veteran's specific exposure scenario. Five of the 99 cases involved unbadged participants and activities that were not covered by a unit dose reconstruction.[1] In 19 of the 99 cases (14 Pacific and five NTS), all for claims filed before 1994, no upper bound was reported. Of the 99 cases, 66 involved primarily Pacific tests and 30 primarily NTS tests, but six people participated in tests at both sites, and a few received comparable doses at the two sites. A neutron dose was reported in three of the 99 cases in the random sample. However, some test participants, not among the 99 cases, received significant neutron doses (Goetz et al., 1981).

In 27 of the 99 cases, a claim or other indication of skin or eye disease was indicated. Beta doses were calculated for nine post-1998 claims, but beta doses were not reported for 18 cases of claims of skin or eye disease before 1998.

The committee's detailed evaluation of methods used in the NTPR program to obtain central and upper-bound estimates of the dose from external exposure based on film-badge data and unit dose reconstructions is discussed below. The conclusions are illustrated with examples from the 99 cases examined by the committee and with examples from additional information submitted to the committee by test participants.

---

[1]The relatively small fraction of such cases reflects the fact that most participants that were expected to be exposed but were not covered by a unit dose reconstruction were issued film badges. However, even if participants were covered by a unit dose reconstruction or film badge, some of the exposures often required additional analysis.

V.B.2 External Gamma Doses Based on Film-Badge Data

V.B.2.1 *Central Estimates*

As discussed in Section IV.B.1, external doses estimated by the NTPR program from film-badge readings are generally biased high compared with estimates based on National Research Council (NRC, 1989) committee recommendations and thus favor the veterans. However, the veteran is not always given the benefit of the doubt regarding allegedly damaged film badges, overlapping mission and permanent badges, and when and how long badges were worn.

The committee found that there was sometimes inconsistency in replacing allegedly suspect film-badge data with a reconstructed dose, particularly for the REDWING test series and for dose assessments before 1995. NTPR program policy regarding REDWING badges changed over time but in early 1995 was clarified to require replacement of all suspect film-badge data with a dose reconstruction if it was feasible (Schaeffer, 1995). As documented in the NRC (1989) report, many of the film badges issued early in the series and worn for more than a few weeks apparently suffered damage that caused the film to appear to record a higher dose—up to several hundred millirem more than actually received. However, later batches of film badges were waterproofed and were not subject to the same problems. Some of the data from the later badges may also have been discarded in favor of reconstructed doses, even for assessments before 1995. Ten of the 99 sample cases involved film-badge data from the REDWING series, and in eight of these the total external dose was based primarily on these data. In 8 of the 10 cases, some film-badge data were determined to be questionable, and the doses were replaced with reconstructed doses; however, in 2 (cases #35 and 42), data were accepted even though the badge was deemed possibly damaged. In three cases (cases #54 in 1995, #82 in 1956, and #55 in 1994), a possibly damaged film-badge reading was accepted but treated as an upper bound. Some of the badges whose data were replaced were rated highly suspect on reanalysis (see case #55), but others were deemed merely questionable (see case #82). However, the fact that case #42 was from 1996 suggests that the policy was not administered uniformly even after clarification.

The committee believes that the present policy of disregarding data for badges rated questionable on reanalysis, as opposed to highly suspect, should be reexamined. Even badges that exhibit slight damage, such as the absence of a distinct filter image, can be analyzed to provide a reasonable dose estimate (NRC, 1989). For example, in case #44 (1998), a potentially damaged badge was reanalyzed, and the originally reported dose of 0.97 R was revised to 0.46 R. However, the analyst replaced the film-badge dose with a reconstructed dose of 0.17 rem. Unless there is clear evidence that a film badge is so highly damaged as to provide a completely unrealistic dose estimate, a policy of accepting the data is more prudent than replacing them with a reconstructed dose. In most cases the data were already recorded in the veterans' medical records, and often they were reported to the veterans in previous NTPR program reports, so the practice of replacing them with lower reconstructed values does not give the veterans the benefit of the doubt as required by law; it also detracts from the credibility of the dose reconstruction process by giving the appearance that the tendency of DTRA is to reduce previously reported doses whenever possible.[2] The committee notes, however, that in most of the REDWING cases examined, the impact on the estimate of total external dose would have been minor.

Instances of uncertainty regarding when a badge was issued and turned in were common (see cases #10, 35, 47, 54, 74, and 97). In many cases, original film-badge records were available, but the fields for date of issue and return had never been filled in. Because reconstructed doses were calculated to account for periods when a person was not badged, incorrect assumptions regarding the period covered by a film-badge dose could have resulted in underestimation of a total dose.

The committee found one case (#32) in which previously reported badge results were in the file but were not used in the dose assessment. Cases in which suspicious data indicated possibly incorrect doses also were found. For example, in case #35, the veteran's film-badge reading was considerably lower than the unit average.

A mission badge was usually assumed to have been worn concurrently with a permanent badge if a permanent badge had been issued, although there is some indication that this was not always true (see discussion of case #47 in Section V.A.2). If the mission-badge dose was lower than the permanent-badge dose, the mission-badge dose was assumed to be included in the permanent-badge dose, and only the dose from the permanent badge was used. If the total dose determined from mission badges exceeded that from the permanent badge, the higher dose was used. It is possible that veterans did not wear mission badges continuously between the time they were issued and the time they were turned in, so in many instances a veteran should have been given the benefit of the doubt and the mission-badge and permanent-badge readings should have been summed.

In many cases, a participant was issued a mission badge but not a permanent badge. His dose for the period not covered by the mission badge was based on a unit dose reconstruction. For example, in case #11, mission badges were issued to a participant who serviced cloud-sampling aircraft; he was not issued a permanent badge. It was assumed that his mission badges accounted for his dose from fallout on the island during the period when the badges were assigned and his reconstructed dose from fallout was modified (on the basis of his unit dose reconstruction) to reflect this, even though he may not have worn the mission badges for the entire period. Therefore, his total dose may well have been underestimated. The committee identified at least eight cases in its sample of 99 in

---

[2]Note that the rationale given by DTRA for accepting the film-badge reading as an estimate of deep equivalent dose, rather than applying a bias factor as recommended by NRC (1989), was supposedly to avoid the appearance of reducing previously reported doses.

which either the mission-badge data were considered to be included in the permanent-badge data or the mission-badge data were assumed to include fallout doses on islands or ships (see cases #10, 11, 16, 32, 38, 44, 92, and 97).

If film badges were generally issued for particular types of mission activities and no data were located, the analyst often assumed that the absence of badge data indicated that the participant did not engage in a mission-related activity and assigned a reconstructed generic dose estimate. The veteran generally is not given the benefit of the doubt even when there is some evidence that he participated in such additional radiation-risk activities on the basis of his general duties or frequency of available mission-badge data. That considerable film-badge data are known to have been lost suggests that the inability to locate such data or extant film does not imply that a permanent or mission badge was not issued or worn during some periods of possible radiation exposure and that the participant therefore could not have been exposed. For example, in case #40, the analyst assumed that the participant was not exposed in high-radiation areas because no film-badge data could be found, even though an earlier assessment had given him the benefit of the doubt and assigned a higher dose (see discussion of this case in Section V.A.2). The participant was not contacted to inquire whether he wore a badge and whether he actually had entered a high-radiation area.

The committee found instances in the 99 sampled cases in which apparently no effort was made to search for film-badge data (for example, see cases #33, 36, and 38). For example, case #38 involved a supply supervisor stationed on Enewetak Island who claimed that he wore a dosimeter. According to his service record, he had previously been assigned a dose of 0.15 rem. That dose might have been based on a film-badge reading, but apparently no effort was made to determine its origin, and this dose increment is assumed to be included in his reconstructed dose rather than being added to it.

The committee found at least 10 instances (cases #21, 22, 27, 38, 40, 81, 87, 93, 94, and 98) in which participants claim to have been issued a film badge but no badge or data could be located. As discussed earlier, film badges and data are known to have been lost and to be no longer available.

For some veterans, some film-badge data were found. That suggests that they did participate in some activities that required badges and that other badge data could have been lost. The committee identified numerous cases in which mission-badge data could have been lost (for example, cases #9, 10, 17, 27, 32, 33, 36, 40, 44, 47, 54, 59, 92, and 99). In case #10, it is clear that the participant did not return all film badges.

Reconstructed doses were usually estimated only for periods for which no film-badge data were found. In its 99 sampled cases, the committee found frequent occurrences of re-evaluation of film-badge data because readings appeared to be anomalously high; many of these, as discussed previously, were for Operation REDWING. The committee found only one example of replacement of an obviously incorrect film-badge reading (zero) with a reconstructed dose (case

#97). There might be a greater tendency to question high readings than to question low readings, inasmuch as damaged film badges were known generally to read high. However, low readings might result from failure to wear badges when they were supposed to be worn or from incorrectly recorded data.

A veteran's estimated dose apparently was not routinely compared with that of others in his unit with similar duties and more complete film-badge records. The committee noted several cases in which such a comparison would have been appropriate (for example, cases #24, 25, 35, 38, 58, 68, and 69). For example, in case #24, the participant's recorded film-badge reading was the lowest of all recorded badge data for his unit.

Doses based on film-badge data are assumed to be definitive for the period covered if such data are available, even though the film-badge data may not agree with a reconstructed dose. Considering that the ratios of the upper-bound dose to the central estimate that the NTPR program provides for "scientific dose reconstructions" are often much lower than the corresponding ratios that it applies to film-badge data, that practice seems contradictory. The committee found that in many cases, the policy does not give the veteran the benefit of the doubt regarding his potential total and upper-bound external dose, particularly if readings from slightly damaged film badges were replaced with reconstructed doses, if possible mission-badge data were not located, if incorrect dates were used for periods when the badges were worn, if badges were not always worn continuously as required, or if incorrect badge data were used. The additional uncertainty in doses based on film-badge data should be reflected in estimated upper bounds, and this is discussed in more detail below.

V.B.2.2 *Upper Bounds*

As discussed in Section IV.B.1, the NTPR program estimates the upper bound of an external dose for a single film-badge reading as recommended in the NRC (1989) report. However, it estimates a slightly lower uncertainty in the sum of multiple film-badge readings than if the most extreme method suggested in the NRC (1989) report had been used. The mean dose estimated by the NTPR program is inflated by about 30% relative to the National Research Council recommendation, because a correction to convert exposure in roentgen to equivalent dose in rem is not applied. Consequently, the upper bound of the sum of multiple-badge data calculated by the NTPR program is still almost always greater than if even the extreme National Research Council recommendation for calculating an upper bound had been followed precisely.[3] Thus, a dose estimate and upper

---

[3]It is interesting to note, however, that even though the committee believes that the modifications to the National Research Council recommendations used in the NTPR program are not unreasonable and are in the veterans' favor, the NTPR program somewhat misleadingly asserts in its communications to veterans that their film-badge results are based entirely on those recommendations.

bound obtained by the NTPR program based primarily on film-badge data should be on the high side if the participant was badged during all periods of potential exposure.

However, the reported film-badge data may not provide an accurate and complete record of a veteran's dose for the period supposedly covered. The upper-bound estimate does not reflect the possibility that the veteran did not wear his badge at all times, that mission badges were issued but not turned in or the data were lost, or that the interval when the film badge was purportedly worn is incorrect. As discussed earlier, the committee found that records indicating when badges were issued and turned in were often ambiguous or had no entries in the date fields. Assumptions had to be made on the basis of reasonable likelihood. However, the possibility that the assumptions were not always correct is not reflected in the upper-bound calculations. Furthermore, no additional uncertainty is assumed to account for the possibility that incorrect data were reported or that a reading was assigned to the wrong individual because of, for example, clerical errors or switching of data. There appears to be considerable evidence that film-badge data were lost, badges were lost, and badges were issued but not always worn. That suggests that for many veterans whose reported doses are based primarily on film-badge data, the upper bound assigned may not truly reflect a credible 95th percentile of the possible dose.

Finally, the NTPR program uses the same uncertainty estimates for a dose estimate based on a cohort badge as for one based on an individual badge. (Recall that "cohort" badging was used when several representative unit members were badged and that the data from such badges were attributed to others in the unit.) The assigned uncertainty clearly should be increased when a dose is based on a cohort badge to reflect the likely variations in doses among the members of the unit. Cohort-badge data were included in the dose estimates in at least six of the 99 cases examined by the committee (cases #6, 9, 15, 26, 39, and 68). In at least one of those (case #15), the committee found that the cohort dose probably considerably underestimated the dose to the participant, who probably entered areas with much higher exposure rates than an average member of the cohort.

In summation, among the 99 random cases reviewed, the committee found a number in which an increased upper-bound estimate of doses based partly or primarily on film-badge data might well be warranted.

V.B.3 Reconstructed External Gamma Doses

V.B.3.1 *Central Estimates*

The committee examined the methods and models used to estimate average external gamma doses assigned to units on the basis of cohort film-badge data or radiation-survey data combined with assumptions about a unit's activities. In many of the unit dose reconstructions, the analysis was thorough and comprehensive. As in a previous National Research Council study (NRC, 1985b), the committee has concluded that the methods used to estimate average doses from external exposure to gamma rays are generally acceptable and, if adequate input data are available, provide credible estimates of the average dose to members of a unit.[4]

Many of the unit dose reconstructions include a comparison of the mean unit dose with available film-badge data. The mean reconstructed doses generally agree fairly well with the mean of the film-badge data. However, the film-badge data are often sparse, and the variability in the film-badge data is often greater than the estimated uncertainty in the mean unit dose. That result is attributed to the fact that many of the film badges were issued to participants who were expected to receive higher doses, such as radiation monitors, so the higher doses do not reflect doses to ordinary participants. However, the estimated uncertainty should still reflect the possibility that not all participants with a potential for high doses had a (surviving) film-badge record.

In many cases, the unit dose estimates are based on sparse survey data and questionable assumptions regarding exposure scenarios, particularly for exposure on some ships, so the dose to some individuals in the unit could substantially exceed the mean. For some smaller units, in which the specific daily activities of individual members could not be precisely determined, daily average doses to members of a subunit were estimated by using a daily weighted average of doses for various radiation-risk activities that was based on previous dose reconstructions, manpower requirements for each activity, and morning-report unit strengths (Ortlieb, 1995).

External dose estimates based on reconstructed doses depend on the validity of assumed exposure scenarios and the inclusion of doses from all possible exposures. However, as discussed below, doses to the most exposed people in many units may not have been realistically estimated in generic dose reconstructions, particularly if the exposure scenario for some did not conform to that assumed for the unit as a whole.

V.B.3.2 Upper Bounds

As described in Section IV.D.2, the unit dose reports generally provide a discussion of the uncertainty in the central or "best" estimate of the dose to a representative member of the unit. As discussed in Section IV.E.2.2, the estimated uncertainty does not reflect the possible upper-bound dose to any individual in the unit but rather the distribution of possible doses about this central estimate. Departures by individuals from the assumed group scenario are not

---

[4]It should be noted that the previous National Research Council review did not address uncertainty in individual dose reconstructions.

considered in estimating uncertainty. This uncertainty analysis is asserted to be "high-sided" because it estimates space-time scenarios in a manner that overestimates exposures, particularly for NTS observer and maneuver units. However, the ratio of the 95th percentile upper-bound dose to the central estimate for external gamma-ray dose based on unit reconstructions is often very low, with the upper bound sometimes only 10 to 20% above the central estimate. Uncertainties of that magnitude are not consistent with uncertainty ranges generally estimated for other dose reconstructions of external radiation exposure based on similar types of data (Henderson and Smale, 1990; Simon et al., 1995). The ratios of reconstructed upper-bound doses to central estimates are often even lower than the ratios based on film-badge data, even though film-badge data, if available, are assumed to provide the most reliable dose estimates and are considered the doses of record.

In many cases, the calculated upper bounds, even though alleged to be "high-sided" estimates, appear to be completely unreasonable. For example, the 95th percentile upper bound estimated for Operation UPSHOT-KNOTHOLE, Shot SIMON maneuver units (Edwards et al., 1985) is only 10% above the "best" estimate (see Table IV.B.1). As discussed in Section IV.B.2, that estimate results from assuming that the uncertainty in the radiation field could be neglected, that the central estimate is itself "high-sided," and that all the uncertainty is due to uncertainty in march speed and stay times. But the resulting total uncertainty estimate is less than the uncertainty in the measurement of exposure rate in the field with available survey meters (Brady and Nelson, 1985).

Similarly, the 95th percentile upper bounds calculated for some Pacific test series unit dose reconstructions also appear to be unreasonably low. For example, the upper bound for doses incurred in boarding or decontaminating target vessels during Operation CROSSROADS is estimated to be only 20% above the mean dose (Weitz et al., 1982), and the upper-bound dose for seamen and island residents from fallout during Operations GREENHOUSE and CASTLE also was generally estimated to be only about 10-20% above the mean dose (Thomas et al., 1982; 1984).

Because uncertainty analyses reported in the unit dose reports are complex and because detailed data and calculations are usually not included in the reports, it was not possible to examine all the specific calculations, assumptions, and supporting data in detail. However, as discussed below, the committee has concluded that the reported upper bounds are not always credible estimates of the 95th percentile dose to all members of the unit.

V.B.3.2.1 *NTS unit dose reconstructions*

For NTS test observers and maneuver troops, the estimated upper bound of an external gamma dose is based on assumptions that the radiation field was well documented, that the times spent by the troops in various locations were fairly well known, and that all participants followed rules (for example, stayed down in trenches for a specified time after a shot and did not venture into areas marked as exceeding mission exposure-rate limits) (Frank et al., 1982; Edwards et al., 1985). The doses were often intended to be "high-sided" by maximizing the time that troops were assumed to have spent at the highest allowed exposure rate (Goetz et al., 1981). That conclusion assumes that there was no error in establishing and marking the lines reflecting the limits and that extrapolations of survey data over time were correctly made with appropriate radioactive-decay factors.

However, monitors with troops at the NTS may have failed to do their job properly, instruments may not have been properly calibrated, and officers may have ignored monitor readings or delayed their response. As discussed in Section V.A, it is clear that procedures were not always adhered to and that errors were made.

The assumption that radiation fields were always well characterized is also questionable. With regard to the Marine Corps maneuvers at Operation UP-SHOT-KNOTHOLE, Shot BADGER (Frank et al., 1982), the unit dose reconstruction report notes that: "a major obstacle to the preparation of this report was the lack of systematic radiological survey data for the area of Brigade operations." For Army maneuver units at UPSHOT-KNOTHOLE (Edwards et al., 1985), isopleths estimated by the US Atomic Energy Commission were used to generate space-time models of exposure rate because original survey data were not available.

Assumptions regarding locations of troops vs time are often asserted to be without error or "high-sided," although the assertions include such words as *likely*, *supposed to*, and *probably* (Frank et al., 1982; Edwards et al., 1985). Finally, departures by individuals from the assumed group scenario are not considered in estimating uncertainty (see Appendix F for discussion of a case in which an actual exposure scenario probably did not correspond to the scenario assumed in a unit dose reconstruction).

V.B.3.2.2 *Pacific test site unit dose reconstructions—exposure to fallout*

Upper-bound estimates of external gamma dose from fallout for participants in many of the earlier Pacific test series assume that participants were exposed at random locations when on deck or outdoors on a contaminated island or ship, rather than at or near a fixed duty station where the external exposure rate may have been higher or lower than the mean. Estimates of the time that veterans may have been indoors or below decks and of the shielding provided by tents and buildings on islands may also have been overestimated for some participants. Furthermore, generic upper-bound estimates are often incorrectly assigned when calculating the dose to a participant who was exposed for only a fraction of the time assumed in the unit dose reconstruction (for example, when a participant was exposed for only a fraction of the interval assumed in the generic dose reconstruction and the dose was prorated with the same fractional error).

The assumption that each exposure outdoors or on deck was at a random location results in too low an upper bound. It is far more likely that a participant's assigned duties placed him repeatedly at particular places on a ship or island, and it is unlikely that he was in entirely different areas on the same day. If, for example, the participant was a member of a decontamination crew, his duty location might more likely be in a high-activity area than in an area with an exposure rate reflected by the mean of all the survey data. Even if a participant was not a crewmember, there is no reason to believe that his topside or island activities would place him in completely random locations during the first few days, when most of the dose would be incurred. For personnel billeted on islands, even though the indoor dose is reduced by an assumed 50% shielding factor, it is not negligible, and the indoor location was likely to be repeated rather than random, although his outdoor locations might have been more varied.

The upper-bound gamma doses from exposure to fallout at the Pacific test sites are generally based on a default 50% coefficient of variation (CV) in the mean exposure rate on an island or ship. However, the 50% CV estimate is itself highly uncertain for most events, and that uncertainty is not considered in estimating upper bounds. The CV estimate includes a component representing the variance due to the measurement itself,[5] so it does not just represent the variability with location.

Even accepting the 50% CV estimate as reasonable, or a modified value for ships with few data or for which better information is available, a more reasonable upper-bound estimate can be obtained by assuming reasonable uncertainty distributions of parameter values and estimating the total uncertainty stochastically with a Monte Carlo calculation. In contrast with the method used in the NTPR program, we make the pessimistic assumption that the participant was exposed at the same indoor and outdoor locations for the entire period of exposure.

The uncertainty due to time topside on a ship (or outdoors on a residence island) versus inside and the shielding factor, uncertainty in the decay rate, and uncertainty in converting free-in-air exposure to dose must also be considered. Let

$$I = \int \{ [E(t_0) \times SF \times IN] + [E(t_0) \times OUT] \} \times t^{-x} \times FBE \, dt.$$

The integral is over the period of exposure t_1 measured from the time of the test to some time T when the veteran left the test site area, and t_0 is the time of the measurements of exposure rate. (If the period of exposure is not continuous, the integral can be written as a sum of integrals.) In the equation, I is the integrated dose, $E(t_0)$ is the mean of the survey exposure rates measured at time t_0, SF is a

---

[5]The uncertainty (precision) in survey meter readings is given as about ±10 to 25% for most instruments. However, the bias is presumed to be small on the basis of an assumption that the instruments were calibrated properly (see Brady and Nelson, 1985).

shielding factor to account for attenuation of gamma radiation when the participant is indoors or below decks, IN is the fraction of time indoors or below decks and OUT the fraction of time on deck or outdoors, FBE is the film-badge-equivalent dose factor (a conversion from free-in-air exposure to dose), and t^{-x} describes the reduction in exposure rate with time due to radioactive decay. A value of 1.2 for the variable x reasonably fits the decay of fallout dose rates for most tests, but a larger or smaller value often gives a better description of the actual decay rate for particular intervals and tests.

Evaluation of the integral given above provides the following expression for the total gamma dose received over the interval from t_1 to T:

$$I = \{[E(t_0) \times SF \times IN] + [E(t_0) \times OUT]\} \times t_0^x \times [(t_1^{1-x} - T^{1-x})/(x-1)] \times FBE.$$

The upper bound or variance in the total dose calculated from the above equation can be estimated by assigning an uncertainty distribution to each variable. An estimate of the total variance and 95th percentile upper bound was obtained with a Monte Carlo simulation by using the following assumptions:

$E(t_0)$ = 1 R h^{-1}, lognormally distributed with a CV of 0.5 (the indoor and outdoor exposure rates are sampled independently in the Monte Carlo simulation).

SF = 0.1 (ships), lognormally distributed with a geometric standard deviation (GSD) of 1.5; SF = 0.5 (islands), normally distributed with an SD of 0.1 (the NTPR program assumes an SD of 0.05).

OUT = 0.4 for ships (as assumed by the NTPR program), lognormally distributed with a GSD of 1.4; OUT = 0.6 for islands (as assumed by the NTPR program), lognormally distributed with a GSD of 1.4.

FBE = 0.7 rem R^{-1}, normally distributed with an SD of 0.05 (consistent with the NTPR program's estimated uncertainty).

x = 1.2, normally distributed with an SD of 0.1 to account for the variability in x from shot to shot as well as with time after a particular shot. An SD of 0.1 is consistent with data presented in various DNA and SAIC reports; for example, Thomas et al. (1984) estimate a 90% confidence interval of ± 0.2 on the basis of observations.

The following compares the normalized results in rem for the 95th percentile upper bound (UB) from this Monte Carlo (MC) analysis for an exposure from $t_0 = t_1 = 9$ h to $T = 120$ h with the approximate upper bounds that would be reported by the NTPR program for similar scenarios:

	Dose	UB-MC	NTPR-UB
Ship	6	13	~7
Island	10	19	~12

Similar comparisons could be expected for other intervals.

From the Monte Carlo analysis, it seems clear that a realistic 95th percentile upper-bound estimate of the dose from a single event could be a factor of 2 or more above the mean dose even if the 50% CV for E_0 is valid, and it could be even higher if the CV is based on little real data.[6]

Even assuming that a participant's location when outdoors or topside varied enough that he was unlikely to have been exposed always at the same exposure rate, thus giving the above analysis of a "high-sided" estimate of his upper-bound dose, it seems clear that the assumption by the NTPR program that he would be exposed completely randomly to the entire distribution of measured exposure rates provides too low an upper-bound estimate. That conclusion is supported by available film-badge results for Operations GREENHOUSE and CASTLE. An SAIC memorandum regarding film-badge data for seamen on the USS *Curtiss* indicates that the highest film-badge readings were a factor of 2–3 above the mean for various exposure periods (reference CIC-67763, available through http://www.osti.gov/opennet/). A similar memorandum regarding film-badge data for servicemen on Parry Island during Shot DOG indicates that the mean was 0.9 R with a range of 0.56-1.4 R (CIC-58845, available through the same Web site). In both cases, the highest film-badge dose was well above the estimated upper bound for reconstructed doses, which was only about a factor of 1.2 above the mean. Data included in the file for case #68 indicate that the upper end of the range of film-badge readings on the USS *Estes* (CASTLE series) is much higher than the upper bound estimated for the reconstructed dose to members of the crew.[7]

Additional exposure to nonbadged crewmembers participating in decontamination activities is not considered in the above analysis. Personnel assigned to decontaminate ships would be likely to receive higher than the average exposure rate for the ship. It is true that weathering and decontamination activities may have reduced the exposure rate somewhat compared with an exposure rate based only on the initial survey data and default decay rate. However, the survey results on ships were often taken during or after the initial decontamination activities, and on-deck activity was generally restricted until decontamination was over.

Finally, the NTPR program does not apply its own uncertainty model correctly. For example, for Operation GREENHOUSE, as discussed earlier, the calculated upper bound is about 20% above the mean for a person exposed over

---

[6]Exposure to multiple events based on additional survey data would reduce the upper bound somewhat depending on the relative fallout levels. However, it is also not clear that the NTPR program properly accounts for the additional variance due to subtracting residual exposure rates from previous events to obtain the appropriate E_0. Any instrument bias in the measurements (assumed to be small relative to the CV) would result in a corresponding bias in the estimated dose.

[7]As discussed in Section IV.B.2, the NTPR program believes that the higher film-badge data reflect doses to personnel whose duties were more likely to result in high radiation exposure and, thus, do not reflect the upper bound for most unit members. The data do, however, call into question whether unit members are being given the benefit of the doubt in estimating the upper-bound dose.

the entire series (several months) to fallout from several tests. However, it is assumed that this 20% factor applies to a person exposed over a shorter period when the dose is calculated by integration with the $t^{-1.2}$ rule, rather than by summing the daily doses and variances for the shorter period (see case #4).[8]

V.B.3.2.3 Pacific test site unit dose reconstructions—exposure on contaminated target ships

The committee believes that upper-bound estimates of external doses received when boarding target ships in Operation CROSSROADS are unreasonably low. The available data used to calculate the mean exposure-rate curves for each ship are not given, but it is asserted that the upper bound of a total dose is less than a factor of 1.2 above the mean. The original unit dose reconstruction (Weitz et al., 1982) assumed a factor of 1.5. However, that was changed in 1986. The rationale was that the reported averages generally fell within a factor of 1.5 of the trend lines for the daily doses (see Figure IV.B.1) and that an uncertainty factor (ratio of upper bound to mean) of 1.2 better represented the standard error of the mean, which is asserted to be a more appropriate estimate of the error for multiple boardings (Schaeffer, 2002b). As was the case with exposure to fallout during later test series, that assertion is based on assuming random locations during each boarding even if the participant was exposed only for one boarding period (Schaeffer, 2002b) rather than allowing for variation across the measured exposure rates on the ship.[9] It is even less likely that a nonbadged participant's location when he was engaging in inspection or decontamination activities aboard a contaminated target ship would be random than for a participant exposed to fallout on an island or ship during later test series. Furthermore, as discussed below, variations in exposure rate with location on target ships were in most cases probably greater than that measured on ships exposed to fallout during later test series.

The committee examined data on the range of exposure rates on target ships. The data suggested large variations with location on contaminated ships (B2, 1946). Table V.B.1 is an excerpt of those data, showing the average and maximum exposure rates measured on a few target ships. The ratio of the maximum exposure rate to the average suggests a CV well above 50%. For most ships, the available data are sparse, and it is not clear how the mean was determined. Extensive survey data are available, however, for one ship, the USS *Salt Lake City* (B2, 1946). For the *Salt Lake City*, the mean exposure rates on deck (topside) ranged from an average of about 0.4 R d^{-1} near the bow to >10 R d^{-1} in other open

---

[8]For the upper-bound factor (95th percentile relative to the mean) to be reduced from the value based on a single exposure with a 50% CV, which corresponds to about 1.9 times the mean, to as low as 1.2 times the mean requires at least 5 days (15 intervals) of random exposure.

[9]Schaeffer (2002b) agrees that a factor of 1.5 "could be justified for a single boarding" but notes that "the standard error of the mean is usually more appropriate."

TABLE V.B.1   Radiation Levels on Selected Target Ships in Roentgen per 24 h (B2, 1946)

Ship[a]	H+day	Reported Mean	Topside Maximum
Parche	8	4	7.5
Pensacola	8	10	50
Salt Lake City	10	3	150
Mugford	8	3	30
Prinz Eugen	9	4	60
Skate	7	4	30
Tuna	7	4	30

[a]*Parche, Skate,* and *Tuna* were submarines, but only *Tuna* was submerged. Other unpublished data suggest that the topside variation in exposure rates was somewhat lower on *Tuna* than on other ships.

areas amidships and up to 150 R d^{-1} for hotspots. Although data for other ships are more limited, data from B2 (1946) and transcripts of radio communications available from the Department of Energy Coordination and Information Center (CIC documents #57001, 57004, 57007, 57020, 57023, 57032, 57033, 57044, 57046, and 57047, which are available through http://www.osti.gov/opennet/) indicate similar variability in exposure rates, excluding hotspots, and suggest that the reported averages and the variations about the means are also very uncertain in that they depend on the number and location of the measurements.[10]

As discussed previously, doses to seamen boarding target ships are based on the average below-decks, topside and amidships daily dose curves constructed for each ship. Average below-deck exposure rates are small fractions of topside values (Weitz et al., 1982). The target-ship dose calculations by the NTPR program generally average topside and below-deck doses instead of assuming a worst-case scenario that the subject was always topside (Figure V.B.1), barring evidence to the contrary (see case #53).

In some cases, a veteran is given the benefit of the doubt and is assumed to have participated in all boardings (see cases #45, 48, and 49). In other cases, however, that assumption was not made. Thus, in cases #45, 48 and 49, the veteran was given a higher dose for service on possibly less contaminated ships than the veteran in case #53 (see discussion of case #53 in Section V.A.2).

The uncertainty in doses from boarding target vessels clearly does not adequately reflect the uncertainty in the average exposure rate or the variability in exposure rate with location. It also does not reflect the uncertainty in the number of boardings and the location when aboard. Limited film-badge data confirm that. In case #53, the veteran was assigned a reconstructed dose from boarding the USS *Skate* after Shot BAKER of 0.51 rem, with an upper bound of 0.6 rem, but film-badge data for 82 members of the *Skate* crew for August indicated a maxi-

---

[10]A memorandum from DNA to the veteran in case #48 states that for some target ships, the highest topside dose rate was as much as 30 times the average.

FIGURE V.B.1   Damaged quarterdeck on USS *Pensacola*.

mum film-badge dose of 1.1 rem (10 other badges were not turned in or were unreadable).

The committee thus finds that the upper-bound estimates of external gamma dose to participants who boarded target ships during Operation CROSSROADS are likely to be considerably underestimated.

An upper bound for being alongside target ships is estimated as a factor of 1.5 above a central estimate on the basis of survey data taken at a distance of 6 ft from the target. However, the data often vary by about a factor of 2 from port versus starboard, and this suggests that the upper bound may also be too low. The contribution to dose from being alongside a target ship was generally a significant fraction of the total dose to most seamen other than those boarding the targets or those who were members of engineering units and thus were exposed to contaminated piping (Weitz et al., 1982).

V.B.3.2.4 *Pacific unit dose reconstructions—summary*

Most of the uncertainty in the calculated external gamma dose to participants on ships and islands is due to variations in the measured exposure rate. That the data are limited suggests that a higher uncertainty should be estimated that allows for the variability in exposure rate with location. Furthermore, the NTPR program assigned default mean exposure rates and corresponding uncertainties (CVs) to islands and ships for which no monitoring data were available, on the basis of the mean and CV estimated for nearby ships for which measurements were available. The committee believes that a higher CV than the mean for nearby ships should be applied on ships for which no data were available to reflect the additional uncertainty in the mean and variance.

The upper-bound calculations for fallout-contaminated ships and islands and the upper bounds for the CROSSROADS target-ship dose calculations are cases in which the NTPR program's uncertainty analysis clearly is flawed. Those exposure scenarios affect a large fraction of the dose assessments that are based primarily on the unit dose reconstructions carried out for the CROSSROADS, CASTLE, GREENHOUSE, and IVY test series in the Pacific. A considerable fraction (about 25%) of the 99 sampled cases examined by the committee involved either exposure to fallout on ships or islands in the Pacific or exposure from boarding CROSSROADS target ships by nonbadged personnel. About 20% of the individual dose assessments carried out by SAIC for DTRA have involved participants in the CROSSROADS tests, and this reflects the large number of participants in that exercise. About 25% of the individual dose assessments have involved participants exposed to fallout on islands or ships at the Pacific test sites during Operations GREENHOUSE, CASTLE, and IVY before the period when all participants were issued film badges.

V.B.4 Summary of Findings on Estimates of External Gamma Dose

V.B.4.1 *Central Estimates*

Estimates of the most likely total external gamma doses to individual participants are usually based on film-badge data, unit dose reconstructions, or both and are generally credible, provided that the assumed exposure scenario is reasonable. However, as discussed above and in Section V.A, it appears that in many cases a plausible set of exposure scenarios for the participant was not fully considered. The unit dose reconstructions estimate the average dose to members of the unit. However, some members of a unit may have had doses well above the average. The committee believes that the dose reconstruction process should give a participant the benefit of the doubt by assuming, without strong evidence to the contrary, that he was a member of the most critically exposed population in the unit. The upper bound assigned to the central estimate should reflect a credible maximum (for example, the 95th percentile) of the dose to such members.

Although most individual dose reconstructions are based on unit dose reconstructions, many also involved some modifications based on unique circumstances or duties. In many cases, the doses for those duties were estimated on the basis of few data regarding exposure rates and time exposed. Examples include workers in laundries where contaminated clothing was washed, small-boat operators that ferried troops between contaminated islands or ships, and workers who prepared trenches for observers in previously contaminated areas. Often, the corresponding estimates were based on assumptions that were intended to "highside" the estimate and were deemed to be upper-bound estimates. Generally, the doses from such activities were relatively low, but it is not clear that the estimates always reflected at least the 95th percentile of possible doses.

V.B.4.2 *Upper Bounds*

Upper-bound estimates of external gamma dose provided by the NTPR program are based primarily on film-badge data or unit dose reconstructions, as discussed above. If a reported dose is based primarily on film-badge data that adequately and reasonably account for all possible external exposures, the reported upper-bound estimates are probably reasonable and even higher than the 95th percentile goal. However, upper bounds based primarily on unit dose reconstructions are, in general, likely to have been underestimated, often substantially.

The upper-bound estimates do not generally include uncertainty due to a possibly incorrect exposure scenario (such as neglect of possible additional exposures or errors in time exposed because of missing or incorrect records). Upper-bound estimates therefore are not credible unless the scenario is correctly specified.

Some of the unit dose reconstructions reviewed by the committee, particularly for smaller units in which the specific daily activities may have varied and were not well known, did not attempt to determine both a central ("best") estimate and an upper bound. Instead, the central estimate is alleged to be "highsided" on the basis of the scenario and exposures assumed (for example, see Ortlieb, 1995), and the estimate is treated as an upper bound when it is combined with other reconstructed or film-badge doses.

Upper bounds estimated from film-badge data and from reconstructed doses are combined in quadrature, assuming that they are uncorrelated, to arrive at an estimate of the upper bound in the total dose. To the extent that the individual upper-bound estimates are credible and all doses and potential uncertainties are included, the upper-bound estimate for the sum is credible, provided that uncertainties in the increments of dose are independent—that is, not correlated—which they may not be because of repetitiveness of behavior and work responsibilities. However, as discussed above, the committee has found a number of instances in which the uncertainty estimates in unit dose reconstructions are not credible and will not adequately reflect the true upper bound (95th percentile) of the dose to an individual participant. The committee has also identified situations in which uncertainty in the exposure scenario, film-badge issuance, and lack of benefit of the doubt suggest a much higher upper bound of the reported doses for individuals, even if they are based primarily on film-badge data.

On the basis of its review, the committee has concluded that reported upper-bound doses of external gamma radiation based primarily on unit dose reconstructions were often markedly underestimated compared with upper bounds that would be obtained if more credible assumptions about parameter values and uncertainties had been used. Of the 50 cases in the 99-case random sample in which a reported upper bound was based partly on a generic (unit) dose reconstruction, the committee has concluded that the upper bounds of about 30 Pacific test-site cases may be underestimated, often by a factor of 2-3 or even more. Reported upper-bound estimates based primarily on film-badge data probably are reasonably "high-sided," provided that the film-badge data accounted for all possible doses. The committee believes that failure to allow for the possibility that the badge was not always worn, that the times worn are incorrect, or that not all badge data have been accounted for makes it likely that many of the reported upper bounds based only or primarily on film-badge data also underestimate a credible upper bound (95th percentile) dose. Of the 25 cases in which the reported upper bound is based almost entirely on film-badge data, several may also warrant a higher value to give the benefit of the doubt to the participant. Thus, the committee has concluded that the estimated upper bounds reported by the NTPR program for most of the 99 cases examined do not represent a credible estimate of the 95th percentile upper bound in the dose from external gamma radiation exposure.

V.B.5 Neutron Dose Estimates

V.B.5.1 *Central Estimates*

Most test participants were not exposed to neutrons, except for observers in trenches at NTS tests and a few cloud-sampling personnel. For most participants who were exposed to neutrons, the doses were very low. However, a small number of volunteer observers in trenches very close to ground zero did receive substantial neutron (and gamma) doses during some NTS tests (Goetz et al., 1981).

Estimates of equivalent dose from exposure to neutrons must take into account the increased biological effectiveness of these radiations compared with gamma rays. As discussed in Section IV.B.3, the NTPR program has assumed a quality factor of 13 or 8.5 to represent this effect. These QFs apply at low doses of neutrons. Kocher et al. (2002) recently surveyed the available data and estimated radiation effectiveness factors (REFs) for neutrons, which represent the biological effectiveness in inducing cancer and other stochastic effects in humans relative to the effectiveness of gamma rays. For acute exposure to fission neutrons at low doses received by participants at atomic tests, a median REF of about 15 for induction of solid tumors would be obtained on the basis of the analysis by

Kocher et al.[11] If it is taken into account that the spectrum of neutrons to which participants were exposed included neutrons of lower energy than fission neutrons, because of scattering in air, and that REFs for lower-energy neutrons are lower than REFs for fission neutrons, the median REF for fission neutrons and solid tumors estimated by Kocher et al. indicates that the QFs adopted by the NTPR program are reasonable. That is, the assumptions about QF yield reasonable central estimates of neutron equivalent doses to participants. However, as discussed below, the committee has concluded that current estimates of upper-bound neutron doses by the NTPR program may be too low.

V.B.5.2 *Upper Bounds*

Reported upper bounds of external dose from exposure to neutrons are based on generally accepted radiation-transport calculations and reasonable corrections to account for shielding by trenches, vehicles, and so on. However, estimated upper bounds of neutron doses do not include any uncertainty in the neutron QF. A review of available data in 1986 by the International Commission on Radiation Units and Measurements (ICRU, 1986) indicated that a credible upper bound of QF could be at least a factor of 5 above a central estimate. Later reviews by the National Council on Radiation Protection and Measurements (NCRP, 1990) and the UK National Radiological Protection Board (Edwards, 1997; 1999) also indicated that there is substantial uncertainty in the biological effectiveness of neutrons.

Kocher et al. (2002) reviewed the available data and estimated probability distributions of REFs for fission neutrons. For induction of solid tumors, the probability distribution of REF is lognormal and has a 95% confidence interval between 2.0 and 3.0; the 50th percentile (median) is 7.7. That REF applies at high acute doses of the reference high-energy gamma rays. Thus, to account for the dependence of the relative biological effectiveness (RBE) of neutrons on the dose and dose rate of the reference gamma rays and to be comparable to the QFs at low doses assumed by the NTPR program, the probability distribution of REF should be multiplied by a factor of about 2.[12] The resulting 95th percentile of the probability distribution of equivalent dose, appropriate for exposure to fission neu-

---

[11]As described in the following section, this value is twice the estimated REF at high acute doses. Kocher et al. (2002) also estimated REFs for induction of leukemia by neutrons. Those REFs are not relevant to most dose reconstructions because many types of leukemia are presumptive diseases under 38 *CFR* 3.309, and a dose reconstruction is not required in evaluating a claim for compensation if a veteran's participation is adequately established.

[12]The RBE of neutrons depends on the dose and dose rate of the reference gamma rays because the dose-response relationship for neutrons generally is linear at any dose and dose rate but the dose-response relationship for gamma rays does not vary linearly with dose and dose rate. In human health-risk assessments, the response at low doses of gamma rays usually is assumed to be about half the observed response at high acute doses.

trons and induction of solid tumors, is about 50, or about a factor 3 above the median.

On the basis of the probability distribution of REF for fission neutrons and solid tumors described above, and taking into account the energy dependence of REF and its uncertainty (Kocher et al., 2002), a credible upper bound (95th percentile) of neutron equivalent dose could be a factor of about 3-5 higher than the QFs of 13 and 8.5 assumed by the NTPR program. Consequently, the upper bound of a combined neutron and gamma dose reported by the NTPR program may not represent a credible upper bound (95th percentile) of the total equivalent dose. For the few participants who were exposed to neutrons,[13] the committee has concluded that the NTPR program should revise the upper-bound estimates of neutron dose to include the uncertainty in biological effectiveness.

V.B.6 Beta Skin and Eye Dose Estimates

V.B.6.1 *Introduction*

Skin cancer was one of the most cited medical issues in the 72 (of 99) sampled individual dose reconstructions that included VA claims. However, for most of the cases involving skin cancer, no beta dose was calculated; beta dose was not routinely calculated in such cases until 1998.

The method for assessing beta dose is discussed in Section IV.B.4. Beta doses from standing on contaminated ground are calculated from upper-bound gamma doses by applying tabulated beta-to-gamma dose ratios that depend on the height above ground, the time after detonation, and whether the shot was at the NTS or in the Pacific. Beta doses from skin contamination and immersion in air or water are calculated by using dose coefficients (beta equivalent-dose rates per unit concentration of radionuclides in the source region). The current method of assessing beta-particle dose from sources outside the body is described in Barss (2000). The Barss report can be considered to have generally documented the method used in 1998 and 1999 and to present the method used after its publication in January 2000.

Of the 99 individual dose-reconstruction cases in the committee's sample, 27 included a claim or other indication of skin or eye disease (cases #2, 4, 9, 12, 17, 18, 19, 20, 25, 28, 29, 35, 38, 39, 40, 54, 55, 64, 65, 66, 70, 71, 87, 88, 93, 96, and 97). Most involved skin cancer and some indicated the type of skin cancer (such as basal cell or melanoma). Three claims or indications were for other skin conditions: rash and spots (case #87), skin disability (case #88), and incurable skin disease (case #54). Three cases were for cataract (cases #19, 39, and 67), and

---

[13]A neutron dose of 0.1-0.3 rem was reported in 3 of the 99 cases in the random sample (cases #37, 55, and 88). However, some test participants, not among the 99 cases, received much higher neutron doses (Goetz et al., 1981).

there was one case of macular degeneration (case #71). In case #64, both skin and eye doses were recorded explicitly, but the file did not indicate the diseases involved. Of the 27 cases, skin or eye dose was recorded explicitly in nine files for which the claim or other indication of skin or eye disease occurred in 1998 or later (cases #9, 12, 25, 39, 40, 64, 66, 96, and 97). The other 18 files—for claims or other indications of skin or eye disease that occurred before 1998—did not provide explicit skin or eye doses. That distribution is consistent with information provided by the NTPR program (Schaeffer, 2002c): skin dose assessments were not performed routinely before 1998. Table V.B.2 summarizes the nine files that state explicit skin or eye doses, and some representative assessments are discussed below.

V.B.6.2 *Summaries of Selected Skin and Eye Dose Assessments*

In case #9, the beta dose calculations appear to have been performed with mathematical software, and the data and calculations were annotated. The beta-to-gamma dose ratio method described in Barss (2000) and Section IV.B.4 was used for the beta dose to the skin and lens of the eye from exposure to contaminated surfaces outside the body. Before the beta-to-gamma dose ratios were applied, the upper bound of each component of the gamma dose was determined by multiplying the estimated gamma dose by an upper-bound factor (ratio of upper bound to central estimate). Upper-bound factors of 1.2–1.6 appear to have been used for gamma doses obtained from film badges and reconstructions. A substantial portion of the beta dose to the upper arms and forearms was ascribed to two 1-min exposures to highly contaminated towlines. The assumed distances were 20 cm for the forearm and 40 cm for the upper arm. No uncertainties were ascribed to the exposure time or the distances. The calculated doses would be very sensitive to errors in the determinations of such small times and distances. Although Barss (2000) includes methods for determining beta dose from standing in descending fallout and from skin contamination, there did not appear to be any consideration of those pathways in this case. That could have been appropriate, but it would have been useful to discuss the reasons for not including them.

In case #64, the file contains no narrative describing the dose assessment or detailed calculations. The gamma dose was determined to be 0.7 rem from ship dose tables, and the neutron dose was determined to be zero on the basis of references. As indicated in Table V.B.2, the upper-bound gamma dose was set to 1.6 rem, but without explanation. There were no beta dose calculations, but the skin and eye doses were also set to 1.6 rem, implying a beta-to-gamma dose ratio of 1. There was no consideration of skin contamination or immersion dose. The veteran performed basic seamanship and watch duties on the USS *Allen M. Sumner* and the USS *Moalem*. If he worked outside on contaminated ships, some beta dose would be expected from contaminated surfaces and possibly from skin contamination or descending fallout.

TABLE V.B.2   Summary of Nine Cases in Committee's Random Sample in Which Skin or Eye Doses Were Reported

Case #	Condition	Claim or Inquiry Date	Assessment Date	External Upper Bound Dose	Skin-dose location
9	Skin cancer	2/25/00	12/13/00	5.2	Head, face, neck Upper arm Forearm Back
12	Skin cancer	8/18/99	10/8/99	3.7	Lower leg
25	Skin cancer	8/11/00	10/6/00	1.5	Face (forehead)
39	Cataract	6/15/00	9/5/00	7.3	N/A
40	Skin cancer	10/19/98	7/15/99	0.2	Face
64	Unknown	12/23/98	1/1/99	1.6	Not stated
66	Skin cancer	12/23/98	1/3/99	1.8	Face, back, arms
96	Skin cancer	4/3/98	8/12/98[c]	4.3	Arm
97	Cataract	3/19/99	4/14/99	1.9	N/A

[a]Methods: A, Barss (2000); B, Lorence et al. (1989) and Finn et al. (1979); C, no method cited.
[b]References were not provided for the beta component of the eye dose, but the methods of Barss (2000) appear to have been used.
[c]Date of transmittal letter; dose-assessment narrative was undated.

Dose (rem)			
Skin	Lens of eye	Method[a]	Comment
30.3 30.1 55.5 8.5	8.7	A	"Date pair/upper bound request" in response to VA inquiry; no narrative dose assessment in file; calculations performed with mathematical software and annotated
77.5	N/A	B	"Upper bound request" in response to VA inquiry; no narrative dose assessment in file; calculations performed with mathematical software and annotated
7.1	N/A	A	"Date pair/upper bound request" in response to VA inquiry; no narrative dose assessment in file; calculations performed with mathematical software and annotated
N/A	25.5	A[b]	"Date pair/upper bound request" in response to VA inquiry; no narrative dose assessment in file; calculations performed with mathematical software and annotated
0.3	N/A	B	Update of previous dose assessment based on request from veteran's family member; beta skin dose specifically requested by JAYCOR; update included detailed narrative and spreadsheet calculations
1.6	1.6	C	"Upper bound request" in response to personal inquiry; no narrative dose assessment in file; external dose determined by reference to previous reports; no indication of how skin and eye doses were assessed
1.8	N/A	C	"Upper bound request" in response to personal inquiry; no narrative dose assessment in file; external dose determined by reference to previous reports; no indication of how skin and eye doses were assessed
35	N/A	B	Radiation dose assessment in response to VA inquiry; this update to the veteran's dose included narrative; two assessments were performed with mathematical software and annotated
N/A	2.4	B	"Upper bound request" in response to personal inquiry; no narrative dose assessment in file; external dose determined by reference to previous reports; eye beta dose calculations performed with mathematical software and annotated

In case #96, the file contained two dose assessments. The calculations in the first assessment appear to have been performed with mathematical software, and the data and calculations were annotated and easy to follow, compared with other assessments. For gamma doses, comparisons were made between the individual calculations for the veteran and the island gamma doses from published reports. The comparisons were well documented, and the two approaches gave results in good agreement. This assessment, which was performed on June 8-15, 1998, preceded publication of Barss (2000), but the method of assessing beta doses, which was based on multiplying gamma doses by beta-to-gamma dose ratios, appears to be fundamentally the same. Although the narrative report cited Lorence et al. (1989) and Finn et al. (1979) for the beta dose assessment, the calculation itself cited the Barss file using the Stiver method, which was probably the documentation of application of the fundamental references to beta dose assessments. The beta-to-gamma dose ratios in the table used in the assessment for the 1-m distance from the source assumed for the arm were identical with those given in Table 2 of the Barss report, which was identified as intended for historical development only. Discussions in the Barss report indicate that the data in Table 2 were based on erroneous assumptions and that using better assumptions lowers the beta-to-gamma dose ratios by a factor of about 2. Revised tables are provided in the Barss report and are identified as the ones to be used. The upper-bound gamma dose was calculated to be 4.1 rem, and the upper-bound skin dose to the arm was assessed to be 95.2 rem.

A second, undated assessment in case #96 was performed by using "new beta/gamma ratios of 7/13/98." The ratios in this table are lower than those used in the first assessment, described above, but are still slightly higher than those later recommended for use by Barss (2000). Although annotated, this calculation was not as easy to follow as in the first assessment. The beta-to-gamma dose ratios used in the second assessment for the 1-m distance from the source assumed for the arm were roughly a factor of 2 lower than those used in the first assessment for the same distance and closer to those later published in Barss (2000). The results of the second assessment were an upper-bound gamma dose of 4.3 rem and an upper-bound skin dose of 35 rem. The skin dose was checked against island gamma doses in published reports and found to be in good agreement (32.4 rem). The dose-assessment narrative reported the upper-bound gamma dose as 4.3 rem and the skin dose as "as much as 35" rem.

V.B.6.3 *Discussion and Conclusions Regarding Estimate of Skin Doses*

From the committee's reviews of the 99 sampled cases, it is not evident that skin or clothing contamination is being considered as a pathway for beta dose to the skin. If skin or clothing contamination was considered and dismissed, the consideration was not documented in any of the seven cases for which the beta-particle components of skin doses were calculated. However, there are examples in which participants had to take multiple showers to decontaminate their bodies (see cases #9 and 26) and situations in which soil, presumably contaminated with fallout, was brushed from their clothing and bodies with brooms (for example, see document submission from veteran Frank Bushey in Appendix C). The committee regards neglect of skin contamination as an important problem in dose reconstructions for maneuver troops and close-in observers at the NTS who filed claims for skin cancer. In the Pacific, "minor radiation burns" were seen on personnel who were below decks on the USS *Phillip* when vents were opened during a period of fallout to reduce intolerably high temperatures (Martin and Rowland, 1982). The committee also notes that a contemporary report (Morgan, 1946) indicated that contamination was found frequently on the clothing and bodies of persons on ships.

Beta-particle doses from standing on contaminated ground are calculated by applying a beta-to-gamma dose ratio to an upper-bound gamma dose. As noted earlier, the committee is concerned that uncertainties in gamma doses may be underestimated in some cases and could lead to underestimates of upper-bound gamma doses and, consequently, to underestimates of beta-particle doses.

The committee also notes that uncertainties are not estimated for the beta-to-gamma dose ratios, although Barss (2000) argues that the ratios are overestimates. However, beta-to-gamma dose ratios depend on the time since detonation and the distance from the source to the exposed tissue. Errors in those quantities may result in substantial underestimation or overestimation of beta-to-gamma dose ratios.

On the basis of the foregoing, the committee found that the beta components of skin doses are questionable. For most of the unit dose reconstructions, beta doses were not calculated, because the method had not been developed. Furthermore, letters to VA by the NTPR program as late as 1997 implied that doses of around 1,000 rem were needed to cause statistical increases in skin cancer (Schaeffer, 1997). The NTPR program also indicated that no evidence suggested that skin cancer was associated with the much lower radiation doses (external or internal) received by participants in atmospheric nuclear testing. That conclusion, however, was apparently based on skin dose calculations that were too low because they did not include a contribution from beta particles and, as discussed below, the assumption that around 1,000 rem was required to induce skin cancer is no longer supported by scientific evidence.

In 1989, NCRP used a linear coefficient, based on recent epidemiological studies, to estimate the cancer risk posed by small radioactive particles on the skin (NCRP, 1989). NCRP noted that skin exposed to ultraviolet (UV) radiation was more susceptible to radiation-induced cancer than UV-protected skin. Most atomic veterans served in Nevada and the South Pacific, places with high solar indexes, so they undoubtedly had UV exposures to some areas of skin.

In 1990, the National Research Council reviewed the relationship between radiation and skin cancer in the fifth report on *Biological Effects of Ionizing Radiation* (BEIR V) (NRC, 1990). On the basis of a study of persons treated for ringworm with radiation, the report noted that tumors began to appear about 20 years after exposure and were not limited to the most heavily irradiated parts of the scalp. Tumors tended to occur more commonly at the margins of the scalp and in neighboring areas of skin that were not covered by hair or clothing. An excess of skin cancers was detected even on the cheeks and the neck, where the doses were estimated to have been only 0.12 and 0.09 Gy (12 and 9 rad), respectively. The distribution of tumors suggested that the carcinogenic effects of X rays were increased by exposure to UV radiation.

In 1991, the ICRP stated in *Publication 59* (ICRP, 1991b) that "although it has traditionally been thought that there was little if any risk of skin cancer below 10 Gy [1,000 rad], there are now several sets of data indicating excess skin cancer following doses of a few grays [a few hundred rad], with one study suggesting risk below 1 Gy [100 rad]. The evidence does not indicate that the risk per unit dose is greater at higher doses than at lower [doses]." The ICRP also noted that risks were greater for UV-irradiated skin.

Thus, by 1991, there was ample indication from authoritative national and international bodies that skin cancer could be caused by doses much lower than 1,000 rad and that UV-exposed skin was particularly sensitive. However, it was not until 1998 that this information began to be incorporated in dose reconstructions for atomic veterans who filed claims for skin cancer.

V.C EVALUATION OF METHODS OF ESTIMATING INTERNAL DOSE

V.C.1 Introduction

Doses due to intakes of radionuclides produced in a nuclear detonation often are considered to be unimportant when compared with doses due to external exposure. That is especially the case in exposure scenarios involving inhalation of fallout particles at locations relatively close to ground zero and shortly after detonation (NRC, 1985b; Levanon and Pernick, 1988; IOM/NRC, 1995). Such an exposure scenario is important for many participants in nuclear tests at the NTS and in the Pacific. The unimportance of the inhalation hazard posed by fallout shortly after a detonation, compared with the hazard posed by external exposure, is attributed to such factors as: the presence of much greater activities of short-lived photon-emitting radionuclides that tend to result in high external doses per unit activity but much lower inhalation doses, compared with the activities of longer-lived radionuclides for which inhalation doses per unit activity often are considerably higher; the dominance of large, essentially nonrespirable particles in fallout relatively close to ground zero; and the insolubility of fallout particles, which can substantially reduce the extent of absorption of inhaled radionuclides into the body. However, there are exposure scenarios for participants at the NTS and in the Pacific in which activities of longer-lived radionuclides compared with shorter-lived radionuclides are much higher than in fresh fallout (for example, when exposures occurred a few weeks or more after a detonation), and internal exposure in these cases can contribute significantly to the total dose received by an organ or tissue of concern; see, for example, dose estimates for occupation forces in Japan given in Table IV.D.1.

This section presents the committee's evaluation of methods used in the NTPR program to estimate internal doses to atomic veterans. Methods of estimating internal dose and the approach to addressing uncertainty are discussed in Sections IV.C and IV.E.4, respectively. Discussions in this section mainly concern methods of estimating doses due to inhalation of radionuclides. Intakes by inhalation are expected to be the most important in determining internal doses to atomic veterans, and only inhalation has been considered routinely in dose reconstructions. Additional discussions and evaluations of the low-level internal dose screen and a bioassay program mentioned in Section IV.C.1 are presented in Sections VI.C and VI.D, respectively.

Data that can be used to estimate inhalation doses to atomic veterans, including data on concentrations of radionuclides in air at times and locations of exposure or amounts of radionuclides excreted in urine or feces, generally are lacking (see Section IV.C.1). Given the lack of relevant data, the basic approach to estimating inhalation doses in the NTPR program has been to use assumptions that are intended to result in substantial overestimates of dose to most participants. In contrast to the approach to assessing external dose from exposure to photons, in which a central (best) estimate and an upper 95th percentile of possible doses are obtained, only a single estimate of inhalation dose, which is intended to be an upper bound (at least a 95th percentile), is obtained in all dose reconstructions for atomic veterans. Thus, the committee's evaluation of methods used in the NTPR program to estimate inhalation doses essentially involves an assessment of whether the methods are likely to yield credible upper bounds (at least a 95th percentile) of possible doses.

This section is divided into four parts. First, we summarize findings of a previous committee of the National Research Council that reviewed methods of estimating inhalation doses to atomic veterans; second, we discuss our own evaluation of methods of estimating inhalation doses; third, we consider the potential importance of ingestion exposures of atomic veterans (as noted in Section IV.C.3, ingestion of radionuclides is rarely taken into account in dose reconstructions); and fourth, we summarize our principal findings from our evaluation of methods of estimating internal doses used in the NTPR program and conclusions based on the findings.

V.C.2 Findings of Previous National Research Council Review

In the middle 1980s, a committee of the National Research Council conducted the first external scientific review of methods used in the NTPR program to estimate doses to atomic veterans (NRC, 1985b). That committee's review of methods of estimating internal doses focused mainly on inhalation doses because, as noted above, ingestion usually was considered to be relatively unimportant and had been included in dose reconstructions only rarely.

At the time of the first National Research Council review, methods of estimating inhalation doses to atomic veterans were largely the same as the methods that have been used since then (Egbert et al., 1985; Barrett et al., 1986). As described in Section IV.C.2, inhalation doses were estimated on the basis of estimates of concentrations of radionuclides in air at locations and times of exposure that were inferred from measurements of external photon exposure with film badges worn by veterans or field instruments, assumed resuspension factors, assumed breathing rates, and other assumptions about the physical and chemical composition of fallout particles or neutron-activated materials in soil.

The 1985 committee generally took a dim view of methods of estimating inhalation doses on the basis of measurements of external photon exposure. The committee stated, for example, that "these methods involve assumptions about relationships between airborne and deposited fallout that are not scientifically valid, and their reliability, even for establishing upper limits of internal radiation doses, is unknown" (NRC, 1985b). Other statements also questioned the credibility and defensibility of the methods. The committee's report did not discuss the basis of the findings in detail. However, the committee's concerns apparently included the methods' insensitivity to the presence of beta- and alpha-emitting radionuclides that are important contributors to inhalation dose, such as ^{90}Sr and plutonium, and the possibility that internal dosimetry models used to estimate doses per unit activity of radionuclides inhaled (inhalation dose coefficients) would not apply to the physical and chemical forms of fallout particles, especially in cases of exposure to large, highly insoluble particles in descending fallout. The committee argued that methods of estimating internal doses to atomic veterans needed to be validated with bioassay testing.

In spite of the 1985 committee's concerns about methods of estimating inhalation doses in the NTPR program, however, it also concluded that the methods "tended to overestimate possible internal doses" and, particularly in cases of inhalation of descending fallout, "probably resulted in large overestimates of radiation exposures" (NRC, 1985b). Those findings also were not discussed in detail, but they apparently were based, at least in part, on the use in dose reconstructions of an assumption that all fallout was in the form of small particles that were respirable when most of the activity in fallout at locations of participant exposure was in the form of large, essentially nonrespirable particles.[14]

---

[14]An option of estimating doses due to inhalation of large, essentially nonrespirable particles (AMAD, 20 μm) was later included in the FIIDOS computer code (Egbert et al., 1985) (see Section IV.C.2.2.1).

The 1985 committee concluded that inhalation exposures had only a "minor impact on total doses expected" (NRC, 1985b). The committee's concerns about methods of estimating inhalation doses thus did not appear to be important with regard to the potential for significant doses to atomic veterans. The view that inhalation doses generally were overestimated in the NTPR program and were unimportant was echoed in a later study (IOM/NRC, 1995).

V.C.3 Evaluation of Methods of Estimating Inhalation Dose

The present committee's evaluation of methods used in the NTPR program to estimate inhalation doses focuses on the question of whether the methods are likely to provide credible upper bounds of possible doses (see Section IV.E.4). The committee's evaluation is divided into three parts. The first part discusses assumptions used in estimating inhalation doses that, in the committee's opinion, tend to result in overestimates of dose. The second part discusses assumptions that, in the committee's opinion, tend to result in substantial underestimates of inhalation doses, and it also considers assumptions that have substantial uncertainty and the importance of that uncertainty in obtaining credible upper bounds of inhalation doses. The third part summarizes the committee's evaluation of methods of estimating inhalation doses used in the NTPR program.

V.C.3.1 *Assumptions Tending to Overestimate Inhalation Dose*

The committee found that several assumptions used to estimate inhalation doses in the NTPR program should tend to result in overestimates of possible doses. In the following discussion, assumptions related to estimating inhalation dose coefficients (equivalent doses to specific organs or tissues per unit activity of radionuclides inhaled) are considered first and are followed by assumptions related to estimating inhalation exposures (intakes of radionuclides in air); these are the two components of models used to estimate inhalation doses (see Section IV.C.2).

[1] **In exposure scenarios in which inhaled particles are assumed to be respirable (that is, when a particle size, AMAD, of 1 μm is used), organ-specific inhalation dose coefficients used in the NTPR program often (but not always) are higher than values for the same particle size currently recommended for use in radiation protection of workers by ICRP.**

An AMAD of 1 μm often is assumed, for example, in scenarios involving suspension of activation products in soil or resuspension of fallout particles that were deposited on the ground or other surfaces, especially when this assumption results in higher estimates of dose than would an assumed particle size of 20 μm (see Section IV.C.2.2.1).

As noted in Section IV.C.2.2, all inhalation dose coefficients used in dose reconstructions were based on dosimetric and biokinetic models described in ICRP *Publication 30* (ICRP, 1979a). Those models represented the state-of-the-art in estimating internal dose when methods of estimating internal doses to atomic veterans (Egbert et al., 1985) were developed.

Beginning in the late 1980s, ICRP developed a new set of dose coefficients for inhalation and ingestion of radionuclides (ICRP, 1989; 1993; 1994a; 1995; 1996a; 1996b) to replace values recommended in ICRP *Publication 30*.[15] The dose coefficients and associated dosimetric and biokinetic models constitute ICRP's current recommendations on methods of calculating dose from intakes of radionuclides for purposes of radiation protection (see also ICRP, 2002).

ICRP's current recommendations on inhalation (and ingestion) dose coefficients incorporate three important changes in methods of calculating internal dose. First, the earlier model used to estimate deposition, retention, translocation, and absorption of inhaled radionuclides and doses to tissues of the respiratory tract (ICRP, 1979a) was replaced by a new and more complex respiratory-tract model (ICRP, 1994b). Second, new biokinetic models to describe the behavior of radionuclides after absorption into blood from the respiratory or gastrointestinal (GI) tract were developed for many chemical elements.[16] Third, when radioactive decay products are produced in the body after intakes of a parent radionuclide, separate biokinetic models are used for the particular chemical elements of concern.[17] In addition, assumed GI-tract absorption fractions and deposition fractions of absorbed activity in different organs or tissues are changed for some radionuclides.

A comparison of dose coefficients for inhalation of radionuclides in respirable form (AMAD, 1 μm) often used in dose reconstructions for atomic veterans with values for workers for the same particle size currently recommended by ICRP (2002) is given in Tables V.C.1 and V.C.2.[18] Radionuclides listed in these tables include selected shorter-lived and longer-lived fission products, activation

---

[15]Dose coefficients given in ICRP *Publication 56* (ICRP, 1989) were superseded by values given in later reports.

[16]The new biokinetic models are physiologically based—that is, translocation and retention are modeled with more realistic representations of physiologic compartments in the body, and cycling among various compartments is taken into account—in contrast to the more empirical approach used previously of modeling retention by fitting of retention or excretion data over time with simple exponential functions (ICRP, 1979a).

[17]The approach to biokinetic modeling of decay products now used by ICRP was incorporated in earlier dose coefficients in the ORNL reports (Killough et al., 1978b) but not in dose coefficients given in ICRP *Publication 30* (ICRP, 1979a) or by Eckerman et al. (1988) (see Section IV.C.2.2).

[18]For purposes of evaluating compliance with dose limits for occupational exposure, ICRP now recommends that a default particle size (AMAD) of 5 μm should be assumed in the absence of information on actual particle sizes (ICRP, 1994a). In the present report, however, current ICRP recommendations based on a particle size of 1 μm are used to be consistent with the assumption for inhalation of respirable particles used in all dose reconstructions.

products, and transuranium radionuclides that often should be among the most important in estimating inhalation doses at various times after a detonation. Not included in the tables is ^{131}I, for which the dose coefficient for the thyroid currently recommended by ICRP (1994a) is about a factor of 2 less than the value used in dose reconstructions, and ^{137}Cs, for which the current dose coefficients for all organs and tissues are slightly lower than the values used in dose reconstructions. The data in these tables illustrate that doses per unit activity inhaled tend to be substantially higher for longer-lived radionuclides than for shorter-lived radionuclides (see Section V.C.1).

The data in Tables V.C.1 and V.C.2 indicate that dose coefficients for inhalation of radionuclides attached to respirable particles (AMAD, 1 μm) used in the NTPR program tend to be higher than values for the same particle size currently recommended by ICRP. That is the case especially for the lung and respiratory lymphatic tissues, but for some radionuclides substantial differences are also

TABLE V.C.1   Comparison of Dose Coefficients for Inhalation of Radionuclides in Respirable Form (AMAD, 1 μm) Used in NTPR Program with Values for Same Particle Size Currently Recommended by ICRP: I. Shorter-Lived Radionuclides

	Dose coefficient (rem μCi^{-1})[a]						
Nuclide[b]	Lung[c]	Lymph tissue[d]	Large intestine[e]	Red bone marrow	Bone surfaces	Liver	Bladder Wall
^{24}Na	4.6E-3	2.5E-2	3.8E-4	1.0E-3	1.2E-3	8.1E-4	5.3E-4
(15.0 h)	(5.2E-4)	(4.1E-4)	(4.4E-4)	(5.6E-4)	(7.8E-4)	(3.7E-4)	(5.6E-4)
^{56}Mn	2.0E-3[f]			3.8E-5[f]	3.0E-5[f]		
(2.6 h)	(1.3E-3)	(8.5E-5)	(3.7E-4)	(3.7E-5)	(3.0E-5)	(4.1E-5)	(2.2E-5)
^{91}Sr	4.3E-3	2.9E-2	2.6E-3	4.1E-4	4.4E-4	4.1E-4	2.6E-4
(9.6 h)	(1.9E-4)	(9.6E-5)	(1.9E-3)	(4.8E-4)	(5.2E-4)	(9.6E-5)	(5.6E-4)
^{93}Y	1.4E-2	1.4E-2	6.4E-3	1.1E-4	1.7E-4	2.3E-4	1.4E-5
(10.2 h)	(5.6E-3)	(1.9E-5)	(7.0E-3)	(6.7E-6)	(4.8E-6)	(6.3E-6)	(5.6E-6)
^{97}Zr	2.1E-2	2.2E-2	1.6E-2	4.0E-4	3.8E-4	3.9E-4	2.4E-4
(16.7 h)	(1.2E-2)	(4.8E-4)	(1.4E-2)	(2.9E-4)	(2.4E-4)	(1.6E-4)	(1.8E-4)
^{143}Ce	1.6E-2	1.9E-2	1.6E-2	1.2E-4	7.7E-5	2.0E-4	7.6E-5
(33.0 h)	(1.5E-2)	(2.5E-4)	(9.3E-3)	(1.6E-4)	(1.9E-4)	(6.3E-4)	(5.6E-5)
^{239}Np	1.0E-2	1.9E-2	1.2E-2	1.8E-4	9.1E-4	2.9E-4	7.3E-5
(2.4 d)	(2.2E-2)	(2.9E-4)	(7.0E-3)	(1.7E-4)	(1.9E-3)	(1.2E-4)	(7.8E-5)

[a]First entry is value from Table 5a of Egbert et al. (1985) based on ORNL reports (Killough et al., 1978a; Dunning et al., 1979) and often used in dose reconstructions for atomic veterans, except as noted; values are based on dosimetric and biokinetic models in ICRP *Publication 30* (ICRP, 1979a). Second entry, in parentheses, is value for AMAD of 1 μm currently recommended for adult workers by ICRP (2002). All values apply to radionuclides in oxide form (Eckerman et al., 1988).
[b]Entry in parentheses is radionuclide half-life.
[c]Dose coefficients for lung are calculated as described in Section V.C.3.1, comment [7].
[d]Lymphatic tissues that drain bronchial and pulmonary regions of lung.
[e]Wall of lower large intestine.
[f]Value from Eckerman et al. (1988).

TABLE V.C.2 Comparison of Dose Coefficients for Inhalation of Radionuclides in Respirable Form (AMAD, 1 μm) Used in NTPR Program with Values for Same Particle Size Currently Recommended by ICRP: II. Longer-Lived Radionuclides

	Dose coefficient (rem μCi^{-1})[a]						
Nuclide[b]	Lung[c]	Lymph tissue[d]	Large intestine[e]	Red bone marrow	Bone surfaces	Liver	Bladder Wall
[60]Co	1.3	1.2E1	2.9E-2	6.4E-2	5.1E-2	1.2E-1	1.1E-2
(5.3 y)	(6.3E-1)	(5.9E-1)	(1.7E-2)	(4.4E-2)	(3.3E-2)	(7.0E-2)	(4.4E-3)
[89]Sr	6.6E-3	7.6E-2	1.4E-2	1.3E-2	2.7E-2	2.2E-3	1.1E-3
(50.5 d)	(7.4E-4)	(6.7E-4)	(1.4E-2)	(1.6E-2)	(2.0E-2)	(6.7E-4)	(2.3E-3)
[90]Sr	9.9E-3	1.4E-1	1.4E-2	1.1	2.2	1.5E-2	7.3E-3
(28.8 y)	(2.3E-3)	(2.2E-3)	(1.9E-2)	(5.9E-1)	(1.4)	(2.2E-3)	(4.8E-3)
[95]Zr	6.9E-2	2.3E-1	1.6E-2	5.5E-3	7.8E-3	4.7E-3	1.6E-3
(64.0 d)	(1.1E-1)	(1.9E-2)	(8.9E-3)	(8.5E-3)	(4.4E-2)	(3.6E-3)	(4.1E-4)
[106]Ru	3.8	3.4E1	1.4E-1	9.4E-3	1.0E-2	1.2E-2	3.9E-3
(373 d)	(1.9)	(3.5E-1)	(7.8E-2)	(3.6E-3)	(3.1E-3)	(4.4E-3)	(2.3E-3)
[144]Ce	2.9	2.2E1	1.3E-1	9.0E-3	1.5E-2	8.1E-2	2.9E-4
(285 d)	(1.4)	(2.1E-1)	(7.0E-2)	(4.1E-3)	(7.0E-3)	(2.0E-2)	(2.7E-4)
[152]Eu	2.7	6.1E1	5.6E-2	1.3E-1	2.0E-1	9.8E-1	1.4E-2
(13.5 y)	(2.1E-1)	(1.2E-1)	(5.6E-2)	(2.4E-1)	(6.7E-1)	(9.6E-1)	(2.3E-2)
[239]Pu[f]	5.8E2	4.1E4	1.1E-1	3.0E2	4.2E3	8.0E2	2.9
	(2.9E2)	(3.0E3)	(1.1)	(3.1E1)	(6.3E2)	(1.3E2)	(1.1)

[a]First entry is value from Table 5a of Egbert et al. (1985) based on ORNL reports (Killough et al., 1978a; Dunning et al., 1979) and often used in dose reconstructions for atomic veterans; values are based on dosimetric and biokinetic models in ICRP *Publication 30* (ICRP, 1979a). Second entry, in parentheses, is value for AMAD of 1 μm currently recommended for adult workers by ICRP (2002). All values apply to radionuclides in oxide form (Eckerman et al., 1988).
[b]Entry in parentheses is radionuclide half-life.
[c]Dose coefficients for lung are calculated as described in Section V.C.3.1, comment [7].
[d]Lymphatic tissues that drain bronchial and pulmonary regions of lung.
[e]Wall of lower large intestine.
[f]Dose coefficients apply to any mixtures of [239]Pu and [240]Pu, which have half-lives of 24,100 and 6,560 y, respectively.

found for other organs and tissues. In some cases, however, the ICRP's current dose coefficient is substantially higher than the value used in the NTPR program. The increase by a factor of 10 for plutonium and the lower large intestine wall is discussed in Section V.C.3.2, comment [3].

Differences in dose coefficients for the lung shown in Tables V.C.1 and V.C.2 are due to a number of factors, including: substantial differences in dose coefficients for respiratory lymphatic tissues combined with a change in how the dose to these tissues is incorporated into the dose to the lung (see comment [7] in this section); differences in dosimetric models for radiosensitive tissues in airways of the respiratory tract; differences in assumptions about deposition fractions of inhaled 1-μm particles in different regions of the respiratory tract, including a separate accounting of depositions resulting from breathing through the mouth or nose in the new respiratory-tract model; and differences in models that describe clearance of radionuclides from the respiratory tract by mechanical transport of particles or absorption into blood, including a separate accounting of the two competing processes in the new model. For inhalation of a long-lived, alpha-emitting radionuclide in insoluble form, such as plutonium, those factors are listed in approximate order of importance. A change in definition of "lung" in the ICRP models also has implications for estimating the probability of causation of lung cancers in atomic veterans; this issue is discussed in comment [7] in this section.

[2] **In exposure scenarios in which inhalation of large particles is assumed (that is, when an AMAD of 20 μm is used), organ-specific dose coefficients used in the NTPR program often are higher than values that would be based on current ICRP recommendations.**

In the NTPR program, an AMAD of 20 μm sometimes is assumed in estimating doses due to inhalation of descending fallout because large particles constitute a substantial fraction of fallout near the location of a detonation (Hicks, 1982; Levanon and Pernick, 1988). That is the case especially when an organ of concern in a dose reconstruction is the thyroid, an organ in the GI tract, or the prostate,[19] and the assumption results in higher estimates of dose than would an assumed particle size of 1 μm. An assumption of large particles is also used in some cases in estimating inhalation doses due to resuspension of deposited fallout. On the basis of dose coefficients used in the NTPR program, an assumption of inhalation of large particles that are mainly deposited in the nose and throat and then swallowed greatly reduces estimates of dose to the lung but increases estimates of dose to organs of the GI tract and the dose to the thyroid from inhalation of ^{131}I compared with an assumption that the inhaled materials are respirable (AMAD, 1 μm) (see Section IV.C.2.2.1).

Large-particle inhalation dose coefficients used in the NTPR program are calculated by assuming that all inhaled particles are deposited in the nose and throat and that 99% of the deposited activity is swallowed, with the remaining 1% absorbed into blood (ICRP, 1979a). Doses to organs of the GI tract result mainly from radionuclides that are swallowed and pass through the body, and these doses are estimated by using dose coefficients for ingestion of radionuclides. Doses to other organs depend on the total activity absorbed into blood from the nose and throat and the GI tract.

In ICRP's current respiratory-tract model (ICRP, 1994b), the fraction of inhaled large particles (AMAD, 20-100 μm) that are deposited in the nose and throat is assumed to be about 0.5, in contrast to the previous value of 1.0 (ICRP,

---

[19]The prostate is not included in the database of dose coefficients used in dose reconstructions, and the bladder wall is used as a surrogate for it.

1979a). Furthermore, about half the particles deposited in the nose and throat are assumed to be expelled by nose-blowing or -wiping (ICRP, 1994b). Thus, dose coefficients for inhalation of large particles used in the NTPR program may be too high by a factor of about 4. As indicated by comparisons in Tables V.C.3 and V.C.4, there usually is little difference between ingestion dose coefficients used in the NTPR program and those currently recommended by ICRP, especially in organs of the GI tract. Again, ingestion dose coefficients are applied to the large fraction of radionuclides attached to large particles that are deposited in the nose and throat and later swallowed. In cases of intakes of ^{131}I, the ingestion dose coefficient for the thyroid currently recommended by ICRP is about 20% less than the value used in the NTPR program. Thus, taking into account differences in the models for deposition and exhalation of large particles in the nose and throat and differences in dose coefficients for ingestion, dose coefficients for inhalation of large particles used in the NTPR program should, in most cases, be higher than values that would be based on current ICRP recommendations.

TABLE V.C.3   Comparison of Dose Coefficients for Ingestion of Radionuclides Used in NTPR Program with Values Currently Recommended by ICRP: I. Shorter-Lived Radionuclides

	Dose coefficient (rem μCi^{-1})[a]						
Nuclide[b]	Kidneys	Pancreas	Large intestine[c]	Red bone marrow	Bone surfaces	Liver	Bladder Wall
[24]Na	1.4E-3	1.5E-3	7.3E-4	1.8E-3	2.2E-3	1.4E-3	1.0E-3
(15.0 h)	(1.1E-3)	(1.4E-3)	(1.5E-3)	(1.4E-3)	(2.0E-3)	(1.1E-3)	(1.6E-3)
[56]Mn			2.0E-3[d]	9.0E-5[d]	3.9E-5[d]		
(2.6 h)	(1.1E-4)	(1.9E-4)	(2.0E-3)	(8.9E-5)	(4.4E-5)	(9.3E-5)	(9.3E-5)
[91]Sr	3.4E-4	3.7E-4	1.6E-2	3.2E-4	2.3E-4	2.6E-4	3.1E-4
(9.6 h)	(2.3E-4)	(2.5E-4)	(1.5E-2)	(5.9E-4)	(5.2E-4)	(1.8E-4)	(7.0E-4)
[93]Y	1.7E-5	2.0E-5	3.3E-2	1.7E-5	6.1E-6	1.2E-5	2.3E-5
(10.2 h)	(1.8E-5)	(1.9E-5)	(3.2E-2)	(1.6E-5)	(7.4E-6)	(1.2E-5)	(2.3E-5)
[97]Zr	4.1E-4	4.0E-4	6.6E-2	4.8E-4	1.7E-4	3.0E-4	6.5E-4
(16.7 h)	(4.1E-4)	(3.7E-4)	(6.7E-2)	(4.4E-4)	(2.0E-4)	(2.9E-4)	(6.7E-4)
[143]Ce	1.0E-4	1.1E-4	4.3E-2	1.9E-4	5.9E-5	7.6E-5	2.1E-4
(33.0 h)	(1.0E-4)	(1.0E-4)	(4.4E-2)	(1.3E-4)	(7.8E-5)	(8.1E-5)	(2.1E-4)
[239]Np	7.8E-5	8.0E-5	2.9E-2	1.7E-4	7.6E-5	6.7E-5	1.8E-4
(2.4 d)	(7.8E-5)	(7.4E-5)	(3.2E-2)	(9.6E-5)	(9.3E-5)	(5.2E-5)	(1.7E-4)

[a]First entry is value from Table 4a of Egbert et al. (1985) based on ORNL reports (Killough et al., 1978a; Dunning et al., 1979) and often used in dose reconstructions for atomic veterans, except as noted; values are based on dosimetric and biokinetic models in ICRP *Publication 30* (ICRP, 1979a). Second entry, in parentheses, is value currently recommended for adult workers by ICRP (1994a; 2002). All values assume GI-tract absorption fraction that applies to radionuclides in oxide form (Eckerman et al., 1988).
[b]Entry in parentheses is radionuclide half-life.
[c]Wall of lower large intestine.
[d]Value from Eckerman et al. (1988) or DOE (1988).

TABLE V.C.4 Comparison of Dose Coefficients for Ingestion of Radionuclides Used in NTPR Program with Values Currently Recommended by ICRP: II. Longer-Lived Radionuclides

	Dose coefficient (rem μCi^{-1})[a]						
Nuclide[b]	Kidneys	Pancreas	Large intestine[c]	Red bone marrow	Bone surfaces	Liver	Bladder Wall
[60]Co	5.7E-3	5.9E-3	4.0E-2	5.4E-3	4.0E-3	6.8E-3	6.2E-3
(5.3 y)	(5.2E-3)	(5.2E-3)	(4.1E-2)	(4.8E-3)	(4.1E-3)	(8.5E-3)	(6.3E-3)
[89]Sr	8.6E-4	8.6E-4	8.7E-2	5.2E-3	1.1E-2	8.6E-4	4.3E-4
(50.5 d)	(7.4E-4)	(7.4E-4)	(8.1E-2)	(1.8E-2)	(2.2E-2)	(7.7E-4)	(2.5E-3)
[90]Sr	6.0E-3	6.0E-3	7.8E-2	4.3E-1	8.6E-1	5.7E-3	3.0E-3
(28.8 y)	(2.4E-3)	(2.4E-3)	(8.1E-2)	(6.7E-1)	(1.5)	(2.4E-3)	(5.6E-3)
[95]Zr	4.2E-4	3.9E-4	2.9E-2	6.6E-4	3.3E-4	3.0E-4	9.0E-4
(64.0 d)	(4.4E-4)	(4.1E-4)	(2.9E-2)	(7.8E-4)	(1.9E-3)	(3.0E-4)	(4.1E-4)
[106]Ru	8.3E-3	8.3E-3	2.6E-1	8.3E-3	9.6E-3	8.3E-3	4.4E-3
(373 d)	(5.6E-3)	(5.6E-3)	(2.6E-1)	(5.6E-3)	(5.6E-3)	(5.6E-3)	(6.3E-3)
[144]Ce	2.4E-4	3.0E-5	2.5E-1	1.4E-4	1.5E-4	7.4E-4	7.3E-5
(285 d)	(7.4E-5)	(7.0E-5)	(2.4E-1)	(7.0E-4)	(1.2E-3)	(3.6E-3)	(1.1E-4)
[152]Eu	1.2E-3	6.6E-4	6.3E-2	9.8E-4	6.9E-4	2.6E-3	1.1E-3
(13.5 y)	(1.2E-3)	(1.2E-3)	(3.7E-2)	(2.2E-3)	(4.1E-3)	(5.9E-3)	(1.6E-3)
[239]Pu[d]	6.3E-2	3.6E-3	2.0E-1	1.9E-1	2.6	4.9E-1	1.8E-3
	(2.5E-3)	(1.0E-3)	(2.0E-1)	(2.9E-2)	(5.9E-1)	(1.3E-1)	(1.0E-3)

[a]First entry is value from Table 4a of Egbert et al. (1985) based on ORNL reports (Killough et al., 1978a; Dunning et al., 1979) and often used in dose reconstructions for atomic veterans; values are based on dosimetric and biokinetic models in ICRP *Publication 30* (ICRP, 1979a). Second entry, in parentheses, is value currently recommended for adult workers by ICRP (1994a; 2002). All values assume GI-tract absorption fraction that applies to radionuclides in oxide form (Eckerman et al., 1988).
[b]Entry in parentheses is radionuclide half-life.
[c]Wall of lower large intestine.
[d]Dose coefficients apply to any mixtures of [239]Pu and [240]Pu, which have half-lives of 24,100 and 6,560 y, respectively.

[3] **An assumption that inhaled particles are respirable (AMAD, 1 μm) should result in large overestimates of dose to the lung if most of the inhaled materials were large particles. Doses to other organs and tissues, except those in the GI tract in many cases, also should be overestimated.**

Some dose reconstructions for atomic veterans assume that inhaled particles were respirable even when a substantial fraction of inhaled material probably consisted of large particles. As noted above and discussed in Section IV.C.2.2.1, an assumption of respirable particles often is used when the organ or tissue of concern is not the thyroid, an organ in the GI tract, or prostate, even in cases of inhalation of mostly large particles in descending fallout. Inhalation of large particles also could be important in other scenarios, such as exposure to fresh fallout that was resuspended by gentle disturbances that did not pulverize fallout particles.

An assumption that inhaled materials were respirable when the materials probably consisted mainly of large particles should result in large overestimates of dose to the lung, because lung doses are proportional to the fraction of inhaled material that is deposited in the bronchial and pulmonary regions and this deposition fraction is small for large particles. In the ICRP's current respiratory-tract model (ICRP, 1994b), the total deposition in all tissues making up the lung is about 15% when inhaled particles are assumed to be respirable (AMAD, 1 μm); but when large particles are inhaled, the fraction deposited in the lung ranges from about 1.5% at an AMAD of 20 μm to less than 0.05% at an AMAD of 100 μm. In the respiratory-tract model used in the NTPR program (ICRP, 1979a), the deposition fraction in tissues making up the lung is assumed to be 33% at an AMAD of 1 μm and zero at an AMAD of 20 μm or greater. Thus, dose to the lung could be overestimated by more than a factor of 10 when inhalation of respirable particles is assumed but most inhaled materials were large particles.

A comparison of inhalation dose coefficients in Tables V.C.1 and V.C.2, which apply to respirable particles, with the corresponding ingestion dose coefficients in Tables V.C.3 and V.C.4, which describe much of the dose from inhalation of large particles, indicates that an assumption of respirable particles when large particles are inhaled also could result in substantial overestimates of dose to organs and tissues other than those in the respiratory and GI tracts. The extent of overestimation depends on the radionuclides inhaled and their relative activities. However, plutonium is an exception (see Section V.C.3.2, comment [3]).

[4] **The extent of absorption of inhaled radionuclides into blood from the respiratory tract or, when swallowed, from the GI tract assumed in dose reconstructions may be overestimated for refractory radionuclides—such as plutonium and isotopes of yttrium, zirconium, and rare-earth elements—in fallout particles, especially when large particles are inhaled.**

In the respiratory-tract model used in the NTPR program (ICRP, 1979a), radionuclides that are not in gas or vapor form are assumed to be attached to surfaces of inhaled particles, from which they can be detached and dissolved in the respiratory and GI tracts, and the solubility of radionuclides is assumed to depend only on their chemical form, independent of the chemical composition of the particles to which they are attached. In fallout particles, however, refractory radionuclides—such as plutonium and isotopes of yttrium, zirconium, and rare-earth elements—are distributed approximately uniformly throughout the particle volume, rather than attached to surfaces (see Section IV.C.2.1.2). Furthermore, fallout particles that contain refractory materials should be highly insoluble, perhaps more so than insoluble chemical forms of radionuclides included in the ICRP models.

Thus, when fallout particles are inhaled, absorption of refractory radionuclides into blood from the respiratory or GI tract before the particles are eliminated from the body may be substantially less than assumed in the NTPR program, especially in cases of inhalation of large fallout particles that contain a relatively high proportion of refractory radionuclides. Reductions in absorption would result in corresponding reductions in doses to organs and tissues other than those in the respiratory and GI tracts.

[5] **Dosimetry models for internal emitters assumed in dose reconstructions may overestimate doses to organs and tissues of the respiratory and GI tracts when large fallout particles that contain refractory radionuclides that emit alpha particles, such as plutonium, are inhaled.**

In dosimetry models for radionuclides in the body used in dose reconstructions (ICRP, 1979a), which provide estimates of doses to target tissues per disintegration of a radionuclide at a site of deposition or transit, emitted radiation is assumed not to be attenuated or absorbed in particles to which radionuclides are attached. That assumption is reasonable when radionuclides are attached to surfaces of small particles. However, doses to organs and tissues of the respiratory and GI tracts could be substantially lower than calculated with ICRP's dosimetry models for internal emitters when large fallout particles that contain refractory radionuclides that emit alpha particles, such as plutonium, are inhaled. In such cases, most of the energy of emitted alpha particles would be absorbed in the fallout particles, rather than surrounding tissues, because of the very short range of alpha particles of a few μm or less. As noted above, this effect could persist during the time that large fallout particles remain in the body, because of their insolubility.

[6] **Dose coefficients for inhalation of radionuclides used in the NTPR program are committed doses; that is, they represent total doses received in specific organs and tissues over a period of 50 years after intake. In organs and tissues other than those in the lung (excluding lymphatic tissues) and GI tract, use of 50-year committed doses can result in substantial overestimates of the dose that could have caused a cancer in an exposed person when an inhaled radionuclide is long-lived and tenaciously retained in the body and the cancer of concern occurred well within 50 years.**

Use of 50-year committed doses from intakes of radionuclides is standard practice in radiation protection of workers. That approach takes into account that an acute intake of a radionuclide can result in a dose that is received over many years (see Section IV.C.2.2).[20]

In dose reconstructions for atomic veterans who file a claim for compensation, the quantity of interest is the dose received in an organ or tissue of concern

---

[20]This practice is intended to ensure that if a worker is exposed continuously over a working life of 50 years at the annual limit on intake, the dose received in any year would not exceed the annual dose limit for occupational exposure.

before a cancer occurred at that site, not the 50-year committed dose used in radiation protection. When an inhaled radionuclide has a half-life or biological half-time in the body of a few years or less, there is little difference between the dose received before the time of occurrence of a cancer, assuming that the cancer did not occur before a minimum latent period after intake, and the 50-year committed dose resulting from a given intake. However, when a radionuclide is long-lived and tenaciously retained in the body, there can be a significant difference between the dose received before a cancer occurred and the 50-year committed dose. In inhalation exposures of atomic veterans, the difference is potentially important mainly for plutonium and, to a lesser extent, ^{90}Sr. The difference between the dose received and the 50-year committed dose is most important for organs and tissues other than those in the respiratory and GI tracts, excluding respiratory lymphatic tissues where long-lived and insoluble radionuclides are assumed to be tenaciously retained (ICRP, 1979a; 1994b).

Consider a hypothetical example in which the disease of concern in an atomic veteran is bone or liver cancer and the dose to bone surfaces or liver was due primarily to inhalation of insoluble plutonium. Suppose that the cancer was diagnosed 35 years after exposure and that the latent period for the cancer is 10 years (Eckerman et al., 1999). In this case, the dose that could have caused the cancer is the dose received within the first 25 years after exposure, and use of 50-year committed doses to bone surfaces or liver would overestimate the dose that could have caused the cancer, mainly because of the biological half-time of plutonium in bone or the liver of several decades (ICRP, 1979a; 1993; 2002). The retention half-time of insoluble plutonium in the lung of a few years when inhaled particles are respirable (AMAD, 1 μm) is less important. Thus, even if inhaled plutonium were rapidly transferred to bone or liver, the 50-year committed dose would overestimate the dose received in the first 25 years by about a factor of 2, and the degree of overestimation would increase somewhat if the inhaled plutonium was respirable and the low rate of absorption of insoluble forms of inhaled plutonium from the respiratory tract into blood is taken into account.[21]

Again, the difference between the 50-year committed dose and the dose received in an organ or tissue is potentially important only if the dose was due mainly to intakes of long-lived radionuclides that are tenaciously retained in the body. Thus, the importance of this difference in dose reconstructions for atomic veterans depends on the activities of particular radionuclides inhaled.

The committee also notes, however, that use of 50-year committed doses from inhalation of long-lived radionuclides that are tenaciously retained in the body, such as plutonium, could result in underestimates of the dose that could

---

[21]The relatively rapid mechanical clearance of some inhaled material to the GI tract in the case of insoluble plutonium in respirable form would not affect the degree of overestimation of dose to a significant extent, because the fraction of ingested insoluble plutonium that is assumed to be absorbed into blood from the GI tract is very low (typically 10^{-4}–10^{-5}).

have caused a veteran's cancer in two situations. The first is illustrated by the example discussed above. In that example, if plutonium is an important contributor to dose and a cancer occurs more than 60 years after exposure, the 50-year committed dose would underestimate the relevant dose. That situation could occur in the future as the population of surviving atomic veterans ages.

The second situation involves cancers for which VA may have assumed that there is no appreciable increase in risk beyond some time after a radiation exposure. For example, studies of the Japanese atomic-bomb survivors indicate that there is little risk of a radiation-induced leukemia beyond about 25 years after exposure (see Section III.E), and a similar assumption may be made for a few other cancers, including lymphoma and multiple myeloma. However, there is an important difference between exposures of the Japanese atomic-bomb survivors and exposures of some atomic veterans that should be taken into account in applying assumptions about decreases in cancer risk at times long after exposure of the veterans. Essentially all of the dose to the atomic-bomb survivors was received at the time of the bombings or shortly thereafter, and there was little exposure due to inhalation of long-lived fission products and plutonium. In contrast, an atomic veteran who inhaled substantial amounts of plutonium and other long-lived radionuclides that are tenaciously retained in the body continued to receive a dose to bone marrow and lymphatic tissues long after the time of intake. Therefore, the practice in the NTPR program of assigning the entire 50-year committed dose from inhalation of plutonium and other long-lived radionuclides to the year of intake, which ignores that the dose is protracted over many decades after an intake, would greatly underestimate the dose that could have caused a veteran's cancer if the risk of that cancer is assumed to be negligible beyond some time after exposure and the veteran's cancer occurred at such a time.

[7]   Dose coefficients for the lung used in the NTPR program could overestimate doses to particular tissues in the respiratory tract where lung cancers occur.

In the respiratory-tract model used by the NTPR program (ICRP, 1979a), dose coefficients for the lung represent the average dose to the tracheobronchial tree, pulmonary region, and pulmonary lymphatic tissues. That is, dose to the lung is calculated as the total energy absorbed in the three regions divided by an assumed total mass of tissue of 1,000 g.

Most lung cancers occur in the bronchial region, which also is the region where most excess lung cancers in the Japanese atomic-bomb survivors have occurred (ICRP, 1994b). In the respiratory-tract model used in dose reconstructions (ICRP, 1979a), calculated doses to the lung overestimate doses to the tracheobronchial tree in cases of inhalation of insoluble (Class Y) longer-lived radionuclides in respirable form (AMAD, 1 μm) by about a factor of 3 because of the influence of the relatively high dose to lymphatic tissues on the average dose in all tissues considered (see Table V.C.2). The difference is smaller when shorter-lived radionuclides are inhaled, because of the smaller influence of doses to lymphatic tissues (see Table V.C.1).

In ICRP's current respiratory-tract model (ICRP, 1994b), dose to the lung is calculated as a weighted average of doses to the bronchial region, bronchiolar region, alveolar-interstitial region, and lymphatic tissues draining these regions, with the dose to lymphatic tissues given a weight of 0.001 and doses to the other three regions each given a weight of 0.333. Thus, the current model gives much less weight to the dose to lymphatic tissues, and the result is that the average dose to all tissues in the bronchial region is about the same as the weighted-average dose to the lung.

[8] Assumptions about resuspension of radionuclides in fallout that was deposited on the ground or suspension of neutron-induced activation products in surface soil used in dose reconstructions should, in some cases, tend to result in overestimates of concentrations in air relative to concentrations on the ground.

Resuspension of fallout deposited on the ground or suspension of activation products in surface soil is important in many scenarios for inhalation exposure of atomic veterans. Resuspension factors normally used in dose reconstructions for these scenarios are discussed in Sections IV.C.2.1.3 and IV.C.2.1.4 and are summarized in Table IV.C.2.

In some exposure scenarios at the NTS or on residence islands in the Pacific, resuspension or suspension of radionuclides on the ground occurred mainly as a result of normal wind stresses or walking and other activities that did not involve vigorous disturbance of surface soil. In those cases, a resuspension factor of 10^{-5} m^{-1} often is assumed in dose reconstructions (see cases #8, 21, 22, 23, 27, 36, 38, 43, 47, 96, and 98), although a lower resuspension factor of 10^{-6} m^{-1}, or even zero, sometimes is assumed when deposited fallout was aged for some time (see cases #31, 58, 63, 78, and 94). The latter assumptions are based on studies that showed that weathering of deposited materials generally reduced the resuspension factor over time (for example, see Anspaugh et al., 1975). As discussed in Sections IV.C.2.1.3 and IV.C.2.1.7, resuspension factors assumed in dose reconstructions are intended to overestimate airborne concentrations of radionuclides relative to concentrations on the ground.

The resuspension factor that should be applied to a particular exposure scenario is a highly uncertain parameter (see Section IV.C.2.1.3), and its value depends on the disturbance that causes resuspension. Data summarized in Table 12.7 of Sehmel (1984) indicate that resuspension factors at a height of 1 m above ground caused by normal wind stresses vary over a range of about 10^{-10} to nearly 10^{-4} m^{-1}, with most of the values being less than 10^{-5} m^{-1}, often by a factor of 10 or more; 1 m is the standard height at which resuspension factors normally are determined. Higher resuspension factors usually apply to freshly deposited materials, although very low values were obtained in some controlled tracer studies.

Most studies of the effects of walking on resuspension, as summarized in Table 12.9 of Sehmel (1984), were conducted in indoor environments, and results of such studies probably are not applicable outdoors. In the few studies outdoors, resuspension factors at a height of 1 m above ground caused by walking were in the range of about 10^{-8} to 10^{-5} m^{-1}.[22]

On the basis of information summarized above, the committee has concluded that resuspension factors used in the NTPR program should tend to overestimate airborne concentrations of radionuclides relative to concentrations on the ground in exposure scenarios in which resuspension or suspension of radionuclides is caused by normal wind stresses, walking, or other actions that do not involve vigorous disturbance of surface soil. That conclusion applies, for example, to resuspension of fresh or aged fallout or suspension of activation products in soil at the NTS and to resuspension of fallout on residence islands in the Pacific under the stated conditions of disturbance. As noted above, resuspension factors of 10^{-5} or 10^{-6} m^{-1} often are assumed in these scenarios. A resuspension factor of 10^{-5} m^{-1} that is normally assumed in scenarios at the NTS involving suspension of activation products in soil also should be a considerable overestimate under the stated conditions of disturbance because suspended materials are part of soil rather than deposited loosely on the ground surface, as is fallout.

The committee also cautions, however, that a resuspension factor of 10^{-5} or 10^{-6} m^{-1} may not be an overestimate during normal, nonvigorous activities on contaminated ships in the Pacific, nor does the conclusion discussed above apply to other exposure scenarios at the NTS in which resuspension is caused by more vigorous disturbances of deposited fallout. Those issues are discussed in the following section.

In summary, the committee has identified several assumptions used in the NTPR program to estimate inhalation dose coefficients and concentrations of radionuclides in air that, in the committee's opinion, should tend to result in overestimates of inhalation doses to atomic veterans; these assumptions are briefly restated in Table V.C.5. The committee also emphasizes, however, that the discussions of these assumptions should not be used to draw conclusions about whether estimates of inhalation doses to atomic veterans provide credible upper bounds without also considering the importance of uncertainties in these assumptions and the importance of other countervailing assumptions used in the NTPR program that may tend to result in underestimates of inhalation doses. Those other issues are discussed in the following section, and the committee's overall evaluation of methods of estimating inhalation doses used in the NTPR program is presented in Sections V.C.3.3, V.C.5, and V.C.6.

---

[22]This range takes into account that the resuspension factor at a height of 1 m caused by non-vigorous disturbances may be about a factor of 30 less than measured values at a height of 0.3 m (Sehmel, 1984).

TABLE V.C.5   Summary of Assumptions Used to Estimate Inhalation Doses in NTPR Program That Should Tend to Result in Overestimates of Dose[a]

Dose coefficients (organ-specific equivalent doses per unit activity of radionuclides inhaled)

- Dose coefficients for respirable particles (AMAD, 1 μm) often are higher than values for same particle size currently recommended for workers by ICRP.
- Dose coefficients for large particles (AMAD, 20 μm) often are higher than values based on current ICRP recommendations.
- Assumption of respirable particles overestimates dose to lung and many other organs when most inhaled materials are large particles.[b]
- Assumed absorption of refractory radionuclides (for example, plutonium and isotopes of yttrium, zirconium, and rare-earth elements) from respiratory or GI tract may be overestimated, especially when large particles are inhaled.
- Dose to respiratory and GI tracts may be overestimated when large particles containing alpha-emitting refractory radionuclides (for example, plutonium) are inhaled.
- Use of 50-year committed doses may overestimate relevant doses from intakes of long-lived radionuclides that are tenaciously retained in the body (for example, plutonium).[c]
- Dose coefficients for the lung may overestimate dose to tissues in respiratory tract where lung cancers occur.[d]

Methods used to estimate inhalation exposures (intakes of radionuclides in air)

- Resuspension factors applied to fallout deposited on ground or to neutron-induced activity in soil may overestimate airborne concentrations in some scenarios.[e]

[a]Assumptions are discussed in detail in Section V.C.3.1.
[b]For most radionuclides, assumption of respirable particles when large particles are inhaled does not overestimate dose to organs of GI tract.
[c]Relevant dose is dose received before disease of concern in exposed person occurs, taking into account latent period between radiation exposure and earliest onset of disease. However, as discussed in Section V.C.3.1, comment [6], use of 50-year committed doses may underestimate relevant doses in some cases.
[d]Most lung cancers, including cancers caused by radiation, occur in bronchial region.
[e]Conclusion applies mainly to scenarios in which resuspension of fallout deposited on the ground or suspension of neutron-induced activity in soil is caused by normal wind stresses or walking and other activities that do not involve vigorous disturbance of surface soil.

V.C.3.2 *Assumptions With Substantial Uncertainty or Tending to Underestimate Inhalation Dose*

The committee also is concerned that some assumptions used to estimate inhalation doses in the NTPR program may not tend to overestimate actual doses and thus may not lead to credible estimates of upper bounds for use in evaluating claims for compensation.

The committee's concerns are of two kinds. The first is that, in some cases, assumptions about scenarios of inhalation exposure or estimates of parameter values probably result in substantial underestimates of possible doses, provided that other assumptions used in estimating inhalation dose are reasonable. The second concern is that uncertainties in assumptions, models, and parameter values used to estimate inhalation doses have not been considered in the NTPR program. As discussed in Section IV.E.4, all estimates of inhalation dose are presented as single values without uncertainty. The justification for that approach is essentially that estimates of inhalation dose are based on assumptions that are sufficiently biased on the high side that the estimates themselves can be considered upper bounds of possible doses. However, the committee is concerned that that may not be the case when uncertainties in assumptions are considered, even when assumed exposure scenarios are reasonable. The two types of concerns are related, in that they both are important in evaluating whether estimates of inhalation doses obtained in the NTPR program are credible upper bounds. As in the previous section, assumptions concerned with estimation of inhalation dose coefficients (organ-specific equivalent doses per unit activity of radionuclides inhaled) are considered first, followed by assumptions concerned with estimation of inhalation exposures (intakes of radionuclides in air).

[1]   **Dose coefficients for inhalation of radionuclides used in the NTPR program are based on dosimetric and biokinetic models that have substantial uncertainty. As a result, credible upper bounds of inhalation dose coefficients may be substantially higher than values used in dose reconstructions, even though the assumed dose coefficients often are higher than values currently recommended by ICRP.**

Dose coefficients for inhalation of radionuclides used in the NTPR program are standard values developed by ICRP. These dose coefficients apply to so-called Reference Man, which is an anatomic, physiologic, and metabolic representation of an average adult (ICRP, 1975). Dose coefficients for Reference Man are assumed to be appropriate for use in radiation protection (that is, in evaluating compliance with dose limits for workers and other requirements).

However, there is substantial uncertainty in inhalation dose coefficients developed by ICRP. The uncertainty results, first, from uncertainty in dosimetric and biokinetic models used to calculate dose coefficients and in data incorporated in the models and, second, from the variability in anatomic, physiologic, and metabolic characteristics among people. Those sources of uncertainty in dose coefficients should be acknowledged and addressed in dose reconstructions for atomic veterans if credible upper bounds of inhalation dose are to be obtained. The approach taken in the NTPR program essentially has been to argue, first, that its methods of estimating inhalation dose, especially assumptions about resuspension factors used to estimate airborne concentrations of radionuclides relative to concentrations on the ground or other surfaces, generally are sufficiently biased on the high side that they compensate for uncertainty in dose coefficients,[23]

---

[23]Oral testimony from J. Klemm, SAIC, and D.M. Schaeffer, NTPR program manager, at open session of committee on October 10–11, 2001.

and, second, that information on uncertainty in dose coefficients was not available for use in dose reconstructions (for example, see Goetz et al., 1987).

The first analysis of uncertainty in dose coefficients was concerned with dose to the thyroid of an adult from ingestion of [131]I (Dunning and Schwarz, 1981); the results also apply to inhalation of [131]I. The analysis showed that when data on the variability in thyroid mass, fractional uptake of absorbed iodine in the thyroid, and retention half-time of iodine in the thyroid are taken into account, the 95th percentile of the dose per unit activity intake in an adult exceeds the median (50th percentile) by a factor of 3. The results of that uncertainty analysis could have been taken into account, but were not, in a later dose reconstruction for a small group of veterans on Rongerik Atoll in the Marshall Islands who were exposed to high levels of fallout after Operation CASTLE, Shot BRAVO and received an estimated equivalent dose of 190 rem to the thyroid from intakes of [131]I (Goetz et al., 1987).

More recently, Bouville et al. (1994) published an analysis of the reliability of dose coefficients for inhalation and ingestion of selected radionuclides; the analysis was adopted in a later report of NCRP (1998). The results of that analysis are summarized in Table V.C.6. An uncertainty factor of 10, for example, means that the 95th percentile of a subjective probability (uncertainty) distribution of the effective dose per unit activity intake by healthy adult males is a factor of 10 higher than the value recommended by ICRP.

Estimates of reliability (uncertainty) in dose coefficients summarized in Table V.C.6 apply to the effective dose, which is a weighted average of equivalent doses to different organs and tissues defined in ICRP *Publication 60* (ICRP, 1991a) and is similar to the effective dose equivalent calculated in some dose reconstructions for atomic veterans (see Section IV.C.2). In specific organs and tissues of concern when a veteran files a claim for compensation, the uncertainty in an inhalation dose coefficient could be larger than the uncertainty in the effective dose coefficient. That is the case, for example, in estimating dose to radiosensitive tissues of the skeleton (bone surfaces and bone marrow) when the important radionuclides inhaled include ^{90}Sr or plutonium (Eckerman et al., 1999). Estimates of dose to the walls of organs of the GI tract from ingestion of alpha-emitting radionuclides (such as plutonium) also have large uncertainty (ICRP, 1979a).

Uncertainty in inhalation dose coefficients of the magnitude indicated in Table V.C.6 clearly is important when methods of estimating inhalation dose used in the NTPR program are intended to provide credible upper bounds. Depending on the particle size of inhaled materials, the organ or tissue of concern, and the important radionuclides inhaled, this source of uncertainty may not be fully compensated by assumptions embodied in dose coefficients used in the NTPR program that should tend to overestimate dose, including that the dose coefficients are higher in many cases than values currently recommended by ICRP.

TABLE V.C.6   Estimated Reliability of Effective Dose Coefficients for Selected Radionuclides, Relative to Values Recommended in ICRP *Publication 30*[a]

Radionuclide	Route of Intake[b]	Uncertainty Factor[c]
High reliability		
^{3}H (HTO)	Ingestion	2
^{14}C (CO_2)	Inhalation	2
^{137}Cs	Inhalation or ingestion	2
^{90}Sr	Inhalation or ingestion	3
^{131}I	Inhalation or ingestion	3
^{140}La	Ingestion	3
Intermediate reliability		
^{140}La, ^{210}Pb	Inhalation	5
^{14}C (CO_2)	Ingestion	5
^{60}Co, ^{144}Ce	Inhalation	5
^{210}Pb, ^{230}Th	Inhalation	5
^{234}U	Inhalation	5
^{55}Fe, ^{95}Nb	Inhalation or ingestion	5
^{140}Ba, ^{226}Ra	Inhalation or ingestion	5
Low reliability		
^{210}Po	Ingestion	10
^{60}Co, ^{210}Pb	Ingestion	10
^{230}Th, ^{234}U	Ingestion	10
^{95}Zr, ^{106}Ru	Inhalation or ingestion	10
^{125}Sb	Inhalation or ingestion	10
^{237}Np, ^{239}Pu	Inhalation or ingestion	10
^{241}Am, ^{244}Cm	Inhalation or ingestion	10
^{144}Ce	Ingestion	10

[a]Estimates given by Bouville et al. (1994) and NCRP (1998). Effective dose is weighted average of equivalent doses to different organs and tissues (ICRP, 1991a).
[b]Inhaled materials are assumed to be attached to respirable particles (AMAD, 1 μm), except that ^{3}H (HTO) and ^{14}C (CO_2) are assumed to be in vapor and gaseous form, respectively.
[c]Ratio of the upper limit of a 90% confidence interval of a subjective probability distribution of the effective dose per unit activity intake by healthy adult males to the effective dose coefficient recommended by ICRP; that is, the 95th percentile of an assumed probability distribution exceeds ICRP's recommended value by the uncertainty factor (see Section V.C.3.2, comment [1]).

[2]   Dose coefficients for inhalation of alpha-emitting radionuclides, such as plutonium, used in dose reconstructions incorporate an assumption that the biological effectiveness of alpha particles is 20 times that of photons and electrons, without uncertainty. However, a credible upper-bound estimate of the biological effectiveness of alpha particles is substantially higher than the assumed value.

Dose coefficients for inhalation of alpha-emitting radionuclides used in dose reconstructions incorporate a standard assumption recommended by ICRP (1977;

1991a) and NCRP (1987b; 1993) that alpha particles are 20 times more effective in inducing stochastic biological effects (cancers and severe hereditary effects) than photons and electrons. That is, in calculating equivalent doses in organs and tissues, absorbed doses of alpha particles are multiplied by a factor of 20. That factor is applied without uncertainty in radiation protection. However, available information on the relative biological effectiveness (RBE) of alpha particles indicates that there is considerable uncertainty in the particular value that should be used to estimate equivalent doses in humans. That uncertainty should be addressed if credible upper bounds of doses from inhalation of alpha-emitting radionuclides, such as plutonium, are to be obtained in dose reconstructions.

A review and analysis of data on the biological effectiveness of alpha particles in inducing lung tumors in various animals was published by ICRP (1980b) at about the time the NTPR program began; the review included information on uncertainty in estimates of biological effectiveness. On the basis of the combined data from studies that used soluble or insoluble chemical forms of alpha-emitting radionuclides, ICRP concluded that the RBE of alpha particles was in the range of about 6-40. However, when only the data from studies using insoluble plutonium oxide were considered, the RBE was in the range of about 10-100. The latter estimate of uncertainty is more relevant to dose reconstructions for atomic veterans in that plutonium and other alpha-emitting radionuclides in fallout are expected to be relatively insoluble.

Later analyses of data on the biological effectiveness of alpha particles were presented by NCRP (1990) and the UK National Radiological Protection Board (Muirhead et al., 1993). The analyses indicate that a central estimate of the RBE of alpha particles obtained from various studies that are relevant to induction of cancer in humans is in the range of about 5-60. That range would be substantially broader if uncertainty in the individual determinations of biological effectiveness were taken into account.

On the basis of analyses by ICRP (1980b), NCRP (1990), and Muirhead et al. (1993) discussed above, Kocher et al. (2002) developed a subjective probability (uncertainty) distribution of the so-called radiation effectiveness factor (REF) for alpha particles in inducing solid tumors in humans; an REF in humans represents data on the RBE of alpha particles in other organisms.[24] Probability distributions of REFs for all radiation types developed by Kocher et al. (2002) will be used in estimating equivalent doses to workers at DOE facilities for the purpose of evaluating claims for compensation for radiation-related diseases.

The probability distribution of REF for alpha particles and solid tumors developed by Kocher et al. (2002) has a 95% confidence interval of 3.4-100; the

---

[24]A probability distribution of REF also was developed for alpha particles and leukemia (Kocher et al., 2002). That distribution is not relevant to most dose reconstructions, because many types of leukemia are presumptive diseases under 38 *CFR* 3.309 and a dose reconstruction is not required in evaluating a claim for compensation when a veteran's participation status is adequately established.

central estimate (50th percentile) is 18.[25] That confidence interval encompasses an estimate of uncertainty by the US Environmental Protection Agency (EPA, 1999). On the basis of the uncertainty estimated by Kocher et al., a credible upper bound of the biological effectiveness of alpha particles, as represented by the 95th percentile, is about 76, or a factor of nearly 4 greater than the value 20 used in dose reconstructions. An uncertainty of such magnitude clearly is important.

[3]   **Dose coefficients for inhalation of plutonium used in dose reconstructions may underestimate doses to organs of the GI tract by a substantial amount in scenarios in which an appreciable fraction of inhaled materials are respirable (AMAD, 1 μm).**

In dose reconstructions for atomic veterans, doses to organs of the GI tract due to inhalation of radionuclides usually are calculated by using dose coefficients for large particles (AMAD, 20 μm). That choice is made because, on the basis of dose coefficients for inhalation and ingestion used in the NTPR program, an assumption of large particles generally results in higher estimates of dose to these organs than an assumption of respirable particles (AMAD, 1 μm); see Section IV.C.2.2.1 and dose coefficients for the wall of the lower large intestine based on ICRP *Publication 30* (ICRP, 1979a) given in Tables V.C.1 through V.C.4.

On the basis of dose coefficients for inhalation and ingestion of radionuclides by adult workers currently recommended by ICRP (1994a; 2002), an assumption of inhalation of large particles still results in higher estimates of dose to organs of the GI tract in most cases (see Tables V.C.1 through V.C.4). However, insoluble plutonium is an exception. In this case, current ingestion dose coefficients for organs of the GI tract are little changed from the values used in dose reconstructions, but inhalation dose coefficients for these organs, assuming an AMAD of 1 μm, are a factor of about 10 higher in the current recommendations. As a result of that increase, the usual assumption of large particles would result in an underestimate of dose to organs of the GI tract from inhalation of insoluble plutonium by a factor of about 5 in scenarios in which inhalation of respirable particles is likely. The importance of this underestimation of dose depends on intakes of plutonium relative to intakes of other radionuclides and the magnitude of the dose.

The increase in dose coefficients for organs of the GI tract from inhalation of insoluble plutonium and the counterintuitive result that the dose to these organs from inhalation is higher than the dose from ingestion of insoluble plutonium is due to a significant change in the biokinetic model for systemic plutonium. In the model used in the NTPR program (ICRP, 1979a), all plutonium absorbed into

---

[25]This probability distribution of REF takes into account a small inverse dose-rate effect, whereby the response per unit dose of alpha particles is assumed to be higher at low dose rates than at the higher dose rates used in radiobiological studies.

blood from the respiratory or GI tract is assumed to be deposited only in bone, liver, or gonads. In ICRP's current model, however, so-called soft tissue compartments are also included, and 14% of all plutonium absorbed into blood is assumed to be deposited uniformly in these tissues, which include all organs of the GI tract (ICRP, 1993). Furthermore, the residence half-time of plutonium in the soft-tissue compartment is many decades. Thus, in the current model, the higher dose to organs of the GI tract from inhalation of insoluble, respirable plutonium than from ingestion is due to three factors: the greater absorption of plutonium from the respiratory tract than from the GI tract, where the absorption fraction is only 10^{-5} (ICRP, 1993; 2002); the low doses to walls of the GI tract during passage of ingested plutonium, because of the short range of alpha particles in matter and short transit time of material in the GI tract (ICRP, 1979a); and deposition of a small but important amount of absorbed plutonium in tissues of GI tract, where it remains for many years.

[4] Methods used in the NTPR program to estimate concentrations of radionuclides in fallout deposited on the ground or other surfaces on the basis of measurements of external photon exposure with film badges or field instruments involve potentially important sources of error and uncertainty that have not been evaluated. An overall bias of the methods toward overestimation or underestimation of concentrations of radionuclides in deposited fallout in all exposure scenarios is difficult to determine from documentation of the methods. However, an assumption of no fractionation of radionuclides in fallout except for removal of noble gases, which has been used in all dose reconstructions, should result in substantial underestimates of concentrations of important refractory radionuclides, such as plutonium, in deposited fallout at locations relatively close to detonations where many participants were exposed, especially at the NTS. In addition, an assumption that the source region is infinite in extent, which is used to relate measured external exposures to concentrations of radionuclides on the ground or other surfaces, is probably not valid in cases of fallout on ships in the Pacific, and concentrations of radionuclides in deposited fallout may be substantially underestimated in these cases.

Scenarios involving inhalation of radionuclides in descending fallout or in fallout that was deposited on the ground or other surfaces and then resuspended in the air are important in dose reconstructions for many participants at the NTS or in the Pacific. In all such scenarios, inhalation doses are estimated in the NTPR program on the basis of estimates of concentrations of radionuclides in deposited fallout. The following discussion concerns methods used in the NTPR program to estimate those concentrations. Assumptions about resuspension factors that are used to estimate concentrations of radionuclides in air breathed by exposed people relative to concentrations on the ground or other surfaces are considered separately in later comments in this section.

Methods used in the NTPR program to estimate concentrations of radionuclides in fallout deposited on the ground or other surfaces are described in Section IV.C.2.1. These estimates depend essentially on two types of data: relative activities of radionuclides in an atmospheric cloud immediately after a detonation, which are estimated from cloud sampling data and calculations of relative activities of fission and activation products produced by the weapon type of concern; and external photon exposures or exposure rates due to deposited fallout, which are estimated from readings of film badges worn by participants or field instruments. All other aspects of the methods involve assumptions about how those data are related to concentrations of radionuclides in deposited fallout. Specifically, on the basis of an assumption about relative activity concentrations of radionuclides on the ground or other surfaces compared with estimated relative activities in the cloud (fractionation of radionuclides) and calculations of external exposure rates per unit concentration of radionuclides on the surface, the desired concentrations on the surface, SA, are estimated as $(SA/I) \times I$, where I is a measured exposure rate due to all radionuclides and (SA/I) is the reciprocal of the calculated exposure rate per unit concentration for the assumed mixture of radionuclides on the surface.

As discussed in Section V.C.2, a previous committee of the National Research Council concluded that methods of estimating radionuclide concentrations in fallout deposited on the ground or other surfaces based on measurements of external photon exposure are not scientifically valid and that their reliability is unknown (NRC, 1985b). The present committee agrees with the previous conclusion that the reliability of the methods is unknown but does not consider the methods to be generally invalid. Relative activities of radionuclides in an atmospheric cloud can be estimated reasonably well from data from cloud sampling and calculations of relative amounts of fission and activation products produced by detonation of a weapon type of concern. If fractionation of radionuclides in the cloud is properly taken into account (see Section IV.C.2.1.2), relative activity concentrations of radionuclides in deposited fallout can be estimated from relative activities in the cloud. Well-established calculation methods can then be used to estimate external exposure rates due to assumed relative activity concentrations of radionuclides in deposited fallout. Finally, measurements of external exposures or exposure rates can be used to normalize calculated exposure rates due to an assumed mixture of radionuclides in deposited fallout to obtain estimates of absolute activity concentrations (Ci m^{-2}) of each radionuclide.

Thus, the committee has concluded that methods used in the NTPR program to estimate concentrations of radionuclides in fallout deposited on the ground or other surfaces are valid, at least in principle, even with regard to estimating concentrations of beta- and alpha-emitting radionuclides, such as ^{90}Sr and plutonium, that can be important contributors to inhalation dose but are not detected by measurements of external photon exposure. However, the reliability of the methods depends on the validity of data and assumptions used to estimate relative activity concentrations of radionuclides in deposited fallout and the validity of assumptions used to calculate external exposure rates per unit concentration of radionuclides on a surface.

The committee is concerned that important sources of error and uncertainty in methods used in the NTPR program to estimate concentrations of radionuclides in fallout deposited on the ground or other surfaces based on measurements of external photon exposure have not been evaluated. Therefore, the reliability of the methods has not been demonstrated, and uncertainty in the calculations has not been quantified. The committee's principal concerns involve two issues: fractionation of radionuclides in an atmospheric cloud, which determines relative activities of radionuclides in deposited fallout at specific locations compared with estimated relative activities in the cloud; and calculation of external photon exposure rates per unit concentration of radionuclides on a surface in cases of fallout on ships in the Pacific.

Methods used in the NTPR program to estimate concentrations of radionuclides in fallout deposited on the ground or other surfaces assume that fractionation of radionuclides in an atmospheric cloud does not occur, except for removal of noble gases (see Section IV.C.2.1.1). However, fractionation is important in determining relative activities of radionuclides in deposited fallout. Although the discussion of fractionation in Section IV.C.2.1.2 is an idealized representation of a complex process (Freiling et al., 1964), neglect of fractionation generally results in underestimates of relative activities of refractory radionuclides—such as plutonium and shorter-lived isotopes of yttrium, zirconium, and rare-earth elements—in fallout at locations relatively close to detonations where considerable resuspension occurred and many participants were exposed and similar overestimates of relative activities of volatile radionuclides, such as ^{131}I and isotopes of strontium. Neglect of fractionation is an important concern, in part, because plutonium probably posed the greatest long-term inhalation hazard at the NTS and in the Pacific.

Data discussed by Hicks (1982) and Freiling et al. (1964) indicate that fractionation typically alters the relative activities of refractory and volatile radionuclides in local fallout, compared with initial activities in an atmospheric cloud, by a factor of about 3-4. In a few shots, however, the effect was as large as a factor of 100 or even more (Freiling et al., 1964). Thus, activities of plutonium and other refractory radionuclides in deposited fallout could be underestimated substantially when fractionation is not taken into account, and activities of volatile radionuclides could be overestimated to a similar extent. That is the case especially at the NTS because all fallout there probably contained a substantial fraction of large particles in which most of the activity of refractory radionuclides but relatively little of the activity of volatile radionuclides was found (Hicks, 1982). Furthermore, there is substantial uncertainty in the extent of fractionation at any shot. There also is some uncertainty in estimates of relative activity concentrations of radionuclides in an atmospheric cloud and in film-badge or field-instrument measurements of external photon exposures from radionuclides deposited on a surface. However, the extent of fractionation and its uncertainty probably is the most important factor affecting the reliability of methods of estimating concentrations of radionuclides in deposited fallout in all exposure scenarios. The committee also notes that the extent of fractionation cannot be estimated reliably from measured external exposures, because relative activities of photon-emitting radionuclides in deposited fallout can vary widely and still give approximately the same reading on a film badge or field instrument and approximately the same dependence of exposure rate on time after a detonation.

In the method used in the NTPR program to calculate external photon exposure rates per unit concentration of radionuclides deposited on the ground or other surfaces, the surface is assumed to be uniformly contaminated and infinite in spatial extent, and the source region is modeled to take into account a small shielding effect of about 0.7 due to ground roughness (Egbert et al., 1985; Barrett et al., 1986). Those assumptions are reasonable in cases of fallout deposited at the NTS or on residence islands in the Pacific because external exposure at a given location was due almost entirely to sources within a few tens of meters and fallout usually was widespread and did not vary irregularly over such small distances. The method of calculation is unlikely to significantly overestimate exposure rates per unit concentration of radionuclides in fallout deposited at the NTS or on residence islands in the Pacific and therefore probably does not significantly underestimate concentrations corresponding to a measured exposure rate.

The assumption of an infinite and uniformly contaminated source region is also used to calculate external exposure rates per unit concentration of radionuclides in fallout deposited on ships in the Pacific (Egbert et al., 1985; Barrett et al., 1986; Goetz et al., 1991). On ships, however, the area of the source region, either on deck or below, is substantially less than the area in an infinite source region that would contribute in an important way to a calculated exposure rate at a given location; as noted above, sources at distances out to a few tens of meters contribute to external exposures due to a surface source of infinite extent. Thus, concentrations of radionuclides in fallout deposited on ships that would result in a given exposure rate are higher than calculated concentrations that would yield the same exposure rate when an infinite source is assumed, and an assumption of an infinite source region results in underestimates of concentrations of radionuclides in fallout on ships. Shielding provided by superstructures on decks of many ships and shielding provided by structures below decks further limit the area of the source region that contributes significantly to a measured exposure rate at a given location, thus increasing the extent of underestimation of concentrations of radionuclides in fallout on ships. On target ships at Operation CROSSROADS, estimation of surface activities of radionuclides that could be resuspended in the air on the basis of measurements of external exposure rates is further complicated by the presence of substantial contamination of hulls and piping inside the hulls.

On the basis of considerations discussed above, the committee has concluded that an assumption of an infinite source region in calculating external exposure rates per unit concentration of radionuclides on a surface probably is not valid in cases of fallout on ships in the Pacific and that concentrations of radionuclides in deposited fallout may be substantially underestimated in these cases. In any event, the reliability of the method used to calculate exposure rates per unit concentration in cases in which fallout is deposited over an area of finite extent has not been evaluated in the NTPR program.

[5] **In estimating inhalation doses in scenarios involving exposure to descending fallout, the resuspension factor used to estimate concentrations of radionuclides in descending fallout based on estimated concentrations on the ground or other surfaces may underestimate airborne concentrations relative to concentrations on the surface when exposure did not occur during the entire period of fallout.**

The method used in the NTPR program to estimate concentrations of radionuclides in descending fallout is based on estimates of concentrations in fallout deposited on the ground or other surfaces and an assumed "effective" resuspension factor of 10^{-4} m^{-1} (see Section IV.C.2.1.5). The resuspension factor is based on an assumption that fallout descended from a height of about 10^4 m (10 km), and the method assumes implicitly that descending fallout is distributed uniformly over that height. When exposure is assumed to occur during only part of the period of descent, inhalation dose is assumed to vary linearly with exposure time (that is, the dose rate is assumed to be constant).

The committee finds that the concept underlying the method of estimating concentrations of radionuclides in descending fallout relative to estimated concentrations on the ground or other surfaces is reasonable, because a concentration in deposited fallout results from descent of an average concentration in air over some height, where the appropriate average concentration in air is inversely proportional to the assumed height. Furthermore, the assumed height is unimportant provided that the assumed average concentration in air yields the correct concentration in deposited fallout. Thus, average concentrations of radionuclides in descending fallout relative to concentrations in deposited fallout should not be underestimated when exposure is assumed to occur during the entire period of descent.

However, that conclusion may not be reasonable in all cases of exposure to descending fallout. Rates of descent of fallout vary greatly with particle size (Sehmel, 1984), as do the relative amounts of refractory and volatile radionuclides (see Section IV.C.2.1.2). As a result, dose rates due to inhalation of descending fallout vary with time during the period of descent. Therefore, in scenarios in which exposure to descending fallout was assumed to occur during only part of the period of descent, average dose rates above (or below) the average dose rate over the entire period of descent could occur. Similarly, on residence islands or ships in the Pacific, doses due to inhalation of descending fallout could be underestimated or overestimated even when a participant was present during the entire period of descent, because inhalation exposure sometimes was assumed to occur during only the fraction of the time that a participant was assumed to spend outdoors (see case #5).

A further difficulty with the method is that, in some cases, fallout probably descended from a height considerably less than 10 km. In such cases, inhalation exposures could be underestimated if the assumed duration of exposure is less than the assumed period of descent from a height of 10 km. That situation could occur, for example, if fallout originated in the stem of a mushroom cloud produced in a detonation, as probably happened at Operation PLUMBBOB, Shot SMOKY (see Appendix F).

[6]   In estimating inhalation doses in scenarios involving resuspension of deposited fallout at the NTS or on islands in the Pacific, the presence of aged fallout that was deposited more than a few months before exposure usually is not taken into account, and the presence of fallout from all prior shots is ignored in some cases. Thus, even in scenarios in which an assumed resuspension factor is a credible upper bound, inhalation doses could be underestimated substantially, especially in cases of exposure at the NTS at times relatively late in the period of aboveground testing.

The possibility of substantial inhalation exposure due to resuspension of aged fallout is an important concern at the NTS and on islands in the Pacific. The concern arises from two considerations. First, concentrations of plutonium and other longer-lived radionuclides, for which doses per unit activity inhaled are the highest and the inhalation hazard thus is the greatest, persisted in aged fallout with little depletion due to decay. Second, at times relatively late in the period of aboveground testing, areas where participants were exposed often were affected by fallout from several prior shots, and concentrations of plutonium and other longer-lived radionuclides on the ground at locations of exposure could be substantially higher than those due to fallout from a single shot. An additional important problem discussed in comment [4] above is that concentrations of plutonium, which probably is the most important longer-lived radionuclide in deposited fallout, may be substantially underestimated by the NTPR program because of neglect of fractionation.

In dose reconstructions for participants on islands in the Pacific, inhalation exposures due to resuspension of deposited fallout are taken into account for periods up to 2,500 h (about 3.5 months) after fallout occurred (Goetz et al., 1991). Thus, fallout from shots that occurred more than 2,500 h before an exposure of concern apparently is not taken into account (see cases #58, 63, and 94). The committee also found evidence that at times of a month or more after fallout was deposited, a resuspension factor of 10^{-6} m^{-1} was used in some cases to estimate airborne concentrations of radionuclides, rather than the standard value of 10^{-5} m^{-1} often used on residence islands (see cases #16, 31, and 78). Use of a time cutoff and a lower resuspension factor for aged fallout was based on an argument that aged fallout is much less susceptible to resuspension than freshly deposited material, although no data are presented to support the argument. As discussed in Section V.C.5, however, the committee does not believe that neglect of resuspension of aged fallout on islands in the Pacific has important consequences for dose reconstructions; that is, potential inhalation doses do not appear to be high.

In dose reconstructions for participants at the NTS, the presence of fallout from previous shots was taken into account in some cases (Barrett et al., 1986); see Section IV.C.2.1.1 and Table IV.C.1 (see also cases #21, 23, 27, 80, and 87). The NTPR program judged the importance of fallout from previous shots on the basis of available data on fallout patterns after each shot at the NTS. An example of an assumed fallout pattern from Operation PLUMBBOB Shot SHASTA at locations of participants at the later PLUMBBOB Shot SMOKY (Goetz et al., 1979) is shown in Figure V.C.1. In all cases except as noted in footnote *a* in Table IV.C.1, fallout from previous shots that was assumed to affect the area at a later shot occurred within 3 months, and fallout that occurred earlier (usually in prior test series) was ignored. To provide a frame of reference for the information in Table IV.C.1, the following discussions, and later discussions in this report, locations of all shots in Operations BUSTER-JANGLE, TUMBLER-SNAPPER, UPSHOT-KNOTHOLE, TEAPOT, and PLUMBBOB at the NTS are shown in Figures V.C.2 through V.C.6.

The committee believes that the presence of prior fallout has been neglected in some cases at the NTS where it is potentially important. Consider first the shots listed in Table IV.C.1 at which fallout from one or more previous shots has been taken into account in dose reconstructions. On the basis of locations of shots and fallout patterns given by Hawthorne (1979), the committee notes the following two examples.

• The area near UPSHOT-KNOTHOLE Shot HARRY also was affected by fallout from TUMBLER-SNAPPER Shot GEORGE (June 1, 1952) because of its nearby location and considerable onsite fallout.
• The area near PLUMBBOB Shots LASSEN, WILSON, OWENS, WHEELER, CHARLESTON, and MORGAN, which were detonated at the same location, also was affected by fallout from BUSTER-JANGLE Shot SUGAR (Nov. 19, 1951) because of its nearby location and considerable onsite fallout. It also is likely that the area near those PLUMBBOB shots was affected by fallout from TUMBLER-SNAPPER Shot GEORGE (June 1, 1952) and TEAPOT Shot APPLE I (March 12, 1955) because of their directions of plume travel and onsite fallout. Finally, onsite fallout from Shot WILSON affected the area near the later PLUMBBOB shots at the same location.

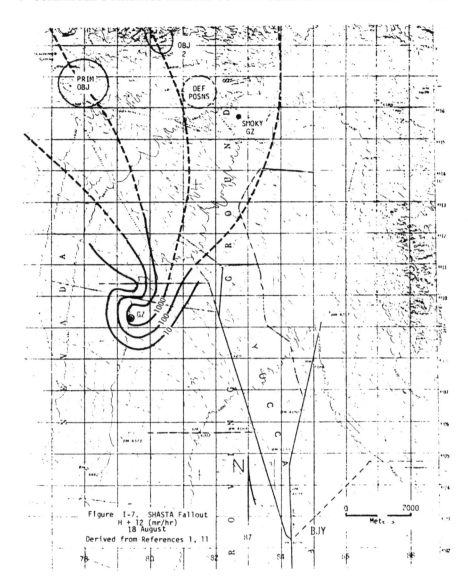

FIGURE V.C.1   Fallout pattern at NTS from Operation PLUMBBOB, Shot SHASTA assumed in dose reconstructions for participants at PLUMBBOB Shot SMOKY (Goetz et al., 1979). Maneuver objectives and defensive positions at Shot SMOKY are shown.

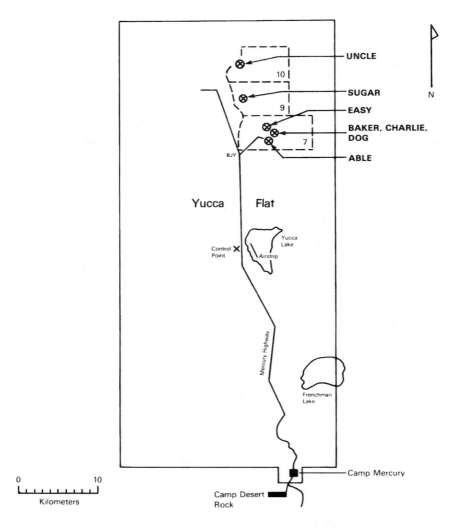

FIGURE V.C.2   Locations of shots in Operation BUSTER-JANGLE in Areas 7, 9, and 10 at NTS (Oct. 22, 1951-Nov. 29, 1951).

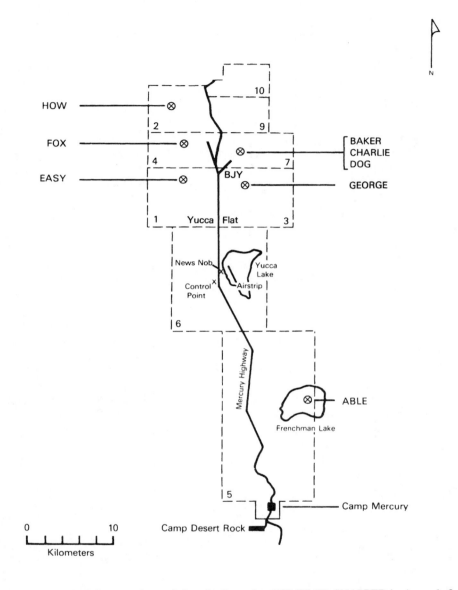

FIGURE V.C.3   Locations of shots in Operation TUMBLER-SNAPPER in Areas 1, 2, 3, 4, 5, and 7 at NTS (Apr. 1, 1952-June 5, 1952).

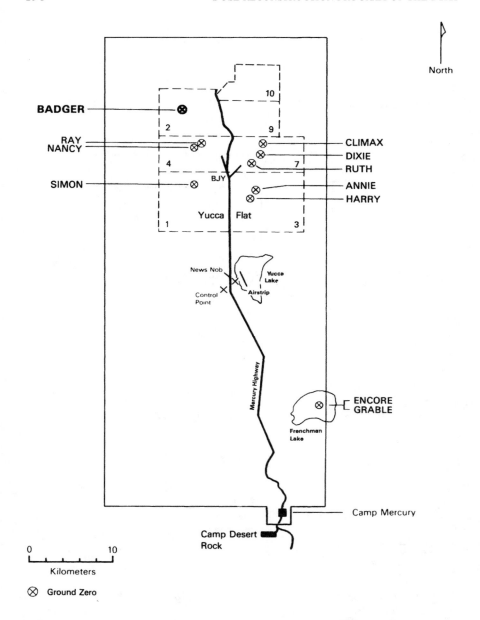

FIGURE V.C.4   Locations of shots in Operation UPSHOT-KNOTHOLE in Areas 1, 2, 3, 4, and 5 (Shots ENCORE and GRABLE) and Area 7 at NTS (Mar. 17, 1953-June 4, 1953).

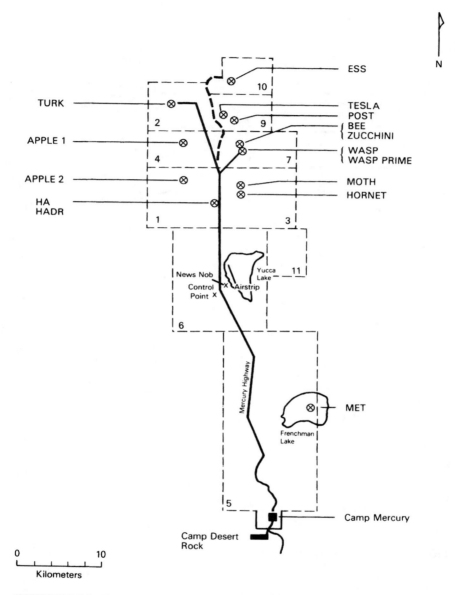

FIGURE V.C.5   Locations of shots in Operation TEAPOT in Areas 1, 2, 3, 4, 5, 7, 9, and 10 at NTS (Feb. 18, 1955-May 15, 1955).

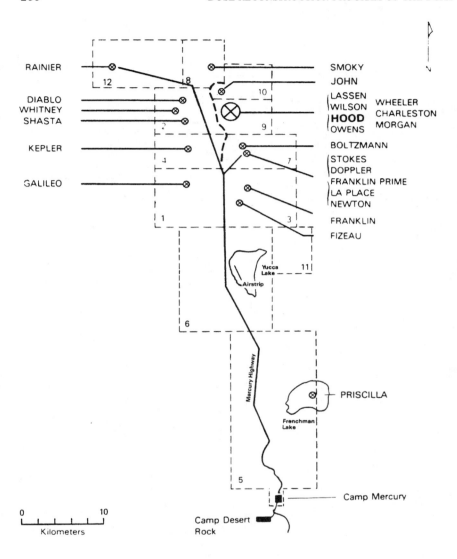

FIGURE V.C.6   Locations of shots in Operation PLUMBBOB in Areas 1, 2, 3, 4, 5, 7, 8, 9, 10, and 12 at NTS (May 20, 1957-Oct. 7, 1957).

In reviewing various documents, including the 99 randomly selected cases of individual dose reconstructions, the committee also encountered cases not included in Table IV.C.1 in which fallout from previous shots apparently impacted the area near a shot of concern but was not considered in dose reconstructions. Those cases are described as follows (examples from the randomly selected dose reconstructions are included in parentheses).

- The area near UPSHOT-KNOTHOLE Shot NANCY (March 4, 1953) was affected by fallout from TUMBLER-SNAPPER Shot FOX (May 25, 1952) because of its detonation at the same location and considerable onsite fallout. It also is likely that this area was affected by fallout from TUMBLER-SNAPPER Shot EASY (May 7, 1952) because of the direction of plume travel and onsite fallout (cases #7, 29, 84, 86, and 87).
- The area near UPSHOT-KNOTHOLE Shot SIMON (Apr. 25, 1953) was affected by fallout from TUMBLER-SNAPPER Shot EASY (May 7, 1952) because of its detonation at the same location and considerable onsite fallout (cases #30, 51, and 81).
- The area near TEAPOT Shot APPLE II (May 5, 1955) was affected by fallout from TUMBLER-SNAPPER Shot EASY (May 7, 1952) and UPSHOT-KNOTHOLE Shot SIMON (Apr. 25, 1953) because of their detonation at the same location and considerable onsite fallout (cases #37, 77, 83, and 90).
- The area near PLUMBBOB Shot HOOD (July 5, 1957) was affected by fallout from prior Shots SUGAR, GEORGE, APPLE I, and WILSON (see second bullet in previous paragraph) and by fallout from Shot BOLTZMANN listed in Table IV.C.1 that was accounted for at other PLUMBBOB shots because Shot HOOD was detonated at the same location as Shots LASSEN, WILSON, OWENS, WHEELER, CHARLESTON, and MORGAN.

That list is not intended to be exhaustive, and other shots presumably could be identified whose ground areas were affected by fallout from previous shots.

The lack of consideration of the impact of fallout from prior shots in the area of PLUMBBOB Shot HOOD noted above seems particularly inexplicable, given that the existence of some prior fallout was considered at all other PLUMBBOB shots at the same location during the same period. That the omission of Shot HOOD from Table IV.C.1 is not an oversight by the NTPR program is indicated by an assumption used in unit dose reconstructions for participant groups in forward areas after the shot that the groups were exposed to suspended neutron-induced radioactive material "in the absence of a fallout field" (see Section 3 and Tables 35 and 37 through 40 of Barrett et al., 1986). An assumption that there were no fission products or plutonium on the ground at the time and location of Shot HOOD is unsupportable, and it clearly results in underestimates of inhalation doses to participant groups in forward areas at that shot, without regard for the particular disturbances that caused resuspension of surface materials. An example analysis of potential inhalation doses in forward areas after detonation of Shot HOOD is given in Appendix E.

More generally, extensive measurements of concentrations of radionuclides in surface soil at the NTS that were made during the 1980s (McArthur and Kordas, 1983; McArthur and Kordas, 1985; McArthur and Mead, 1987; McArthur and Mead, 1988; McArthur and Mead, 1989; McArthur, 1991; IT and DRI, 1995) indicate that by the times of later test series, large areas of the NTS had received substantial fallout. The extent of substantial fallout is indicated, for example, by the distribution of ^{137}Cs shown in Figure V.C.7. Thus, without the need to consider locations of particular shots and associated directions of fallout patterns and the extent of fallout, as has been done in dose reconstructions (Barrett et al., 1986) and in the committee's evaluation as given above, it is virtually certain that participants, including maneuver troops and observers, who engaged in activities in any of several areas, especially in the northeast quadrant of the NTS where most shots were detonated, during later periods of atomic testing received inhalation doses due to resuspension of previously deposited fallout. The magnitude of possible doses depends, of course, on the particular locations and times of exposure, activities of the participants, and the nature of the disturbances that caused resuspension.

Neglect of resuspension of previously deposited fallout in many dose reconstructions for participants at the NTS perhaps was based on an assumption that the resuspension factor would decrease substantially over time, as indicated by data obtained at the site (Anspaugh et al., 1975). However, fallout in many areas of the NTS increased over time as more shots affected the areas, and there undoubtedly were scenarios in which the resuspension factor did not decrease substantially over time. An example is exposures during assaults or marches behind armored vehicles (see case #88). A high resuspension factor of 10^{-3} m^{-1} normally is assumed during such activities (see Table IV.C.2), and the vigorous action of vehicle treads most likely resuspended aged and fresh fallout about equally. It also is not obvious that a pronounced decrease in the resuspension factor over periods of a few years would apply to resuspension caused by walking or other light activities, especially if large groups of participants were involved, because measured resuspension over long periods at the NTS (Anspaugh et al., 1975) probably was caused mainly by wind stresses rather than human activities.

On the basis of considerations discussed above, neglect of prior fallout clearly is a potentially important source of underestimation of upper bounds of inhalation doses to many participants at the NTS, especially if participants were exposed in forward areas during later periods of atomic testing. The presence of prior fallout is especially important in scenarios in which higher resuspension factors of 10^{-3} or 10^{-4} m^{-1} are assumed (see Table IV.C.2), given that plutonium probably was the principal long-term inhalation hazard and that concentrations of plutonium in fallout at the NTS are substantially underestimated in dose reconstructions because of neglect of fractionation (see comment [4] above).

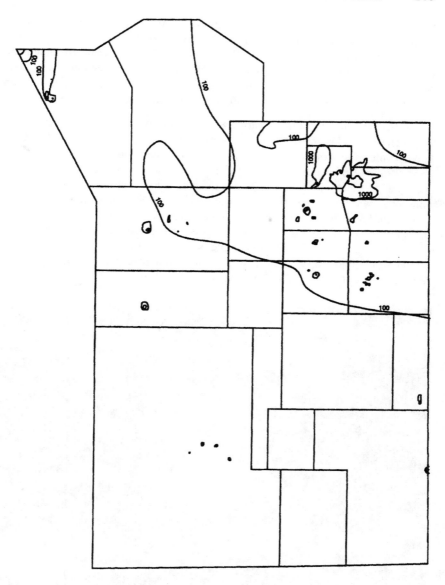

FIGURE V.C.7   Distribution of concentrations of ^{137}Cs in surface soil at NTS as of January 1, 1990 (McArthur, 1991). Isopleths represent concentrations of 100, 1,000, and 10,000 nCi m^{-2}; concentrations at end of period of atomic testing were about a factor of 2 higher.

[7]   **In dose reconstructions for participants at the NTS, resuspension of previously deposited fallout by the blast wave produced by a detonation generally has been ignored. Neglect of effects of a blast wave on resuspension could result in underestimation of upper bounds of airborne concentrations of radionuclides in previously deposited fallout relative to concentrations on the ground by a factor of about 100 or more in some exposure scenarios.**

When a nuclear weapon is detonated above ground, a blast wave is produced in which the wind speed close to the location of the detonation can reach several hundred miles per hour (1 mph = 1.6 km h^{-1}) and the wind speed at distances of about 1 mile (1.6 km) can be about 180 mph (Glasstone and Dolan, 1977). Wind speeds of such magnitude can result in extensive resuspension of radionuclides in fallout that was deposited after previous shots. Depending on the height and yield of a detonation, a blast wave can produce a dense cloud of dust at distances up to about 6 miles (10 km). The effect of a blast wave is evident in photographs taken after detonations at the NTS; an example is shown in Figure V.C.8.

The potential importance of a blast wave on resuspension of previously deposited fallout at the NTS is indicated by the following considerations. First,

FIGURE V.C.8   Photograph taken shortly after detonation of Operation PLUMBBOB, Shot PRISCILLA showing formation of dust cloud along the ground by blast wave produced by the detonation.

the resuspension factor associated with a blast wave should be substantially higher than values that apply to other, less vigorous disturbances, such as walking, that often are considered in dose reconstructions. If the height of the dust cloud caused by a blast wave is assumed to be about 100 m and it is further assumed that all radionuclides on the ground surface are resuspended by a blast wave, the resuspension factor is $1/(100 \text{ m})$, or 10^{-2} m^{-1}. Thus, even if only 10% of the radionuclides in deposited fallout were resuspended by a blast wave, the resuspension factor would be 10^{-3} m^{-1}. A credible upper bound in the range of 10^{-2}-10^{-3} m^{-1} seems reasonable when one considers that values as high as 10^{-3}-10^{-4} m^{-1} caused by vehicular and pedestrian traffic have been reported (Sehmel, 1984). The committee also notes that a resuspension factor of 10^{-3} m^{-1} is assumed in some exposure scenarios that involved vigorous disturbances of surface soil (see Table IV.C.2) and that a value of 10^{-2} m^{-1} was assumed in an unusual scenario involving short-term exposures during a localized dust storm (see Section IV.C.2.1.3). A credible upper bound of the resuspension factor associated with a blast wave should be at least as high as the value assumed by the NTPR program in these other cases of unusually high resuspension.

Second, a substantial fraction of materials resuspended by a blast wave should be in the form of small, respirable particles. Although fresh fallout at the NTS consisted primarily of large, essentially nonrespirable particles as a result of fractionation, particle sizes probably were reduced by wind stresses and other natural disturbances (NCRP, 1999), and a blast wave itself should tend to pulverize larger particles on the ground.

Third, smaller resuspended particles have low fall velocities. At the low wind speeds that occurred at ground level at the time of most detonations at the NTS (Hawthorne, 1979), the deposition velocity of small particles is expected to be about $10^{-2} \text{ cm s}^{-1}$ (Sehmel, 1984). That estimate agrees with a study at the NTS in which deposition velocities of respirable particles in the range of 3×10^{-1} to $3 \times 10^{-3} \text{ cm s}^{-1}$ were inferred (Luna et al., 1969). If the deposition velocity is assumed to be $10^{-2} \text{ cm s}^{-1}$ and the height of the dust cloud is assumed to be 100 m, the time required for dust to settle is 10^6 s, or approximately 10 days. That estimate is comparable to measured half-times of several tens of days for settling of resuspended radionuclides at the NTS, as summarized in Table 12.8 of Sehmel (1984). Thus, when resuspension was caused by a blast wave, most of the resuspended material that was respirable probably remained airborne during periods of possible inhalation exposure in forward areas after a detonation.

Fourth, since weapons were detonated at times of calm winds or low wind speeds at ground level (Hawthorne, 1979), and the dust cloud caused by a blast wave often covered a wide area, it is reasonable to presume that the cloud usually was not completely blown away from forward areas where exposures occurred for up to a few hours after a detonation.

A blast wave undoubtedly occurred at most aboveground shots at the NTS, the exceptions being detonations at high altitudes, safety shots, and misfires. The extent of the resulting dust cloud presumably depended, for example, on the height of the detonation above ground and yield, but there were some blast-wave effects at most shots. There is little doubt that many participants who engaged in activities in forward areas after a shot encountered a dust cloud caused by the blast wave. Reports of activities of participant groups and individual dose reconstructions reviewed by the committee sometimes referred to high dust levels (see cases #21, 27, and 77), and levels sometimes were so high that a planned activity was delayed or canceled (see case #21 and Appendixes E and F). The presence of high dust levels in areas where participant groups engaged in activities is also indicated by the routine procedure after many operations of using brooms to brush accumulated soil from participants' clothing, as shown in Figure V.C.9, even when participants had engaged in activities, such as walking, that should not have caused extensive resuspension (see Goetz et al., 1981; U.S. Army, 1957).

FIGURE V.C.9   Illustration of procedure for routine decontamination of participants (maneuver troops) after operations in forward areas at NTS.

Thus, neglect of the effects of a blast wave on resuspension at the NTS could result in underestimation of upper bounds of airborne concentrations of radionuclides in previously deposited fallout relative to concentrations on the ground by a factor of about 100 or more in some scenarios. That is especially the case when a resuspension factor of 10^{-5} m^{-1} or less is assumed in dose reconstructions, as in scenarios in which resuspension after detonations was assumed to be caused by walking or other light activities. In addition, the frequent neglect of aged fallout that contained substantial amounts of important long-lived radionuclides, such as plutonium, and the likelihood discussed in comment [4] that concentrations of plutonium in fallout deposited at the NTS have been underestimated substantially, because of neglect of fractionation, could increase the extent of underestimation of upper bounds of airborne concentrations of radionuclides to which participants in forward areas were exposed by another factor of perhaps as much as 10 in the worst cases. Therefore, regardless of uncertainties in estimating inhalation doses due to blast-wave effects, it is virtually certain that inhalation doses to many participants in forward areas at the NTS have been greatly underestimated by the NTPR program.

An example analysis to investigate potential inhalation doses due to blast-wave effects in forward areas at Operation PLUMBBOB, Shot HOOD is presented in Appendix E. The results of that analysis indicate that upper bounds of equivalent doses to some organs well above 1 rem are plausible in some cases and, therefore, that blast-wave effects at the NTS are potentially important. In its review of randomly selected dose reconstructions for individual veterans, the committee encountered many cases in which consideration of blast-wave effects could be important for obtaining credible upper bounds of dose to participants at the NTS from all exposure pathways combined.[26]

[8] **In several individual dose reconstructions reviewed by the committee, an internal dose of zero was assigned even though a substantial external dose was estimated and inhalation exposure was plausible. Regardless of the magnitude of possible inhalation doses, assigning a zero dose does not conform to the stated policy that the veteran will be given the benefit of the doubt in estimating dose, and it does not provide a credible upper bound.**

In its review of 99 randomly selected individual dose reconstructions, the committee found four cases (cases #7, 29, 89, and 99) in which a veteran filed a claim for compensation for cancer in an internal organ, the veteran was a confirmed member of a participant group that engaged in activities in the forward area at one or more shots at the NTS, the veteran was assigned an external dose of

---

[26]In about 20% of the 99 randomly selected cases, a participant was at the NTS and engaged in activities within a few hours after a detonation in a forward area that probably was contaminated by fallout from previous shots and could have been affected by the blast wave; see, for example, cases #7, 21, 27, 29, 30, 37, 51, 55, 77, 81, 84, 86, 87, 88, 89, 90, and 99.

about 1–4 rem, and the group's activities probably took place in an area of fallout from prior shots, but the veteran was not assigned an internal dose. Indeed, in one of those cases (case #99), the presence of fallout from previous shots was acknowledged but not taken into account in estimating dose. In another case of a participant who served on boarding parties on contaminated ships at Operation CROSSROADS (case #49), an internal dose was calculated, but the assigned dose was zero. An assignment of no internal dose in such cases is difficult to understand.

The committee also encountered cases in which a veteran filed a claim for compensation for an unspecified disease or a dose reconstruction was requested without the filing of a claim and the veteran was assigned an external dose of about 1–6 rem, but the veteran was assigned no internal dose (see cases #1, 28, 30, 34, 52, 54, 55, 56, 64, 72, 74, 77, and 92). In some of those cases, the potential for a substantial internal dose was clear—for example, when a veteran was a member of a radiation-safety team or other group that engaged in activities in contaminated forward areas after a detonation at the NTS (cases #1, 30, and 77), or a veteran spent considerable time on residence islands in the Pacific (cases #52, 54, 55, and 56). It was not possible to receive external doses of a few rem on the ground at the NTS or in the Pacific without any internal exposure.

In addition, estimates of inhalation dose in some unit dose reconstructions for participant groups almost certainly are much too low. One such case mentioned in the previous comment and discussed in Appendix E involves exposures at Operation PLUMBBOB, Shot HOOD.

[9] Resuspension factors used in the NTPR program to estimate inhalation doses from exposure to fallout deposited on ships may not represent credible upper bounds of actual resuspension factors in many cases.

Resuspension factors normally used to estimate inhalation doses to participants from exposure to fallout deposited on ships in the Pacific are summarized in Table IV.C.3. During normal activities (that is, excluding decontamination and ammunition loading and unloading), the assumed resuspension factor is 10^{-5} or 10^{-6} m^{-1}, with the higher value applied to exposures below decks and the lower value applied to exposures on deck. Assumed resuspension factors on ships were based on a review of data obtained in indoor and outdoor environments (Phillips et al., 1985). The committee believes, however, that credible upper bounds of resuspension factors during normal activities on ships could be substantially higher.

In indoor environments, data reviewed by Sehmel (1984) but not considered by Phillips et al. (1985) indicate that a credible upper bound of the resuspension factor could be 10^{-2} m^{-1} or higher, and values above 10^{-4} m^{-1} are not uncommon. Such high values are reported even in cases in which the stress that caused the resuspension—such as walking, changing clothes, or several people moving in a room—did not involve vigorous activity, and high values occurred in both ventilated and unventilated rooms.

The committee believes that resuspension factors during walking and other normal activities on decks of ships could be substantially higher than values observed under similar conditions on land. Fallout particles probably do not adhere to smooth deck surfaces to nearly the extent that they do in surface soil; this may be an important reason for the high resuspension factors observed indoors. The committee also recognizes, however, that the possibility of high resuspension factors for fallout deposited on decks of ships may be mitigated by several factors, including the generally damp conditions on ships in the Pacific due to the high humidity, frequent rains, and periodic swabbing of decks; the propensity for loose particles that could be moved by the wind to accumulate in nooks, crannies, and cracks, where resuspension is less likely; and the small area of contamination on a ship, which increases natural dilution and the chance of the winds blowing resuspended material away from locations of participants before exposure occurs, compared with exposure to large areas of contamination on land. Thus, although assumed resuspension factors summarized in Table IV.C.3 may not be credible upper bounds on ships in the Pacific, it seems likely that the possible degree of underestimation of resuspension is greater below decks than on deck.

[10] **In dose reconstructions for participants on residence islands in the Pacific, exposures to descending or resuspended fallout during the fraction of the time spent indoors normally are ignored; that is, concentrations of radionuclides in indoor air are assumed to be zero. That assumption does not provide a credible upper bound of possible inhalation doses indoors. Inhalation doses during the fraction of the time spent below decks on ships in the Pacific also may be underestimated in some cases.**

The committee's review of dose reconstructions for veterans who served on residence islands in the Pacific, as included in the 99 randomly selected cases, indicates that inhalation of descending or resuspended fallout normally is assumed not to occur during the fraction of the time spent indoors (see cases #16, 43, and 78). The indoor exposure time on residence islands normally is assumed to be 40%.

The committee agrees that concentrations of radionuclides on residence islands in the Pacific probably were lower in indoor air than outdoors. Furthermore, an assumption that a person was exposed outdoors for 100% of the time while on a residence island would not increase estimated inhalation doses by a large amount. However, an assumption of no inhalation exposure indoors is unreasonable, given that windows and doors of buildings on residence islands presumably were open much of the time to promote ventilation, and it does not provide a credible upper bound of possible inhalation doses in these cases.

In some dose reconstructions for participants who served on ships in the Pacific, the inhalation dose during the fraction of the time spent below decks also is assumed to be zero (see cases #6, 24, 25, and 44). Especially in cases of exposure during known periods of fallout on ships, this assumption takes into account that forced-air ventilation systems often were turned off in an effort to minimize contamination below decks. However, it is unreasonable to assume that ventilation systems were always turned off during periods of substantial fallout and that there was no contamination below decks. At Operation CASTLE, for example, forced-air ventilation systems were sometimes left on during periods of fallout to maintain tolerable temperatures below decks (Martin and Rowland, 1982). Thus, inhalation doses to participants during periods spent below decks on ships probably were underestimated in some cases.

In summary, the committee has identified several assumptions used in the NTPR program to estimate inhalation dose coefficients and concentrations of radionuclides in air that have substantial uncertainty that has not been taken into account in dose reconstructions, and the committee has also identified several assumptions that should tend to result in underestimates of inhalation doses to atomic veterans; these assumptions are briefly restated in Table V.C.7. Additional concerns about situations in which use of 50-year committed doses and assignment of committed doses to the year of intake could result in underestimates of the dose that could have caused a veteran's cancer are discussed in Section V.C.3.1, comment [6]. The committee also emphasizes, however, that the discussions of assumptions summarized in Table V.C.7 should not be used to draw conclusions about whether estimates of inhalation dose to atomic veterans in particular scenarios provide credible upper bounds without consideration of the importance of assumptions discussed in the previous section that should tend to result in overestimates of inhalation doses. The committee's overall evaluation of methods of estimating inhalation doses used in the NTPR program is presented in the following section and in Sections V.C.5 and V.C.6.

V.C.3.3 *Evaluation of Methods of Estimating Inhalation Dose*

The committee found that several assumptions used in dose reconstructions for atomic veterans should tend to result in overestimates of inhalation doses to participants at the NTS or in the Pacific. Those assumptions are discussed in Section V.C.3.1 and summarized in Table V.C.5. Nearly all the assumptions concern dose coefficients for inhalation of radionuclides used in all dose reconstructions, but one assumption concerns resuspension factors that are applied in estimating airborne concentrations of radionuclides in some exposure scenarios.

The committee also found, however, that several assumptions used in dose reconstructions for atomic veterans have substantial uncertainty that has not been taken into account in the NTPR program and that several other assumptions should tend to result in underestimates of inhalation doses to participants at the NTS or in the Pacific. Those assumptions are discussed in Section V.C.3.2 and summarized in Table V.C.7. Most of the assumptions concern methods used to estimate airborne concentrations of radionuclides in various scenarios, but im-

TABLE V.C.7   Summary of Assumptions Used to Estimate Inhalation Doses in NTPR Program That Have Substantial Uncertainty That Is Not Taken into Account or Should Tend to Result in Underestimates of Dose

Dose coefficients (organ-specific equivalent doses per unit activity of radionuclides inhaled)[a]

- Uncertainties in dose coefficients due to uncertainties in dosimetric and biokinetic models are not taken into account.
- Uncertainty in dose coefficients for alpha-emitting radionuclides due to uncertainty in biological effectiveness of alpha particles is not taken into account.
- Dose coefficients for organs of GI tract from inhalation of plutonium may be underestimated when inhaled materials are respirable (AMAD, 1 μm).

Methods used to estimate inhalation exposures (intakes of radionuclides in air)

- Sources of error and uncertainty in methods of estimating radionuclide concentrations in deposited fallout based on measured external photon exposures have not been evaluated, and reliability of methods is unknown. The assumption of no fractionation (except for removal of noble gases) should result in substantial underestimates of concentrations of refractory radionuclides (such as plutonium), and the method of calculating external exposure rates per unit concentration of radionuclides on a surface probably is not valid for fallout deposited on ships in Pacific and should result in underestimates of concentrations in these cases.
- Resuspension factor used to estimate radionuclide concentrations in descending fallout may result in underestimates of exposure when exposure did not occur during entire period of fallout.
- Presence of fallout deposited more than a few months before exposure usually is ignored, especially late in period of atomic testing at NTS, when buildup of plutonium and longer-lived fission products from many prior shots was extensive.
- Effect of blast wave from detonations at NTS on resuspension of substantial fraction of previously deposited fallout over large areas generally is ignored.
- In some dose reconstructions for veterans who filed claim for compensation for cancer in internal organs and received substantial external dose, inhalation dose of zero was assigned even though inhalation exposure almost certainly occurred.
- Resuspension factors applied to fallout deposited on ships in Pacific, especially below decks, may be too low.
- Inhalation dose during time spent indoors on residence islands in Pacific is assumed to be zero; some inhalation doses below decks on ships also may be underestimated.

[a]Additional discussions of situations in which use of 50-year committed doses from inhalation and assignment of committed doses to the year of intake could result in underestimates of the dose that could have caused a veteran's cancer are given in Section V.C.3.1, comment [6].

portant sources of uncertainty in dose coefficients also have not been taken into account.

The basic question in evaluating methods used in the NTPR program to estimate inhalation doses is whether the methods provide credible upper bounds of doses from this intake pathway (see Section IV.E.4). The question is difficult to answer in general terms, especially in the more important cases of exposure to fallout. Participants were exposed to airborne radionuclides in descending or resuspended fallout under a wide variety of conditions, especially at the NTS, and the importance of different sources of overestimation or underestimation of inhalation dose depends on the conditions of exposure. Therefore, estimates of inhalation dose obtained in dose reconstructions, which are intended to be upper bounds, probably provide credible upper bounds in some cases; they almost certainly do not in other cases, and it is difficult to determine one way or the other in the rest.

Another complicating factor is that the committee could not fully evaluate methods used by the NTPR program to estimate concentrations of radionuclides in fallout deposited on the ground or other surfaces on the basis of assumptions about the composition of fallout and external photon exposures measured with film badges or field instruments. Those methods are important because estimated concentrations in fallout are used to calculate inhalation dose in most cases. Therefore, although some assumptions embodied in the methods of estimating concentrations of radionuclides in fallout are likely to be overpredictive or underpredictive, it often is difficult to judge whether the net effect of all such assumptions is that estimated inhalation doses from exposure to descending or resuspended fallout tend to be overestimates or underestimates and by how much.

An example of a scenario in which credible upper bounds of inhalation dose probably are obtained in dose reconstructions involves exposure to descending fallout throughout the period of descent. Such exposures occurred, for example, on residence islands in the Pacific (see cases #3, 5, 8, 16, 22, 32, 38, 43, 47, 58, 60, 63, 78, and 96). Suppose that a participant who was exposed mainly to descending fallout filed a claim for compensation for a cancer in an internal organ other than the lung or an organ in the GI tract (such as the kidney). Several assumptions used in such a case should result in substantial overestimates of inhalation dose. Assumed dose coefficients for those organs usually apply to a particle size (AMAD) of 1 μm, even though most particles in descending fallout presumably were large and essentially nonrespirable, and dose coefficients for inhalation of respirable particles used by the NTPR program often are at least a factor of 2 higher than values for inhalation of large particles based on current ICRP recommendations. Because fractionation of radionuclides in fallout is ignored in all dose reconstructions, doses due to inhalation of volatile radionuclides attached to particle surfaces should be overestimated by a factor of about 3 or more. Doses due to inhalation of refractory radionuclides may also be overestimated substantially, even though their amounts in fallout probably are underestimated by a factor of about 3 or more because of neglect of fractionation; refractory radionuclides are dispersed mainly throughout the volume of large and highly insoluble fallout particles and therefore may be absorbed from the respiratory and GI tracts into blood to only a small extent before the particles are eliminated from the body. If the radionuclide composition in an atmospheric cloud is reasonably well characterized on the basis of cloud sampling data, the several factors that should tend to result in overestimates of inhalation dose probably are sufficient to compensate for uncertainties in all dose coefficients amounting to a factor of about 3-10 about a central estimate, owing to uncertainties in dosimetric and biokinetic models and the uncertainty in the biological effectiveness of alpha particles of a factor of about 4. Estimates of inhalation dose to the lung should also be credible upper bounds when descending fallout is assumed to consist of small (1-μm) particles. These discussions and summaries in Appendix B of estimated upper-bound doses for the specific cases identified above indicate that upper bounds of organ equivalent doses in scenarios involving exposure to descending fallout usually were low (that is, less than 1 rem).

A clear example of when estimates of inhalation dose obtained in dose reconstructions almost certainly do not provide credible upper bounds involves a scenario for exposure to resuspended fallout at the NTS. Participants who engaged in activities in forward areas within a few hours after a shot almost certainly were exposed to previously deposited fallout that was resuspended to a large extent by the blast wave produced by the detonation. However, effects of a blast wave have been ignored in all dose reconstructions, so the upper bound of the resuspension factor probably has been underestimated by more than a factor of 100 in scenarios in which resuspension is assumed to be caused by walking or other light activities. In addition, plutonium probably was the most important inhalation hazard in previously deposited fallout and, as noted above, concentrations of plutonium in fallout at the NTS probably are underestimated by a factor of about 3 or more because of neglect of fractionation. Furthermore, fallout that occurred more than a few months before a shot of concern generally has been ignored, but many prior shots contributed to fallout at the NTS toward the end of the period of aboveground testing. Therefore, unless concentrations of plutonium in fallout are overestimated by the NTPR program by substantially more than a factor of 100—which seems highly unlikely considering the interest in measuring plutonium in cloud samples—biases in other assumptions that tend to result in overestimates of inhalation dose almost certainly are not sufficient to compensate for neglect of blast-wave effects in all dose reconstructions at the NTS. Furthermore, as noted in Section V.C.3.2, comment [7], upper bounds of organ equivalent doses in this scenario could be substantially above 1 rem in some cases.

As an example of how it can be difficult to determine whether estimated inhalation doses are credible upper bounds, consider a scenario in which participants walked or engaged in other light activity in an area contaminated by fallout. This type of scenario occurred before some shots at the NTS (see cases #1, 7, 23, and 87). Suppose that lung cancer is the disease of concern, and consider the dose to the lung from plutonium only. The dose coefficient for the lung for inhalation of respirable particles (AMAD, 1 μm) used in dose reconstructions is based on ICRP *Publication 30* (see Table V.C.2), the assumed breathing rate is $1.2 \ \text{m}^3 \, \text{h}^{-1}$, and the assumed resuspension factor, which is intended to be an upper bound, often is $10^{-5} \, \text{m}^{-1}$. An intended upper bound of the inhalation dose (rem h^{-1}) per unit activity concentration of plutonium on the ground used in dose reconstructions is proportional to the product of those three factors, or 7×10^{-3}. This factor is applied to an estimated concentration of plutonium on the ground, which is based on an assumption of no fractionation in fallout.

Now, consider the effects of bias and uncertainty on the estimate of dose from inhalation of plutonium given above. We make the following assumptions:

- When inhalation of respirable particles (AMAD, 1 μm) is assumed, the central estimate of the dose coefficient for the lung should be reduced by a factor of 2 to conform to current ICRP recommendations (see Table V.C.2).
- The uncertainty in the dose coefficient due to uncertainties in dosimetric and biokinetic models is represented by a lognormal probability distribution with a 90% confidence interval that spans a factor of 10 above and below the central estimate (see Table V.C.6).
- The uncertainty in the biological effectiveness of alpha particles relative to photons and electrons is described by a lognormal probability distribution with a 50th percentile at 18 and a 97.5th percentile at 100 (see Section V.C.3.2, comment [2]).
- Only a fraction of resuspended plutonium is in respirable form, with the dose to the lung from inhalation of large particles assumed to be essentially zero, and the uncertainty in this fraction is described by a uniform probability distribution over the range of 0.2-0.8.
- On the basis of a review of available data (EPA, 1997), the uncertainty in the breathing rate that applies during light activity is described by a lognormal probability distribution with a 90% confidence interval of 0.6–1.4 m^3 h^{-1}.[27]
- Fractionation increases the concentration of plutonium on the ground compared with the concentration assumed in dose reconstructions by an uncertain factor that is described by a lognormal probability distribution with a 90% confidence interval of 2-4.
- The uncertainty in the resuspension factor under conditions of walking or other light activity is described by a lognormal probability distribution with a 90% confidence interval of 10^{-8}-10^{-5} m^{-1} (see Section V.C.3.1, comment [8]).

By multiplying those probability distributions with Latin Hypercube sampling techniques and the Crystal Ball® 2000 software (Decisioneering, 2001),[28] we obtain the following results:

- The central estimate (50th percentile) of the probability (uncertainty) distribution of the result is 1×10^{-4}.

---

[27]The committee notes that the NTPR program generally assumes a single breathing rate, with no uncertainty, in a given scenario (see Section IV.C.2). Assumed breathing rates probably underestimate upper bounds when uncertainties in breathing rates for various activities (EPA, 1997) are considered. However, neglect of uncertainties in breathing rates is not an important concern, because they clearly are small compared with uncertainties in other parameters used in calculating inhalation doses to atomic veterans, including dose coefficients and resuspension factors.

[28]Crystal Ball® 2000 is licensed by Decisioneering, Inc., 1515 Arapahoe St., Suite 1311, Denver, Colorado 80202.

- The estimate of 7×10^{-3} based on input parameters assumed in dose reconstructions, which is intended by the NTPR program to be an upper bound (at least a 95th percentile), lies between the 75th and 80th percentiles of the probability distribution based on assumed uncertainties in the parameters.
- The 95th percentile of the probability distribution of the result is 9×10^{-3}, which is a factor of about 1.3 greater than the presumed upper bound obtained in dose reconstructions.

The analysis described above is not intended to be definitive and should not be taken as such. Fractionation of plutonium may be misrepresented in the analysis (for example, the assumed upper bound of the probability distribution describing fractionation could be too low), and other uncertainties, such as uncertainties in the duration of exposure and the nature of work activities, would need to be considered. Nonetheless, the results suggest that an estimate of inhalation dose based on estimates of input parameters used in dose reconstructions is more likely than not to overestimate actual doses in the assumed scenario. In contrast, that estimate is slightly less than an estimated upper bound (95th percentile) of a probability distribution of inhalation dose based on assumed uncertainties in input parameters. Thus, when possible errors in estimating concentrations of plutonium in fallout and the need to include uncertain contributions to dose from resuspension of other radionuclides in fallout are considered, it is difficult to draw a definitive conclusion about whether estimates of inhalation dose obtained by the NTPR program for this scenario provide credible upper bounds. However, as indicated by discussions in Appendix E.5 and summaries in Appendix B of estimated doses for the specific cases identified at the beginning of this example, upper bounds of organ equivalent doses in scenarios in which resuspension of previous fallout was caused only by walking or other light activity almost certainly were low (substantially less than 1 rem).

Further summary discussions of the committee's evaluation of methods used to estimate inhalation doses to atomic veterans are given in Sections V.C.5 and V.C.6.

V.C.4 Evaluation of Potential Ingestion Doses

Ingestion of radionuclides is rarely considered in dose reconstructions for atomic veterans, and ingestion doses are not estimated in any randomly selected cases reviewed by the committee (for example, see Section IV.C.3).[29] Thus, with

---

[29]Case #3 includes a statement that ingestion intakes would be minimized during routine flights through radioactive clouds because operational standards prohibited drinking or eating when contamination was present. In case #58, consumption of contaminated food and water during operations on a residence island in the Pacific is mentioned, but the analysis assumes that there was no potential for ingestion exposure.

rare exceptions (see Goetz et al., 1987), ingestion is considered to be unimportant compared with inhalation and external exposure.

As noted in Section I.D, atomic veterans have expressed concern about ingestion doses they might have received. This section discusses the potential importance of ingestion exposures of atomic veterans. The potential importance of ingestion is assessed on the basis of assumed exposure scenarios at the NTS and in the Pacific.

V.C.4.1 *Example Analysis of Potential Ingestion Doses at the NTS*

Two scenarios of ingestion exposure at the NTS are considered. The first involves ingestion of contaminated soil that was transferred to the hands and then swallowed. This scenario could occur when participants engaged in such activities as digging trenches (see case #87) or installing or removing displays or electronic equipment in contaminated areas (see case #1). This scenario is used to investigate possible doses due to ingestion of longer-lived radionuclides in fallout.

A bounding estimate of potential doses due to ingestion of contaminated soil at the NTS is obtained on the basis of the following assumptions. Data given by McArthur (1991) and companion reports (see Section V.C.3.2, comment [5]) indicate that concentrations of plutonium in surface soil at the end of the period of atomic testing exceeded 500 pCi g^{-1} at a few locations and that the highest concentrations of ^{90}Sr and ^{137}Cs exceeded about 500 and 200 pCi g^{-1}, respectively. On the basis of an estimate that an adult ingests soil at 20 mg h^{-1} while gardening (EPA, 1997), we assume that a soil-ingestion rate of 100 mg h^{-1} is a credible upper bound for a participant who worked in contaminated soil at the NTS.[30] On the basis of those assumptions, the intake rate would be about 50 pCi h^{-1} for plutonium and ^{90}Sr and 20 pCi h^{-1} for ^{137}Cs. If ingestion dose coefficients for workers currently recommended by ICRP (1994a; 2002) are assumed, a central estimate of the dose to any organ would not exceed 0.1 mrem h^{-1}. Taking into account uncertainties in ingestion dose coefficients (see Table V.C.6), a credible upper bound would not exceed about 1 mrem h^{-1}. Thus, an assumption of reasonable exposure times would give total doses to any organ of no more than about 10 mrem. The assumed radionuclide concentrations in soil are at the upper end of measured concentrations at the NTS and the assumed soil-ingestion rate should be an overestimate, so it is reasonable to conclude that ingestion of radionuclides in soil is not an important concern at the NTS. That conclusion also takes into account that concentrations of shorter-lived radionuclides presumably were not high when digging and other such activities were undertaken.

---

[30]This assumption should overestimate intakes of contaminated soil because fallout was confined to the top layer of soil, but such activities as digging would result in some intakes of uncontaminated soil from deeper layers.

A second credible exposure scenario that is used to investigate doses due to ingestion of shorter-lived radionuclides in fallout involves participants at the NTS who consumed contaminated milk that was produced near St. George, Utah. A bounding estimate of potential ingestion doses in this scenario is obtained from estimates of ingestion doses to adults who lived near St. George. After Operation UPSHOT-KNOTHOLE, Shot HARRY, which resulted in unusually high levels of fallout near St. George, central estimates of absorbed doses to an adult due to ingestion are 4.3 rad to the thyroid; 0.12 and 0.3 rad to the upper and lower large intestine wall, respectively, and 0.07 rad or less to all other organs (Ng et al., 1990).[31] Doses to participants at the NTS due to consumption of contaminated milk obtained from St. George would have been far less, because doses at St. George due to fallout from Shot HARRY were considerably higher than doses due to fallout from other shots and the amount of contaminated milk consumed by participants at the NTS would have been a small fraction of the total diet of locally grown foods consumed by residents near St. George. Thus, it is reasonable to conclude that ingestion of shorter-lived radionuclides also is not an important concern at the NTS.

V.C.4.2 *Example Analysis of Potential Ingestion Doses in the Pacific*

As in the previous section, two scenarios for ingestion exposure on residence islands in the Pacific are considered here. The first scenario involves ingestion of locally produced terrestrial and aquatic foodstuffs that were contaminated by fallout. Although participants on residence islands consumed mainly imported foods, such items as coconut milk and seafood obtained from lagoons presumably were consumed on occasion. This scenario is used to investigate possible doses due to ingestion of longer-lived radionuclides in fallout.

A bounding estimate of potential doses due to ingestion of locally produced foodstuffs is based on assessments of dose to native Marshall Islanders today. Doses to natives should be far higher than potential doses to atomic-test participants because a far greater fraction of foodstuffs consumed by native populations is obtained locally. An assessment of ingestion doses to residents of northern Marshall Islands (Robison et al., 1997b) indicates that the highest annual effective dose to a resident of Enewetak Atoll is about 0.03 rem. The atoll was the location of residence islands for participants. The estimated dose to native residents assumes that only local foods are consumed. Ingestion doses result mainly from intakes of ^{137}Cs because of its high accumulation in terrestrial and aquatic foodstuffs in environments with low concentrations of potassium (Whicker and Schultz, 1982), with minor contributions from intakes of ^{90}Sr and insignificant

---

[31]Because ingestion doses at St. George were due primarily to intakes of shorter-lived, beta- and gamma-emitting radionuclides, equivalent doses in rem are essentially the same as absorbed doses in rad.

contributions from plutonium; doses to any organ are about the same as the effective dose. An assessment of ingestion doses to Marshall Islanders who might resettle on Bikini Atoll (Robison et al., 1997a) indicates that annual equivalent doses to different organs would be about 0.3-0.4 rem if 20% of the caloric content of the diet is obtained from local foods. Contamination on Bikini Atoll is substantially higher than on islands where participants resided (Robison et al., 1997b). Finally, an assessment of exposures of residents of Bikini, Enewetak, Rongelap, and Utirik Atolls using whole-body counting (Sun et al., 1997a) indicated that annual internal doses to all organs from intakes of ^{137}Cs were less than 0.02 rem.

Concentrations of ^{137}Cs and ^{90}Sr on residence islands during the period of atomic testing were higher, by a factor of 2-3, than concentrations when the dose assessments described above were performed because of radioactive decay. Nonetheless, the results indicate that potential doses to participants on residence islands in the Pacific due to ingestion of longer-lived radionuclides were very low, perhaps a few mrem or less, given that only a small fraction of the caloric content of a participant's diet would have been obtained from local foods and relatively few participants spent more than a few months on residence islands.

A second credible exposure scenario used to investigate potential doses due to ingestion of shorter-lived radionuclides in fallout in the Pacific involves participants who ingested fallout particles that were deposited directly on food and water as they were being consumed. This scenario is investigated using the following analysis. Service personnel on Rongerik Atoll received high doses after Operation CASTLE, Shot BRAVO. A dose reconstruction for them gave the following results (Goetz et al., 1987). First, estimated external doses are about 30-50 rem. Second, estimated internal doses are 190 rem to the thyroid, 76 rem to the lower large intestine wall, 44 rem to the upper large intestine wall, 13 rem to the small intestine wall and lung, and from 0.6 to about 6 rem to all other internal organs. Those doses were due mainly to shorter-lived radionuclides in fallout that occurred within 1 day of detonation. Furthermore, on the basis of the consideration that most of the fallout on Rongerik was in the form of large particles, the analysts concluded that internal doses were dominated by ingestion that resulted from deposition of fallout on foods while they were being consumed and that intakes by inhalation were relatively unimportant.[32]

Dose reconstructions reviewed by the committee indicate that estimated external doses to most participants who were stationed on residence islands are less than 1 rem. If the ratios of internal to external doses to personnel on Rongerik Atoll given above are assumed to apply to participants on residence islands and if

---

[32]It should be noted that as a result of an assumption that ingestion intakes were dominant, estimated internal doses to some organs and tissues are similar to estimated external doses. That result differs from the expectation that when ingestion is unimportant, doses from inhalation of descending fallout normally should be substantially less than external doses (see Section V.C.1).

internal doses on Rongerik are assumed to be due entirely to ingestion, ingestion doses on residence islands would not have exceeded a few rem to the thyroid, about 1-2 rem to the large intestine walls, and a fraction of a rem to all other internal organs. Those results should greatly overestimate ingestion doses on residence islands if precautions about eating were taken during known periods of fallout.

V.C.4.3 *Summary of Evaluation of Ingestion Doses*

The committee acknowledges the concerns of atomic veterans about doses they may have received from ingestion of radionuclides. However, on the basis of an analysis of bounding scenarios for ingestion exposure of participants at the NTS and on residence islands in the Pacific, doses to specific organs and tissues due to ingestion of radionuclides probably were low, especially compared with doses from external exposure, except in rare circumstances. Ingestion doses to most participants probably were around a few mrem or less. Doses of that magnitude are unimportant, so neglect of intakes of radionuclides by ingestion in dose reconstructions for atomic veterans does not appear in most cases to be an important concern with regard to evaluating claims for compensation for radiation-related diseases.

A conclusion that ingestion doses to most atomic veterans were very low may seem unreasonable, especially at the NTS, given the considerable attention that has been paid to ingestion exposures of the US population due to fallout from atmospheric weapons tests at the NTS (NCI, 1997; IOM/NRC, 1999). However, even in the population of the US, doses to most organs and tissues due to ingestion of radionuclides in fallout were substantially less than doses from external exposure (Anspaugh and Church, 1986; Anspaugh et al., 1990; Till et al., 1995; Whicker et al., 1996; UNSCEAR, 2000). The one exception was doses to the thyroid from ingestion of ^{131}I in milk, but even in such cases the principal concern was doses to infants and children who consumed large quantities milk, and doses to adults who drank milk were substantially less. Therefore, ingestion doses normally would have been a concern at the NTS only if participants drank large quantities of milk that had been contaminated by high levels of fallout from recent atmospheric tests. That situation is not known to have occurred at the NTS, and it generally was not a concern in the Pacific, given the absence of sources of milk near the locations of Pacific tests.

V.C.5 Summary of Principal Findings Related to Estimation of Internal Dose

The committee's evaluation of methods used in the NTPR program to estimate internal doses to atomic veterans focused on methods of estimating inhalation dose. As discussed above, the committee has concluded that internal exposures of most participants were due mainly to inhalation and that intakes by ingestion usually were insignificant.

The committee recognizes that estimation of inhalation doses to atomic veterans is difficult. Given the lack of data on airborne concentrations of radionuclides at locations and times of exposure and data on amounts of radionuclides excreted in urine or feces, inhalation doses can be estimated only by using indirect methods that involve substantial uncertainty. It also is likely that in some exposure scenarios, such as those involving exposure to suspended neutron-activation products in soil at the NTS or exposure to descending fallout at the NTS or on residence islands in the Pacific, inhalation doses were inconsequential compared with external doses that could be monitored with film badges or field instruments. In scenarios in which inhalation doses should be much lower than external doses, uncertainties in methods used by the NTPR program to estimate inhalation dose are unlikely to be important.

The committee's detailed evaluation of methods used in the NTPR program to estimate inhalation doses to atomic veterans is given in Section V.C.3 and summarized in Tables V.C.5 and V.C.7. In some respects, the methods should tend to overestimate inhalation doses. In other respects, however, the methods involve substantial uncertainty or they should tend to underestimate inhalation doses to such an extent that it is often difficult to determine whether estimated doses to atomic veterans are credible upper bounds, as intended by the NTPR program. Furthermore, the committee has identified exposure scenarios in which neglect of resuspension of previously deposited fallout by the blast wave produced in most detonations at the NTS almost certainly has resulted in underestimation of upper bounds of inhalation doses by a factor of at least 100. Such scenarios are important because thousands of participants at the NTS could have been exposed to substantial airborne concentrations of fallout that was resuspended by a blast wave. The committee also identified other cases in which an inhalation dose of zero was assigned to an organ in which a veteran's cancer occurred but there is little doubt that there was some inhalation exposure.

On the whole, the committee has concluded that methods used in the NTPR program to estimate inhalation doses to atomic veterans have important shortcomings that center around three issues.

[1] Most estimates of inhalation dose to participants at the NTS and in the Pacific depend on estimates of concentrations of radionuclides deposited on the ground or other surfaces or distributed over a depth in surface soil at locations and times of exposure. Those estimates are based, in part, on measurements of external photon exposure with film badges worn by participants or field instruments, combined with calculations of external exposure rates per unit concentration of radionuclides on the surface. However, especially in scenarios involving exposure to descending or resuspended fallout, the reliability of the methods of estimating concentrations of radionuclides

that are important contributors to inhalation dose has not been demonstrated and therefore is unknown.

Methods of estimating inhalation doses based, in part, on measured external photon exposures were criticized by a previous committee of the National Research Council (NRC, 1985b). The essence of the criticism was that the methods lacked scientific credibility and that their reliability is therefore unknown. Similarly, on the basis of an evaluation of simultaneous measurements of airborne concentrations and ground deposition at the same locations near the NTS during periods of atomic testing, Cederwall et al. (1990) concluded that the relationship between airborne and surface concentrations of fallout is too complex to be treated adequately by simple approaches, such as use of a deposition velocity. The present committee shares those concerns about the reliability of methods used in the NTPR program to estimate concentrations of radionuclides that are potentially important contributors to inhalation dose.

The previous National Research Council committee suggested that urinanalysis should be used to assess the validity of methods used in the NTPR program to estimate internal dose (NRC, 1985b). The present committee also believes that some indication of reliability is essential if estimates of inhalation dose are to be considered credible. However, because of experience with a bioassay program that was recently undertaken to assess internal exposures to plutonium and difficulties with the use of present-day measurements to estimate intakes that occurred many years ago, as discussed in Section VI.D, the committee believes that urinanalysis is not likely to provide useful information on the reliability of methods used to estimate inhalation doses to atomic veterans.

A potentially more fruitful approach would be to compare estimated radionuclide concentrations in deposited fallout or in neutron-activated soil used in the NTPR program with measurements that were made at the NTS or in the Pacific after the period of atomic testing ended. As noted in Section V.C.3.2, comment [6], radionuclide concentrations in surface soil over portions of the NTS that were affected by fallout were measured extensively during the 1980s. Important constituents of fallout on which data were obtained are ^{241}Am, ^{238}Pu, ^{239,240}Pu, ^{60}Co, ^{90}Sr, and ^{137}Cs. Although data on shorter-lived radionuclides in fallout are lacking, measurements of longer-lived constituents and knowledge of the relative activities of different fission and activation products that were produced in each shot presumably could be used to assess the reliability of estimated concentrations of all radionuclides in deposited fallout that are used in dose reconstructions. An illustration of the importance of those data is provided by an analysis presented in Appendix E. In addition, later measurements of ^{152,154,155}Eu in surface soil could be used to assess the reliability of estimated concentrations of activation products at the NTS.

Similarly, concentrations of radionuclides in fallout deposited on residence islands in the Pacific have been estimated in many studies, some of which began during the period of atomic testing (see, for example, Wilson et al., 1975; Robison et al., 1997b; Simon and Graham, 1997; Donaldson et al., 1997). Those data could be used to assess the reliability of estimated concentrations of radionuclides in fallout deposited on residence islands that are used in dose reconstructions and the potential importance of inhalation doses, and they may also be useful in assessing the reliability of estimated concentrations of radionuclides in fallout on ships. The potential importance of the data is illustrated by the following example. On residence islands at Enewetak Atoll, the total deposition of plutonium reported by Wilson et al. (1975) is about 0.3-25 nCi m^{-2}. If we assume that those data define a 90% confidence interval of plutonium concentrations and use the same assumptions about uncertainties in parameter values as in the example analysis of a scenario involving resuspension caused by walking or other light activities discussed in Section V.C.3.3—except that an assumption about fractionation is not needed when concentrations of plutonium on the ground are measured—we find that a central estimate of inhalation dose to the lung is about 10^{-4} mrem h^{-1}, and an upper bound (95th percentile) of a probability (uncertainty) distribution is about 0.02 mrem h^{-1}. Those results indicate that inhalation doses due to resuspension of longer-lived radionuclides in fallout deposited on residence islands in the Pacific are unlikely to be important in most cases. That conclusion is supported by later assessments of doses to native Marshall Islanders from inhalation of plutonium (Robison et al., 1997b; Sun et al., 1997b). Knowledge of amounts of shorter-lived radionuclides in fallout relative to plutonium could be used to infer possible inhalation doses due to resuspension of all radionuclides deposited on residence islands.

The committee is particularly concerned about two assumptions used in the NTPR program to estimate concentrations of radionuclides in fallout deposited on the ground or other surfaces. The first is an assumption of no fractionation of radionuclides in fallout except for removal of noble gases. That assumption almost certainly results in substantial underestimates of concentrations of refractory radionuclides (such as plutonium) in fallout at the NTS and in the Pacific. An assumption of no fractionation is especially important at the NTS because accumulation of fallout plutonium during the period of atomic testing presented an important inhalation hazard to thousands of participants who engaged in activities in forward areas. The second is an assumption, used to calculate external exposure rates per unit concentration of radionuclides in deposited fallout, that the source region is a surface of infinite extent. That assumption is reasonable at the NTS and on residence islands in the Pacific, but it probably results in underestimates of concentrations of radionuclides in fallout deposited on ships.

Estimates of concentrations of radionuclides on the ground or other surfaces used in dose reconstructions are of crucial importance because calculated inhalation doses in most scenarios depend on those estimates. The committee is not aware of any efforts by the NTPR program to assess the reliability of those estimates at the NTS or in the Pacific. If a key element of a method on which estimates of dose depend has unknown reliability, all estimates of dose based on the method are called into question unless it can be demonstrated by other means that the method as a whole most likely results in substantial overestimates of dose. The committee does not believe that it has been demonstrated that the method as a whole tends to overestimate inhalation doses.

[2] **An important deficiency in dose reconstructions for many participants at the NTS is the lack of consideration of resuspension of previously deposited fallout by the blast wave produced in aboveground detonations. When combined with the frequent neglect of aged fallout that accumulated at the NTS during the period of atomic testing and the general neglect of fractionation in fallout, neglect of resuspension caused by a blast wave could result in underestimates of upper bounds of inhalation doses by a factor of at least 100 in some scenarios in which participants engaged in activities in forward areas within a few hours after a shot, and perhaps by a factor of as much as 1,000 in the worst cases.**

The issue of neglect of resuspension caused by the blast wave produced in a detonation in all dose reconstructions at the NTS and the possible degree of underestimation of upper bounds of inhalation dose in some scenarios due to neglect of blast-wave effects are discussed in Section V.C.3.2, comment [7]. The committee believes that neglect of effects of a blast wave on inhalation exposures of participants in forward areas after detonations at the NTS, combined with the frequent neglect of aged fallout that accumulated during the period of atomic testing at the NTS and neglect of fractionation in fallout, is an important deficiency for which there is no apparent explanation. The potential importance of resuspension caused by a blast wave on inhalation doses is demonstrated by an analysis in Appendix E. Neglect of blast-wave effects is important not only because of the likelihood of large underestimates of inhalation dose but also because thousands of participants at the NTS (maneuver troops and close-in observers) probably were exposed to fallout that was resuspended by a blast wave, and credible upper bounds of doses to organs of concern could have exceeded 1 rem in many cases.

[3] **Dose coefficients for inhalation of radionuclides (equivalent doses to specific organs and tissues per unit activity intake) have substantial uncertainty that has not been taken into account in the NTPR program. In the worst cases, such as the dose coefficient for the lung from inhalation of plutonium, a credible upper bound of a dose coefficient based on current ICRP recommendations and a full accounting of uncertainty is more than a factor of 10 higher than values used in dose reconstructions for atomic veterans.**

Dose coefficients for inhalation of radionuclides are uncertain because of uncertainty in the associated dosimetric and biokinetic models and in the biological effectiveness of alpha particles. Evaluations of those uncertainties have been available for use in dose reconstructions at least since 1994 (see Section V.C.3.2, comments [1] and [2]). The conclusion that the upper bound of a dose coefficient for inhalation could be underestimated by a factor of more than 10 in the worst cases takes into account the presumed bias of most dose coefficients used in dose reconstructions to overestimate dose when a particle size (AMAD) of 1 μm is assumed (see Tables V.C.1 and V.C.2).

The substantial uncertainty in dose coefficients is important because it affects all calculations of inhalation dose to participants. Uncertainty in dose coefficients should be acknowledged and taken into account in the NTPR program if credible upper bounds of inhalation doses to atomic veterans are to be obtained.

V.C.6 Conclusions on Credibility of Estimated Upper Bounds of Inhalation Dose

All estimates of inhalation dose to atomic veterans obtained in the NTPR program are reported as single values without uncertainty, and those estimates are intended to provide upper bounds of possible inhalation doses. Thus, the key question in evaluating methods of estimating inhalation doses used in dose reconstructions is whether the methods provide credible upper bounds. If they do, estimates of inhalation dose to atomic veterans are appropriate for use in evaluating claims for compensation for radiation-related diseases. However, if estimates of inhalation dose are substantially less than credible upper bounds, the veterans are not given the benefit of the doubt and, depending on the magnitude of possible doses from all exposure pathways, their claims for compensation may not be evaluated fairly; that is, a veteran's claim could be denied even though a credible upper-bound estimate of dose, taking all exposure pathways and uncertainties into account, would qualify the veteran for compensation.

As discussed in Section V.C.3.3, the committee does not believe that the question of whether estimates of inhalation dose obtained in the NTPR program are credible upper bounds can be given a single answer that applies to all exposure scenarios for participants at the NTS and in the Pacific. However, partly on the basis of conclusions obtained in previous reviews by committees of the National Research Council (see Section V.C.2), the NTPR program has often claimed that its methods of calculating inhalation dose provide overestimates of dose (the doses are "high-sided"), the implication being that the claim applies generally (see, for example, Schaeffer, 2001b). Therefore, the question is whether the methods of estimating inhalation doses provide credible upper bounds in all or nearly all cases.

The present committee's review of methods of estimating inhalation dose used in the NTPR program has been considerably more extensive than previous reviews by other committees of the National Research Council. The present committee considered many issues involved in estimating inhalation doses that were not evidently considered in previous reviews. Furthermore, the present committee had access to documentation of methods that was not available when the first review was conducted in 1985; and for the first time, extensive and detailed evaluations of dose reconstructions for individual veterans who filed a claim for compensation or who requested information on their doses were conducted.

On the basis of its review, the present committee has reached a different conclusion about methods of estimating inhalation dose used in the NTPR program from the one based on previous reviews. Its conclusion is summarized as follows: **Methods used in the NTPR program to estimate inhalation doses to atomic veterans do not consistently provide credible upper bounds. Furthermore, the extent of underestimation of upper bounds is a factor of at least 100 in important scenarios involving maneuver troops and close-in observers at the NTS who were exposed to old fallout that was resuspended by the blast wave produced in a detonation.**

There are some important scenarios in which estimates of inhalation dose obtained in dose reconstructions probably are credible upper bounds, as intended by the NTPR program. An example of such a scenario discussed in Section V.C.3.3 is exposure to descending fallout throughout the period of descent on residence islands in the Pacific when cancer in an internal organ other than an organ in the GI tract is the disease of concern, although an unequivocal conclusion is difficult even in this scenario because of the unknown reliability of methods used by the NTPR program to estimate concentrations of radionuclides in descending fallout. It also seems likely that estimates of inhalation dose in scenarios at the NTS involving suspension of neutron-activation products in surface soil are credible upper bounds, given that assumed resuspension factors are likely to be considerable overestimates for radioactive materials that are fixed in soil. Estimates of inhalation doses to occupation forces in Japan discussed in Section IV.D also should be credible upper bounds if they are based on an assumption that exposure occurred only at locations of highest fallout.

However, the types of exposure scenarios for which estimates of inhalation dose obtained in dose reconstructions probably are credible upper bounds are somewhat limited. In many frequently occurring scenarios, such as scenarios of exposure to previously deposited fallout in forward areas at the NTS, the committee believes that uncertainties in assumptions used to estimate inhalation dose are sufficiently important that doses estimated by the NTPR program may not be credible upper bounds even if some parameter values used in the calculations, especially resuspension factors, are credible upper bounds. Even in scenarios involving exposure to descending fallout, exposure during the entire period of fallout probably was a rare occurrence at the NTS, in which case concentrations of radionuclides in air could be underestimated, depending on when exposure occurred; and the committee again notes that concentrations of radionuclides in fallout that descended on ships in the Pacific may be underestimated. Furthermore, in some dose reconstructions, it is evident to the committee that upper bounds of inhalation doses to atomic veterans have been underestimated by large factors. The most obvious cases involve exposure scenarios for participants in forward areas at the NTS, including maneuver troops and close-in observers, in which resuspension of substantial amounts of previously deposited fallout by the blast wave produced in a detonation has been ignored even though exposure to relatively high concentrations of resuspended radionuclides caused by the blast wave almost certainly occurred. For example, when the NTPR program has assumed that resuspension of previously deposited fallout was caused by walking or other light activity in cases in which blast-wave effects probably occurred but were ignored, the committee believes that upper bounds of inhalation doses are underestimated by a factor of at least 100, and perhaps by a factor of as much as 1,000 in the worst cases. Furthermore, in such cases, upper bounds of equivalent doses to some organs and tissues could have been substantially above 1 rem.

Of paramount importance is the issue of whether deficiencies in methods of estimating inhalation dose identified by the committee could have affected decisions about compensation of atomic veterans. The committee believes that possible underestimation of upper bounds of inhalation doses by the NTPR program is unlikely to be important for most participants in the Pacific or occupation forces in Japan. Inhalation doses to most of those participants probably were too low for possible underestimation of upper bounds to have affected decisions about compensation. The committee also believes that neglect of possible ingestion doses in dose reconstructions is unlikely to be important for most participants at any site. However, the neglect of blast-wave effects, combined with the frequent neglect of aged fallout that accumulated during the period of atomic testing at the NTS and neglect of fractionation in fallout, is an important concern for thousands of participants who were exposed in forward areas at the NTS shortly after a detonation. On the basis of an example analysis of the effects of a blast wave on inhalation doses (see Appendix E) and screening doses that have been used in evaluating claims for compensation (see Section III.E), use of credible upper bounds of inhalation doses in scenarios involving resuspension by a blast wave could have changed decisions not to grant compensation in some cases, depending on the disease of concern (for example, lung cancer in a nonsmoker).

The question of the importance of deficiencies in methods of estimating inhalation doses in the NTPR program with respect to evaluating claims for compensation for radiation-related diseases is discussed further in Sections VI.F and VII.C.

V.D DOSE RECONSTRUCTION FOR OCCUPATION FORCES IN JAPAN

As discussed in Section IV.D, the upper-bound external dose for the 195,000 troops who participated in the occupation of Japan or were prisoners of war at or near Hiroshima or Nagasaki was estimated, on the basis of very pessimistic assumptions, to be always less than 1 rem, even though the likely dose to most participants was at least a factor of 10 lower (McRaney and McGahan, 1980). The dose from ingestion of contaminated food or water or inhalation of resuspended debris was also found to be insignificant. The highest possible dose is for a participant who was present throughout the entire operation and spent 8 h d^{-1} at the location of highest exposure rates. However, most troops were rotated, troops were billeted well away from contaminated areas, and the highest exposure rates occurred over an area of only about 0.1 km^2. In examining a sample of 12 cases, the committee found that detailed calculations of worst-case upper-bound doses were carried out for most of the veterans, and the calculations included both internal and external doses. In those cases, the calculated upper bound was considerably less than the overall generic upper-bound value of 1 rem. The one exception was a person with a calculated upper-bound dose of 0.62 rem. At the other extreme, three veterans were given an upper-bound dose of zero because they did not have an opportunity to be close to contaminated sites (for example, they remained on board a ship in the Nagasaki harbor). In one case, the veteran was in a different part of Japan.

The committee concurs with the assessment by the NTPR program that the dose to even the most exposed of the occupation troops in Japan from both internal and external exposure was probably well below 1 rem.

V.E COMMITTEE EVALUATION OF METHOD OF ESTIMATING UNCERTAINTY IN DOSE AND UPPER BOUNDS

As stated in Section II.A, dose reconstruction is an inexact science. Uncertainties in quantifying dose arise from uncertainties in the various components that must be brought together to calculate a dose: in reconstruction of the activity scenario, in characterization of the radiation environment through time and space, in parameters assumed for calculations (such as resuspension factors and decay factors for radiation fields), in characterization of the mixture of radionuclides produced by a particular detonation, and in quantifying exposures through various routes (such as inhalation, ingestion, and dermal exposure).

Clearly, uncertainties in the dose assigned to an atomic veteran are highly relevant to the adjudication process, particularly for diseases not categorized as "presumptive," that is, diseases whose probability of causation is evaluated, because those uncertainties can inform the decision regarding the merits of a claim for service-connected disability. According to 32 *CFR* 218.3, which describes the approach to dose reconstruction used in the NTPR program: "Due to the range of activities, times, geometries, shielding, and weapon characteristics, as well as the normal spread in the available data pertaining to the radiation environment, an uncertainty analysis is performed. This analysis quantifies the uncertainties due to time/space variations, group size, and available data. Due to the large amounts of data, an automated (computer-assisted) procedure is often used to facilitate the data-handling and the dose integration and to investigate the sensitivity to variations in the parameters used."

However, the committee did not see evidence in the case files that this kind of thorough uncertainty analysis was often done, although Monte Carlo methods can bring together sources of uncertainty in this way. The standard operating procedures (SOPs) document provided to the committee (DTRA, 1997) provides almost no information about how uncertainty is quantified by the NTPR program, and this complicated the committee's review of methods used. The unit dose reports do provide uncertainty estimates, but they are usually estimates of the uncertainty in the average unit dose and, as discussed earlier, they may not provide a credible estimate of the uncertainty in the dose to the most exposed individuals in the unit. Furthermore, they often provide little detail regarding the specific method used, the exact correlations assumed or neglected, and the specific data used to calculate the upper bounds. Often, the reports acknowledge that the procedures used to combine various sources of uncertainty are based on approximate methods.

The NTPR program's intention with an upper-bound calculation is to provide at least a 95th percentile of the dose, that is, a dose that is intended to ensure that we can be at least 95% confident that the true dose is lower. Upper bounds estimated from film-badge data and from reconstructed gamma and neutron doses are combined in quadrature, assuming that they are uncorrelated, to arrive at an estimate of the upper bound in the total external dose. To the extent that the individual upper-bound estimates are credible and all doses and potential uncertainties are included, the upper-bound estimate for this sum is credible, provided that uncertainties in the increments of dose are independent—that is, not correlated—which they may not be because of repetitiveness of behavior and work responsibilities. If the components being summed are positively correlated, then the quadrature method will systematically underestimate the upper bound for the aggregated dose. Another problem arises in the context of combining uncertainties across different types of radiation. In recent years, after it became routine to report an upper bound for the external gamma plus neutron dose to VA, the sum of the estimated upper bounds of the gamma and neutron doses and the estimated "high-sided" internal organ dose has been used as the dose of record in evaluating probability of causation of a veteran's claimed disease in the adjudication process. Summing upper bounds of external and internal doses would generally result in an overestimate of the upper bound of the total organ dose. However, as discussed earlier in this chapter, the committee found that in many cases the estimated upper bounds for external gamma and neutron dose were not credible and the "high-sided" estimates of internal and beta skin doses may not always reflect the 95th percentile dose (that is, a credible 95th percentile could be considerably higher).

To the extent that the external gamma-plus-neutron dose upper bounds and inhalation dose estimates are reasonable estimates of at least 95th percentile or higher doses, the VA practice of summing the reported upper-bound external dose and the "high-sided" inhalation dose will result in a high-sided estimate of the 95th percentile upper bound of the total organ dose.[33] Although external and inhalation dose estimates are sometimes correlated to some extent, such as when both are based on the same exposure-rate measurement, most of the pertinent uncertainties involved are independent of each other. The estimated beta skin dose calculated by the NTPR program is directly related to the reported upper bound in gamma external dose. Thus, summing the reported beta dose estimate with the reported upper-bound gamma dose estimate will result in a credible estimate of the upper bound of the skin dose when the beta and gamma dose estimates both are credible upper bounds.

The committee acknowledges that calculation of an upper-bound dose is itself an uncertain process. Furthermore, it is not clear how one ought to quantify effects of uncertainties in an activity scenario. For example, for external radiation exposure, NTPR program policy guidelines sometimes seem to target a best or even "high-sided" central estimate together with a 95th percentile upper-bound dose, and at other times seem to opt for only a "high-sided" estimate, in accordance with the benefit-of-the-doubt provision. For internal dose, the policy of the NTPR program is to provide a "high-sided" estimate that supposedly incorporates benefit of the doubt with respect to the exposure scenario. However, as discussed elsewhere in this chapter, the committee has concluded that assumed exposure scenarios often did not give the veteran the benefit of the doubt.

V.F SUMMARY OF COMMITTEE FINDINGS REGARDING DOSE AND UNCERTAINTY ESTIMATES BY NTPR PROGRAM

The central ("best") estimates of external gamma and neutron doses to participants obtained by the NTPR program based on film-badge data and/or unit dose reconstructions are generally credible, provided that the assumed exposure scenario is reasonable. However, the committee has documented numerous examples in which the NTPR program has failed to establish the participant's exposure scenario adequately; that is, plausible scenarios could be developed, on the basis of available information, that would have resulted in higher estimates of dose.

The committee finds that estimates of uncertainty in external dose obtained by the NTPR program in unit dose reconstructions often are not credible and do not adequately reflect the upper bound (95th percentile) in the external dose to an individual participant, because deviations in individual exposure scenarios from the assumed group exposure scenario are not considered. Furthermore, the committee has identified a number of situations in which uncertainty in film-badge issuance dates, interpretation of data, and failure to give the veteran the benefit of

---

[33]The equivalent dose to any specific organ from external gamma irradiation differs little from the reported whole-body dose because of the high penetrating power of the energetic photons emitted in detonations and by radionuclides in fallout and in activation products.

the doubt suggest the possibility of a much higher credible upper bound of the dose to an individual than reported by the NTPR program, even when the dose is based primarily on film-badge data. Upper-bound estimates of external dose should include consideration of the possibility of incorrect exposure scenarios, possibly missing or erroneous film-badge data, the impact of limited survey data, and other such factors. To give the veteran the required benefit of the doubt, some method should be devised to increase upper-bound estimates of external dose **when there is reason to believe that any of those events may have occurred**.

The committee has concluded that, contrary to claims by the NTPR program, calculated internal doses from inhalation are not always "high-sided." The committee has identified scenarios for which the method used by the NTPR program to estimate inhalation dose probably provides credible upper bounds (95th percentiles of possible doses or above). However, the committee has also identified important scenarios for which estimates of inhalation dose obtained by the NTPR program probably underestimate upper bounds by as much as a factor of 100 or more. Furthermore, organ equivalent doses could be substantial in some of those cases.

The committee found that beta doses to the skin and lens of the eye, although claimed by the NTPR program to be "high-sided," may not represent a credible estimate of the 95th percentile beta dose. Furthermore, beta doses from direct contamination of skin or clothing apparently have not been considered in dose reconstructions in any cases in which a veteran filed a claim for skin cancer.

The committee believes that upper bounds of neutron doses reported by the NTPR program are not credible, because of neglect of the uncertainty in the biological effectiveness of neutrons. When neutron doses were important, estimated upper bounds of the combined gamma-plus-neutron doses obtained by the NTPR program may be low by as much as a factor of 5.

The committee thus has concluded that the external gamma and neutron dose upper bounds and "high-sided" internal and beta skin and eye doses reported by the NTPR program often do not represent a credible estimate of the 95th percentile upper bound of the possible dose to an individual participant.

As discussed in Section III.E, VA uses the sum of the reported external-dose upper bound and organ internal dose to evaluate the probability of causation of a claimed radiation-related disease. By using the upper-bound dose estimate to evaluate probability of causation, rather than the best (central) estimate, VA intends to give the veteran the benefit of the doubt. However, to the extent that the reported doses do not provide credible estimates of 95th percentile upper bounds of organ total equivalent doses, evaluations of probability of causation may be less favorable to the veteran than intended. Implications of the committee's findings with regard to evaluating claims for compensation are discussed further in Section VI.F.

VI Findings Related to Other Issues

VI.A QUALITY ASSURANCE AND DOCUMENTATION

The NTPR program provided written and oral information to the committee describing its quality assurance (QA) practices, including standard operating procedures (SOPs) and documentation requirements. That information, which is summarized in Section IV.G, consisted of a document on SOPs (DTRA, 1997), letters in response to the committee's inquiries (see Appendix D), and oral discussions at open meetings of the committee. The following discussion summarizes the committee's findings related to QA and documentation of procedures and individual dose reconstructions.

According to DTRA (Schaeffer, 2001a), the SOPs (DTRA, 1997) and 32 *CFR* Part 218 serve as the only written guidelines and procedures for the conduct of dose reconstructions. However, the SOPs really constitute a statement of approach and general principles followed by the NTPR program rather than a manual documenting the procedures used to reconstruct doses. Furthermore, the committee found that the SOPs are incomplete, are out of date, and contain no references to supplement the limited text. Many of the methods used to estimate doses are not discussed at all, nor are the methods that are used to estimate upper bounds. Details of the reconstruction methods are neither discussed nor referenced in the SOPs.[1]

---

[1]Some procedures are reported in unit dose reports, specialized reports (for example, Banks et al., 1959; Barss, 2000; Egbert et al., 1985; Barrett et al., 1986; Goetz et al., 1991), or unpublished SAIC internal memoranda (for example, Ortlieb, 1991; 1995; Flor, 1992; Schaeffer, 1995).

The SOPs provided to the committee (DTRA, 1997) contain a provision for periodic review and updating. The version provided to the committee had not been modified in several years, even though significant changes had occurred in the program. An example of an important change that should have triggered an update of the 1997 SOPs is the routine assessment of beta dose to skin that began in 1998. Documentation of procedures for determining beta dose to skin is important because the number of claims filed for skin cancer under the nonpresumptive regulation has increased dramatically over the last 4 years.

The committee was unable to locate formal documentation detailing when particular procedures were implemented or revised,[2] although those changes might have been documented in internal memoranda not reviewed by the committee. Examples include implementation of large-particle inhalation dose coefficients, revision of the upper-bound gamma dose for target-ship boarding parties at Operation CROSSROADS, and the upper-bound analysis for sums of film-badge readings.

QA is not discussed in any detail in the SOPs, nor are many important procedures, such as details on how film-badge uncertainties are calculated. Because of the evolution of the program and the lack of formal documentation of changes in policy available either publicly or in the individual files, it is difficult for a reviewer to evaluate a given dose reconstruction to ensure that up-to-date approved and consistent procedures were used. Some of the case files contained no narrative discussions of the dose assessments.

Dose reconstruction memoranda prepared by the NTPR program supposedly contain references to all methods used. However, references in dose memoranda are often internal NTPR program memoranda.[3] A veteran reviewing such a dose report would have no ready access to the referenced documents. Dose reconstructions by different analysts for similar scenarios often referenced different internal memoranda or reports for the same method.

---

[2]For example, in a letter to DTRA, the committee requested documentation regarding the use of a lower upper-bound estimate of external dose for participants who boarded target ships during Operation CROSSROADS than given in the unit dose report (Weitz et al., 1982). The reply (Schaeffer, 2002b) stated that the upper bound had been revised, but no specific documentation as to when and why was supplied. Similarly, the committee requested specific information as to how the upper bound for sums of film-badge readings was calculated. In reply, the committee was told that the NTPR program followed recommendations of the NRC (1989) and supplied an unannotated copy of the computer program used to calculate upper bounds. Comments in the program listing clearly indicated that the method recommended by the NRC had been slightly modified. However, no additional documentation justifying the modifications was supplied.

[3]The committee found a number of instances in which the references cited in a dose memorandum were not the appropriate or correct source of the information used to calculate the dose. Often, the cited reference contained only a reference to another document that discussed the procedure.

VI.A.1 Quality Assurance

As noted in Section IV.G.2, the statement of work in the DTRA solicitation for NTPR program support required that the contractor provide QA monitoring for the program in database management, dose assessment, and veteran assistance. In response to the solicitation, JAYCOR and SAIC submitted a technical proposal that specified QA measures for the three program tasks. The committee did not see the JAYCOR-SAIC proposal or any detailed QA procedures, but it did receive a letter from the NTPR program that described the QA program (Schaeffer, 2002e); this description is included in Appendix D and summarized in Section IV.G.2.

The committee noted that its sample of 99 case files contained little evidence of uniform application of basic QA measures. Dose calculations often were not signed, dated, or initialed by the analyst. Many of the typed assessments included typed initials of analysts and dates, but several did not. In files containing several recalculations of dose, the lack of dates made it difficult to determine which was the most recent. Although dose assessments are supposed to be reviewed before release to VA or to the veteran (DTRA, 1997), the files generally contained no documentation to show that the reviews occurred or who performed them. Many files did, however, contain logs in which activities associated with the files (but not final reviews) were noted and dated. Dose assessments were transmitted to VA or the veteran by a letter signed by the DTRA program manager, thus indicating final managerial approval.

The committee also found several examples of poor quality control that resulted in errors in the calculation or reporting of external dose. In case #2, the reported dose failed to account for a film-badge exposure during an earlier test series. The participant also was assumed to be present during Operation GREEN-HOUSE for a shorter time than indicated by his service record. In case #84, the dose memorandum references an incorrect unit dose report. In case #87, the dose memorandum assigns a dose of 0.4 rem but references an SAIC memorandum that indicates that the dose was 0.8 rem. In case #88, the dose memorandum and a letter in the file from the NTPR program to the veteran give the dose as 1.0 rem, but the SAIC database lists the total external dose as 1.8 rem, with no upper bound; according to the referenced exposure scenario, if the veteran is given the benefit of the doubt and is assumed to have been in an armored personnel carrier rather than a tank during a maneuver, his dose would have been 1.7 rem with an upper bound of 2.7 rem.

Published reports of the NTPR program did not indicate that they had been subjected to peer review. Reports are often published without documentation of their review process, and the committee did not see any written description of a process by which the NTPR program reviews its documents. As is the case with some of the committee's other concerns, peer reviews may have occurred informally or even formally, but they could not be verified by the committee, because neither the reviews nor any indications that they had occurred were recorded in documents available to the committee. Some reports contained erroneous statements, which suggested to the committee that effective peer review had not occurred.[4]

VI.A.2 Documentation of Dose Calculations

The SOPs (DTRA, 1997) list the documentation to be supplied to a veteran in connection with a dose assessment. On the basis of the committee's reviews of the sample of 99 case files, the documentation was generally supplied. However, detailed reviews of the calculations used for dose reconstructions[5] were central to the committee's work. Those reviews were hampered by the uneven and generally poor documentation found in the individual files. Many calculations were recorded by hand and were illegible. Variables often were not defined, and committee members had difficulty in understanding the calculations. One file (case #77) even had important information recorded on an unsigned adhesive note that appeared on the side of a page in the record.

Most of the records for the 99 cases in the committee's sample were considered to be inadequately documented, but we identified 28 case files in which the calculations were particularly poorly documented (cases #1, 3, 4, 6, 8, 9, 18, 19, 21, 26, 37, 38, 40, 43, 47, 49, 53, 57, 59, 60, 65, 68, 73, 77, 78, 82, 85, and 98). Comments on a few selected examples follow.

Case #3: Dose calculations were handwritten and scribbled although one part of the work was dated and signed. No upper-bound calculation was evident, although this assessment was performed in 1997.

Case #6: The handwritten dose calculations were unclear, hard to read, and difficult to interpret. The dose summary sheet had a typed date and initials, but the dose calculations themselves were not dated or initialed.

Case #18: The handwritten dose calculations had insufficient detail. Although the typed dose assessment was dated and included the analyst's name, the dose calculations themselves were undated and the analyst unidentified.

---

[4]In Barss (2000), for example, the linear energy transfer (LET) of photons was compared with the LET of beta particles, but LET is a quantity that applies only to charged particles, such as betas, and not to photons. Another statement in the report claimed that "although only a small fraction of the associated mixed fission product external gamma dose is actually deposited in the basal cell layer (of the skin), it is assumed that the entire whole body external gamma dose was deposited in the basal cell layer for upper bounding." That statement illustrates a misunderstanding or misuse of the quantity "dose." The committee is of the opinion that those inexact statements would have been identified in a peer review.

[5]Examples include calculations for combining film-badge data, applying unit dose reconstructions, assessing internal dose, and assigning uncertainty or upper bounds.

Case #53: The file contained a hard-to-read handwritten excerpt, regarding doses on the USS *Skate*, that was attributed to "the back-up file." The hard-to-read handwritten calculations of daily target-ship doses and upper-bound dose calculations were not dated, and the analyst was not identified.

Case #57: The calculations in the file were mostly handwritten, with some use of mathematical software. No source was given for a handwritten table concerning generic doses and their "error factors" for ship-related activities. The handwritten calculations were difficult to read, and what could be read was difficult to interpret.

Case #77: The file had an important inference about the impossibility of the veteran's being involved in a dose-producing activity (Chemical Biological Radiological Team training) attached as an adhesive note (see discussion of this case in Section V.A.2).

Case #78: A dose of 150 mrem was assigned to account for the dose received during island "sweeps" in connection with Operation GREENHOUSE. There were neither details on how these "sweep" doses had been calculated nor a citation. Internal dose calculations were handwritten and difficult to follow. The calculations were not dated or initialed by the analyst.

VI.A.3 Summary

In summary, the committee did not see a comprehensive manual of SOPs that is complete and current and covers all methods. The committee saw no manuals that document when changes in procedures occurred, that contain copies of computer programs, that document a formal quality assurance and quality control (QA-QC) program, or that contain copies of all internal NTPR program procedures memoranda referenced in dose reports. Although the NTPR program (Schaeffer, 2002e) cited the Environmental Protection Agency's QA guidelines concerning the importance of SOPs,[6] and the committee agrees with this approach, we did not see sufficient evidence of its implementation in the case files and other documents reviewed. The lack of a comprehensive document explaining the dose reconstruction methods was also an important concern to the National Research Council committee that reviewed the program in 1985 (NRC, 1985b).

Assessment documents in the case files often are undated and show no authorship. Some assessments are difficult to follow, even for scientists, because internal documentation is inadequate. In fairness, others are good, but the dispar-

---

[6]These guidelines (EPA, 2001) state that: "The development and use of SOPs are an integral part of a successful quality system as it provides individuals with the information to perform the job properly, and facilitates consistency in the quality and integrity of a product or end result."

ity makes it clear that there are not uniformly applied standards of quality. The committee was told that all assessments are reviewed, but there is not even simple documentation of reviews in the files (for example, date of review and signature of the reviewing authority), much less memoranda or other written documentation of the substance of the reviews.

A comprehensive program explaining QA-QC objectives and procedures should be developed and documented. The program should ensure that all doses are checked and correctly calculated, that consistent and up-to-date-methods are used, that all dose estimates are fully and adequately documented, and that backup records and calculations in files are complete, legible, annotated, and dated. All dose assessments should be documented clearly, and the veteran's file that contains details of the calculations, backup material, and so on, should be complete, legible, and comprehensible. Any references not readily available, such as internal memoranda, should be included in the file. All entries in the file should be typed or clearly written and dated, and the author should be clearly identified. Methods of validating film-badge doses, such as routine comparisons with generic dose reconstructions or with data on other veterans who performed the same type of activity, should be part of such a program. Some QA-QC should be performed by outside experts, possibly under the direction of an advisory board.

VI.B COMMUNICATION WITH ATOMIC VETERANS

As discussed in Section III.B, communication with an atomic veteran concerning compensation decisions is the primary responsibility of the VA Regional Office (VARO), which receives dose reports and other correspondence from DTRA. But the response time has often been lengthy, particularly before the recent effort to respond in a more timely manner. In case #32, for example, the NTPR program wrote to the VARO on October 20, 1995, apologizing at the outset for a delay in follow-up of an inquiry first filed on February 6, 1995. It appears to have taken that long for the veteran's participation in an atmospheric test to have been confirmed and a radiation dose assessment completed. In our sample of cases, it was not uncommon for at least 6 months to pass between the VARO inquiry and the NTPR report on the results of a dose reconstruction.

When a veteran feels aggrieved, he is owed accurate and responsive communication by the government. In any effort of this magnitude and complexity, there will be errors in communication, delays, failures to respond appropriately, and questionable judgments about suitable content in a response. But there were enough lapses within the committee's sample cases to cause concern. In some cases, intervention by members of Congress who inquired on behalf of constituent veterans or family members seemed to speed the process or to revive an investigation years after it appeared to have concluded. In a few other instances, an attorney was retained to pursue the matter on behalf of a veteran. On the whole, veterans and their families have needed patience and determination to resolve their concerns.

Over the years, as represented in our sample cases, written communications with veterans or their families were courteous, and doses were reported accurately. When much time had elapsed between letters, which was often the case, suitable apologies were expressed. When veterans described serious health problems, regret and sympathy usually were offered.

Nonetheless, veterans often had reason to find the process frustrating. In one instance (case #68), the VARO received a letter from DNA reporting a dose to the skin, but the claim was for cancer of the kidney. The Veterans Benefits Administration identified the error about 6 weeks later.

The same file exemplifies a problem in many of the cases in which the veteran was deceased. As part of information gathering, a letter from the VARO to the widow asks 20 questions of a highly detailed nature that the widow is unlikely to know how to answer, such as:

• "Is there personal history of smoking? If yes, provide the starting date, what was smoked (cigarettes, pipes, cigars, etc.), how much (number of packs, cigars, or pipes, etc., daily), and the date stopped (if applicable)."
• "Give the organization or unit attached to (ship, tank group company or squadron) and the rank at the time of the test? If different for more than one test, provide this information for each test."
• "Provide a detailed description of duties and other activities during the entire period of each test."
• "How far away was ground zero at the time of each test shot?"; and "Was there any direct contact with contaminated materials? If yes, describe the specific circumstances."

Although the widow is informed that she may write "I do not know" for any question, it is hard to understand the point of this exercise. The respondent is asked to provide such specific information that the reliability of her responses leaves wide scope for doubt. The committee recognizes that there are legal requirements for service-connected benefits, but the failure to modify the questions out of consideration for the widow's circumstances seems excessively formalistic.

Obtaining accurate detailed information from deceased veterans' families is sometimes hampered by veterans' unwillingness to discuss their nuclear-test-related activities when they were alive. In at least one instance among those sampled by the committee (case #32), the veteran's family reported his reluctance to violate his sense of patriotic duty by discussing the atomic-testing program because he had been told, for reasons of national security, not to discuss his role with anyone. That sometimes made a veteran reluctant to apply for additional benefits when he became ill. As a result, only an incomplete account of his service activities was available to his surviving family. When one veteran was deposed by VARO counsel concerning a claim, he refused on security grounds to disclose all details of a rocket test shot (case #36).

Attitudes about the process and the candor of VA and DTRA varied. Although expressions of patriotism characterized many of the veterans' approaches to the dose reconstruction process, there were also those in our sample whose mistrust was demonstrated in remarks they made about the circumstances of their exposure. It is striking, for example, that several veterans alleged that they were used as a "human guinea pig" or a "human monitor" or that they were in an "experiment." It is not clear that anyone in authority could dissuade veterans of such views, but delays and bureaucracy surely do not lessen such suspicions.

More recent direct communication between veterans and DTRA evinces an improved effort to tailor responses to the veteran's circumstances. Yet in at least one exchange (case #60), the veteran asked several specific questions concerning bioassay (see Section VI.D), sampling criteria in the comparable U.S. population that did not participate in atomic testing, and the name and address of the agency conducting the sampling, but the response did not address the second and third questions.

Over the years, various attempts have been made to provide veterans with general information on what was being learned about their radiation exposures. For example, a letter from the Navy to a veteran (case #49) includes the following statement, which is found in a number of such letters in the early to middle 1980s: "An important finding to date is that radiation exposures to the participants were generally quite low." The letter also refers to "the consensus of the medical community" that "the risk of any adverse health effect from exposures such as experienced by nearly all test participants is very, very slight." It is not clear what "consensus" the letter intended to identify. About a year later, a letter from the NTPR program to another veteran (case #58) reports the conclusions of a 1985 National Research Council report concerning mortality of test participants (NRC, 1985a) and includes a press release of the report. The committee found such generic statements regarding the low incidence of cancer among veterans and low probability of radiation as a causal agent for cancer possibly confusing to laypersons.

Although attempts to provide general information and reassurance on the significance of radiation doses and risks are laudable, the committee finds the way in which scientific views have been used in these communications to be somewhat troubling. Appearing in the context of an individual request for record review, dose reconstruction, and potential compensation for a disease suffered by a veteran, the communications have a defensive quality that is at best unpersuasive for the veteran's case and at worse an overt attempt to "expert" the veteran into retreat. Veterans and their families may, and apparently often do, regard the source of this information, which is the authority responsible for analyzing radiation exposure, as operating under a conflict of interest.

Letters sent to veterans often were ambiguous in their presentation of the risk and the extent of scientific agreement. A letter sent in 1996 (case #15) stated that "the doses received by over ninety-nine percent of the test participants were less than the current Federal guidelines for radiation workers which permit external exposures of 5 rem gamma per year." The next paragraph begins by saying that "While medical science has no proof that exposure to low levels of ionizing radiation is hazardous to health, it is generally assumed by scientists that even low levels of exposure carry some slight risk." Quite apart from questions about the accuracy of the latter assertion, the two statements taken together may be confusing to a layperson.

This language appeared in letters from the NTPR program to atomic veterans for years. An earlier version found in the file for case #63 sought more aggressively to downplay the significance of radiation exposure. It included the statement that: "In sum, the studies described above have not revealed a basis for concern by participants of the atmospheric nuclear tests, or by the veterans of the Hiroshima and Nagasaki occupation forces over an increased risk of adverse health effects due to radiation." The magnitude of the beta dose to skin and the associated risks were downplayed for years after it became clear that there was a stochastic risk of skin cancer at doses experienced by some participants (see Section V.B.6).

In some cases, veterans seem to have received a more tailored response when they communicated directly with the NTPR program and bypassed the VARO. Sometimes, however, the tailored responses unfolded over a long period. Case #9 exemplifies the bureaucratic odyssey that sometimes characterizes multiple dose assessments and the tenacity required by the veteran. After a 1978 application for service-connected disability benefits, a 1980 letter from a Navy NTPR official to the veteran offered a medical examination and reported a "careful search of dosimetry data" at the test shot, which was said to have revealed an exposure within then-prevalent occupational standards. After what seems to have been a series of further contacts, a 1992 letter included then-common language that "radiation exposures to the participants were generally quite low" and that "the consensus of the medical and scientific community is that the risk of any adverse health effect from exposures such as experienced by nearly all test participants is very, very slight." DTRA completed the latest dose report on this veteran on June 20, 2000.

The long period of several dose assessments often resulted in letters concerning revised dose reconstructions that could as easily have undermined a veteran's confidence in the process as reinforced it. One instance is a letter in 1995 (case #56) that reports a revision in an estimate provided 10 years earlier.

On the whole, communication with veterans, although courteous and accurate with respect to reporting doses, has sometimes not been timely, has varied in detail, has conveyed mixed messages, and, considering the time and effort involved, has failed to inspire confidence that the process was fair, orderly, and expeditious.

VI.C THE LOW-LEVEL INTERNAL DOSE SCREEN

VI.C.1 Introduction

As mentioned in Section I.D, one of the key issues of concern to the atomic veterans has been the application of a low-level internal dose screen to eliminate the need for a more detailed estimation of internal dose in many dose reconstructions. In an attempt to provide some insight into this issue, the committee undertook a thorough evaluation of the low-level internal dose screen and provides the following discussion about its use in dose reconstructions.

The methods applied in the NTPR program to estimate internal doses to atomic veterans are discussed in Section IV.C, and the committee's evaluation of the methods is presented in Section V.C. As emphasized previously, intakes of radionuclides by inhalation are expected to be more important in most cases than intakes by ingestion or absorption through the skin or open wounds.

Early in the NTPR program, it was recognized that the task of performing detailed calculations of inhalation dose to all participants in the atomic-testing program could be overwhelming, given that data needed to estimate intakes of radionuclides, including measured concentrations of radionuclides in air at times of exposure or amounts of radionuclides excreted in urine or feces, generally were lacking. It also was believed that inhalation doses usually would be insignificant compared with external doses. Inhalation doses would be insignificant, for example, when most of the airborne radionuclides to which a participant was exposed were shorter-lived, photon-emitting fission or activation products. Those types of radionuclides produce relatively high external doses per unit activity concentration in the environment because of the long distances of travel of higher-energy photons in air, but relatively low internal doses because of their short residence times in the body and the low absorption in body tissues of higher-energy photons emitted by radionuclides in the body.

On the basis of those considerations, a method was developed that could be used to quickly evaluate potential inhalation doses to groups of participants who were exposed under similar conditions and to eliminate from further consideration groups whose members most likely did not receive a significant inhalation dose. Such a process is referred to as screening. A method of evaluating the potential importance of inhalation exposures, referred to as the low-level internal dose screen, was first developed and applied to participant groups at the NTS (Barrett et al., 1986). The same screening method was later applied to participant groups in the Pacific (Goetz et al., 1991).

This section discusses the general requirements of a screening method and the low-level internal dose screen developed in the NTPR program. Particular attention is paid to concerns that the atomic veterans have expressed about use of the internal dose screen in dose reconstructions, especially when a veteran files a claim for compensation for a radiation-related disease.

VI.C.2 General Requirements of Screening Methods

Any method of screening that is used to draw conclusions about the significance of potential radiation doses, or lack thereof, incorporates two basic elements:

- A model to estimate dose in an assumed exposure scenario.
- A dose criterion to define a level of exposure below which there is no concern.

If the dose estimated with the model is below the dose criterion, potential doses in the assumed scenario are considered to be so low that the associated health risks are insignificant.

Given the basic elements listed above, a method of screening must satisfy two conditions to meet its intended purpose:

- The model used to estimate dose must tend to overestimate doses for any exposure conditions that could be encountered.
- The dose criterion must correspond to a dose that clearly is insignificant with regard to potential health risks to an exposed person.

It also is helpful if the model used to estimate dose is simple and transparent, so that others can understand that it is likely to overestimate dose.

VI.C.3 Assumptions Used in Internal Dose Screen

The methods used to estimate inhalation doses to participant groups for purposes of screening are the same as those described in Section IV.C.2. That is, the same four basic exposure scenarios and the same methods of estimating inhalation dose in each scenario were used in screening. For example, when a participant group was assumed to be exposed to resuspended fallout that had been deposited on the ground, the inhalation dose to an organ or tissue of concern is estimated on the basis of (1) estimates of the relative activities of radionuclides produced in shots of concern combined with an assumption of no fractionation of radionuclides in fallout except for the removal of noble gases, which are used to estimate the relative activities of radionuclides in fallout deposited on the ground; (2) measurements of external photon exposure or exposure rate due to the deposited fallout with film badges or field instruments combined with calculations of the exposure rate per unit activity concentration for the assumed mixture of radionuclides, which are used to estimate the absolute activity concentrations of radionuclides on the ground; and (3) an assumed resuspension factor, which is applied to the estimated concentrations of radionuclides on the ground to obtain an estimate of the concentrations in air.

The dose criterion used in screening of potential inhalation exposures of participant groups is a 50-year committed equivalent dose of 0.15 rem to bone, where "bone" denotes the entire volume (mass) of bone (Barrett et al., 1986). Although the NTPR program recognized that selection of the particular organ and value of the dose criterion to be used in screening is arbitrary, bone was selected because of its importance as a site of deposition of many radionuclides, including long-lived alpha emitters, such as plutonium, and the dose criterion was set at 1% of the dose limit for bone in standards for occupational exposure that had been recommended by the National Council on Radiation Protection and Measurements (NCRP, 1971).

VI.C.4 Use of Internal Dose Screen in Dose Reconstructions for Participant Groups

In the low-level internal dose screen, the dose criterion of 0.15 rem to bone and the model to estimate inhalation dose in an assumed scenario for a participant group have been used in a dose reconstruction for that group in the following way (Barrett et al., 1986). As discussed in the previous section, the model essentially relates the inhalation dose to an organ or tissue of concern (bone in this case) in an assumed scenario to a measurement of external photon exposure.

Consider the scenario involving inhalation exposure to deposited fallout that was resuspended in the air (this discussion also applies to suspension of neutron-induced activity in soil). If D_{FB} denotes the total external photon exposure indicated by a participant's film badge worn during the time of exposure to resuspended fallout,[7] this exposure is given by

$$D_{FB} = 0.7 \times I \times T, \qquad (VI.C-1)$$

where I denotes the average external exposure rate over the time of exposure and T is the duration of exposure. The factor 0.7 takes into account that the external exposure registered by a film badge attached to a participant's body is less than the exposure that would be registered by the badge in isolation, because of the shielding provided by the body. Combining Equation VI.C-1 with Equation IV.C-3 in Section IV.C.2.1.1, the external exposure registered by a film badge worn by a participant corresponding to a 50-year committed equivalent dose of 0.15 rem to bone is given by

$$D_{FB} = \frac{0.15 \times 0.7}{(SA/I) \times K \times BR \times DF}, \qquad (VI.C-2)$$

where SA/I is the reciprocal of the calculated exposure rate per unit concentration of the assumed mixture of radionuclides on the ground, K is the assumed resus-

---

[7]In cases where film-badge data are not available and external exposure rates are based on readings of field instruments, the total exposure that would be indicated by a film badge is estimated by assuming that the exposure rate varies as $t^{-1.2}$, where t is the time after detonation (in hours), and integrating the exposure rate over the time of exposure.

pension factor, *BR* is the assumed breathing rate, and *DF* is the inhalation dose coefficient for bone and the assumed mixture of radionuclides.

Thus, if the film-badge dose calculated with Equation VI.C-2 is greater than the estimated external dose to individuals in the participant group of concern, as obtained from measurements with film badges or field instruments, the committed equivalent dose to bone presumably was less than the screening criterion of 0.15 rem, and the inhalation dose is judged to be insignificant.[8]

The low-level internal dose screen was applied to many participant groups at the NTS (Barrett et al., 1986) and in the Pacific (Goetz et al., 1991). On the basis of an assumed exposure scenario at a particular test, a code was assigned that indicates the estimated external dose in rem that would be required to yield a committed equivalent dose of 0.15 rem to bone from inhalation. For example, at the NTS, an assigned screening code of IIC2 indicates that members of the participant group were assumed to be exposed to resuspended fallout for up to 9 h after a detonation under conditions in which the resuspension factor was assumed to be 10^{-5} m^{-1} and the external dose corresponding to a committed dose of 0.15 rem to bone, calculated with Equation VI.C-2, was 130 rem. That is, if the estimated external dose to members of the group is less than 130 rem, the committed dose to bone due to inhalation presumably was less than 0.15 rem. Thus, in this scenario, the external dose was expected to be about a factor of 1,000 greater than the inhalation dose.

The documented uses of the low-level internal dose screen (Barrett et al., 1986; Goetz et al., 1991) described above apply to dose reconstructions for participant groups (generic or unit dose reconstructions). In unit dose reconstructions, all members of a group who engaged in similar activities are assigned the same dose. Of greater interest to the committee is the question of whether the low-level internal dose screen is used in dose reconstructions when a veteran files a claim for compensation for a nonpresumptive radiation-related disease and a dose estimate for that person is required. That issue is discussed in the committee's comments on use of the low-level internal dose screen in the following section.

VI.C.5 Discussion of Low-Level Internal Dose Screen

The committee is aware that the atomic veterans have expressed concerns over use of the low-level internal dose screen in dose reconstructions. The essence of the veterans' concerns appears to be that the screening method is a means to avoid having to estimate inhalation doses and that when the screen was used, significant inhalation doses to participants were not taken into account in dose reconstructions. More specific concerns apparently include that:

---

[8]If *D* denotes the estimated external dose to the participant group of concern, the estimated inhalation dose to bone is linearly proportional to the external dose and is given by $D \times (0.15/D_{FB})$, where D_{FB} is estimated with Equation VI.C-2.

• The resuspension factors assumed in calculating airborne concentrations of resuspended fallout are too low under conditions of exposure that were encountered by some participants.

• The use of a committed equivalent dose of 0.15 rem to bone for purposes of screening is inappropriate because most of the mass of bone is not believed to be radiosensitive (the radiosensitive tissues of the skeleton are the much smaller mass of endosteal cells that lie on bone surfaces), and doses to other organs could be substantially higher when the dose to bone is 0.15 rem.

• The screening codes assigned to some participant groups were based on an incorrect exposure scenario.

With regard to the last point, the veterans have expressed concern, for example, that some participant groups were assumed to be exposed only to suspended neutron-induced activity in soil in the absence of a fallout field, which results in very low inhalation doses relative to external doses, in cases where fallout from prior shots also was present, thus greatly increasing potential doses from inhalation of resuspended plutonium and longer-lived fission products relative to external doses.

The committee has carefully considered the veterans' concerns about the low-level internal dose screen. Questions about whether assumed exposure scenarios are correct and whether the models and parameter values used to estimate inhalation dose for purposes of screening might underestimate actual doses to participant groups are important because, as emphasized in Section VI.C.3, the same scenarios and models also are used in dose reconstructions for individual veterans. Indeed, the veterans' concerns about assumed scenarios and methods of estimating inhalation dose used in screening are shared by the committee (see, for example, Section V.C.3.2 and Table V.C.7).

The committee also agrees that the dose criterion used in screening is not the most suitable choice. It does not seem logical to use an organ (bone) that is not radiosensitive. Furthermore, depending on the radionuclides inhaled, a calculated dose of 0.15 rem to bone can correspond to substantially higher doses to other organs and tissues.[9] Thus, the dose criterion used for purposes of screening of inhalation doses may not correspond to doses to all organs and tissues that would be considered insignificant.

The committee emphasizes, however, that there is nothing inherently wrong with the use of screening to eliminate unimportant radionuclides or exposure pathways from further consideration in a dose reconstruction, provided that the method of screening meets the two conditions described in Section VI.C.2. It is

---

[9]Consider, for example, inhalation of plutonium in respirable and insoluble form. In the database of dose coefficients used in some dose reconstructions (Dunning et al., 1979), doses to the lung, respiratory lymph nodes, bone surfaces, and liver exceed doses to bone by a factor of 1.6, 115, 16, and 2.2, respectively.

not sensible to expend resources to estimate doses that can be shown by use of simple and transparent methods to be below a minimal level of concern. Screening of radionuclides or exposure pathways has been used, for example, in dose reconstructions for members of the public who were exposed to radionuclides released from DOE sites (see Section I.C.1) or to fallout from testing of nuclear weapons at the NTS (for example, see Ng et al. [1990]).

In light of the veterans' concerns about the low-level internal dose screen, the committee gave careful consideration to the issue of whether the screen was used in dose reconstructions for individual veterans, especially when they filed claims for compensation. In response to specific inquiries on this matter, the committee received written assurance that the generic internal dose screen developed in 1986 "had no impact on individual organ doses from intake of fallout for VA claims" (Schaeffer, 2001a) and, later, that "DTRA/SAIC does not and has not used internal dose screening factors to evaluate inhalation doses to individuals' organs or tissues" (Schaeffer, 2002b).

Based on its review of the 99 randomly selected dose reconstructions for individual veterans, the committee has inferred that the low-level internal dose screen, meaning the screening codes assigned to participant groups at the NTS or in the Pacific (Barrett et al., 1986; Goetz et al., 1991), as discussed above, has not been used in dose reconstructions for veterans who filed claims for compensation. That conclusion may not be immediately evident from an examination of the documentation of an individual veteran's dose reconstruction. For example, in some cases, a statement that the veteran's unit passed (or did not pass) screening is included in the assessment of inhalation dose; the dose to the organ or tissue of concern in evaluating the veteran's claim for compensation is reported as less than 0.15 rem, which is the dose criterion used in screening, rather than the actual estimate; or a dose of less than 0.15 rem to bone is assigned even though the veteran did not claim a bone disease (see cases #8, 18, 21, 23, 27, 32, 36, 38, 41, 47, 59, 63, 68, 73, 76, 78, 81, 94, and 98). In all such cases, however, the committee found that the dose reconstruction for the veteran included an assessment of inhalation dose that did not rely on the screening code that had previously been assigned to the veteran's unit. That is, a separate calculation of inhalation dose to the veteran for the assumed conditions of exposure was performed in all cases.

It also came to the committee's attention, however, that the letter from DTRA to VA documenting the dose reconstruction for a veteran who files a claim for compensation (see Section III.B.2, Figure III.B.1) sometimes included a statement that a report on the low-level internal dose screen (Barrett et al., 1986; Goetz et al., 1991) indicates that the claimant's dose to bone and the organ in which the claimant's cancer occurred are both less than the screening criterion of 0.15 rem; a copy of this letter also was sent to the veteran or veteran's representative (for example, surviving spouse if the veteran was deceased). Such statements give the appearance that the low-level internal dose screen sometimes has been used to estimate inhalation doses when a veteran filed a claim for compensation, contrary to the assertions by DTRA noted above.[10]

On the basis of the committee's efforts to understand the low-level internal dose screen and its use in dose reconstructions, the committee appreciates the confusion that can result when the NTPR program states that the internal dose screen is not used in dose reconstructions for veterans who file claims for compensation but there appears to be evidence to the contrary in the documentation of dose reconstructions or in letters sent to veterans that summarize the results. However, any such confusion can be resolved if one recognizes that the methods of estimating inhalation dose that were used to develop the screening codes for participant groups at various tests are the same as the methods used to estimate inhalation doses in individual dose reconstructions (see Section VI.C.3).

In judging the adequacy of an assessment of inhalation dose in the dose reconstruction for an individual veteran, the most relevant questions concern whether the assumed exposure scenario for that person is reasonable, according to knowledge of his activities and the radiation environment in which those activities took place, and whether the models, parameter values, and other assumptions used to estimate inhalation dose in the assumed scenario provide credible upper-bound estimates. If credible upper-bound estimates of inhalation dose have been obtained for a veteran, it is of secondary concern whether the documentation of the dose reconstruction or the letter sent to the veteran summarizing the results includes statements or reported doses that indicate that the low-level internal dose screen may have been used to estimate inhalation dose or may have influenced the calculation. If the screening code assigned to the veteran's unit in a low-level internal dose screen report was based on reasonable assumptions, the inhalation dose based on the screening code and the assigned external dose for the unit should provide a credible upper bound. However, if the assumptions used in developing the screening code for the veteran's unit are incorrect and do not describe the conditions of exposure of the veteran, the inhalation dose based on the screening code and the assigned external dose will not provide a credible upper bound.[11]

---

[10]The NTPR program also has acknowledged that there were instances when a veteran filed a claim for compensation but an individual dose reconstruction was not performed (Schaeffer, 2002d). That situation occurred when the exposure scenario for the veteran was judged to be adequately described by assumptions used in an existing dose reconstruction for the veteran's unit in which the low-level internal dose screen had been applied. In such cases, JAYCOR apparently reported doses estimated in the unit dose reconstruction but an individual dose reconstruction was not performed by SAIC. Thus, in a sense, the low-level internal dose screen was used to estimate inhalation doses to a few veterans who filed claims for compensation.

[11]A clear example discussed in Section V.C.3.2, comment [6] and [7], and Appendix E involves inhalation exposures of participants in forward areas after Operation PLUMBBOB Shot HOOD. The screening code assigned to several units at Shot HOOD was based on an assumption that the only radionuclides inhaled were neutron-activation products in soil (Barrett et al., 1986), even though substantial amounts of plutonium and longer-lived fission products undoubtedly were present.

It is important to bear in mind that the credibility of a dose reconstruction does not depend on how the analysis is described. Rather, the credibility of a dose reconstruction depends only on the assumptions used in calculating dose, whether the dose was calculated correctly on the basis of the assumptions, and whether the calculated dose was reported correctly. The description of a dose reconstruction may indicate that there are problems in communicating to a veteran how the analysis was performed, and poor communication certainly can affect a veteran's belief about the credibility of an analysis. However, the issue of communication to the veteran is separate from the issue of the credibility of the dose reconstruction itself.

VI.D BIOASSAY PROGRAM TO ASSESS INTERNAL EXPOSURES TO PLUTONIUM

VI.D.1 Description of Plutonium Bioassay Program

In the late 1990s, the NTPR program undertook an effort to determine whether plutonium bioassay testing of atomic veterans, specifically measurement of the amounts of plutonium in urine samples, could be used to estimate intakes that occurred during participation in the weapons-testing program (Schaeffer, 2002f). If intakes of plutonium could be estimated on the basis of bioassay testing, the estimated intakes could be used to investigate the reliability of the methods of estimating inhalation doses to participants discussed in Section IV.C.2, which are based on data other than measured concentrations of radionuclides in air or amounts of radionuclides excreted in urine or feces shortly after exposures occurred.

Bioassay testing for the purpose of assessing the reliability of methods of internal dose estimation was recommended in a previous review by a committee of the National Research Council (NRC, 1985b) discussed in Section V.C.2. Plutonium was selected as the radionuclide to be studied because it is an important component of fallout, it presents a significant inhalation hazard, and it is tenaciously retained in the body after an intake (a substantial fraction of plutonium absorbed into blood is retained for the rest of life, even if intakes occurred at an early age), and because there appeared to be suitable methods of analysis that could detect very low concentrations of plutonium in urine.

It is the committee's understanding that data on the amounts of plutonium in urine samples provided by atomic veterans that have been collected so far have not been used in assessing internal dose in any dose reconstructions. The committee found no evidence to the contrary in its review of dose reconstructions for individual veterans.

VI.D.2 Discussion of Plutonium Bioassay Program

On the basis of information on results obtained in the plutonium bioassay testing program (Schaeffer, 2002f), the committee has concluded that it will be difficult to obtain reliable estimates of internal doses to participants during periods of atomic testing. The measured activities of plutonium in urine are sufficiently low in most cases that background levels of plutonium in urine due to global fallout can obscure the signal derived from participation in atmospheric tests. There also are problems with the method of bioassay itself at these low levels, and analyses of split samples revealed poor agreement in pairs of measurements.

Furthermore, even if activities of plutonium in urine could be measured reliably, there is substantial uncertainty in applying assumed biokinetic models for the behavior of plutonium in the body and rates of excretion (which may vary among individuals) to obtain estimates of intakes that occurred many years ago. That uncertainty would need to be evaluated and taken into account in assessing the extent to which recent bioassay data could be used to assess the reliability of methods that have been used by the NTPR program to estimate intakes of plutonium by atomic veterans.

VI.E RETROACTIVE RECALCULATIONS OF DOSES AND RE-EVALUATIONS OF PRIOR COMPENSATION DECISIONS

An important challenge throughout the more than 20-year existence of the NTPR program has been the need to adapt to changes in applicable laws and regulations and to incorporate improvements in the scientific foundations of dose reconstruction and methods of estimating radiation risks and probability of causation of radiation-related diseases. For example, one issue is how to incorporate improvements in science into the methods of dose reconstruction and the decision process used to evaluate claims for compensation while striving to treat all claims equitably.

The committee was concerned about what is done within the NTPR program to re-evaluate dose reconstructions when a claim for compensation was denied but it is thought that later changes in laws, regulations, or methods of reconstructing doses or estimating probability of causation might have affected the outcome if they had been in place at the time of the claim. For example, suppose that a veteran filed a claim for compensation for kidney cancer before this form of cancer was declared to be presumptive in 38 *CFR* 3.309 (see Section I.B.4), but the claim was denied because of a low estimate of dose in the veteran's dose reconstruction (see case #99). The question in this case is: Would the veteran's claim have been reopened after kidney cancer was declared to be a presumptive disease and would compensation have been granted from the time the presumptive law was passed, assuming that the veteran's participation in the atomic-testing program is adequately established? Or suppose that a veteran filed a claim for skin cancer before 1998, when such claims were routinely denied without assessment of possible skin doses on the basis of the presumption that doses of around 1,000 rem were required to cause skin cancer and participant doses of this magnitude were not credible (see Section V.B.6.3). Would the veteran's claim have been reopened and an assessment of dose to the skin performed for use in re-evaluating the previous compensation decision after the method for estimating beta dose to the skin was developed and claims for compensation for skin cancer were granted in some cases when the estimated equivalent dose to skin was as low as a few tens of rem (see case #9)?

In response to a verbal inquiry on the issue of retroactive recalculations of dose and re-evaluations of prior compensation decisions in cases in which claims had been denied, the committee was informed that VA generally does not take the initiative to reopen cases when a change in law, regulations, or methods of reconstructing doses or estimating probability of causation of a radiation-related disease could have affected the compensation decision. Nor does VA or the NTPR program inform individual claimants about changes that could have af-fected their denied claims. Rather, the NTPR program disseminates such infor-mation, for example, in public announcements or statements submitted to veterans organizations, such as the National Association of Atomic Veterans, for publica-tion in a newsletter. Once the information is disseminated, the responsibility lies with the veterans or their representatives to request that a prior claim be reopened and re-evaluated.

The committee has identified several specific issues that could be important with regard to retroactive recalculations of doses and re-evaluations of prior compensation decisions. The issues are summarized as follows (all examples apply to claims for compensation for nonpresumptive diseases under 38 *CFR* 3.311):

- Re-evaluation of claims that were denied before the disease of concern was declared to be presumptive in 38 *CFR* 3.309.
- Re-evaluation of claims for skin cancer that were denied before develop-ment of the method of estimating beta doses to skin and before the large reduction in the skin dose that could be judged to be at least as likely as not to cause skin cancer.
- Re-evaluation of claims that were denied when the wrong disease was identified in the dose reconstruction (that is, the dose to the wrong organ or tissue was estimated).[12]
- Retroactive calculation of upper bounds of external dose in claims that were denied before such upper bounds were calculated routinely (that is, in cases in which only a central estimate of external dose was obtained in the dose reconstruction).

---

[12]The committee encountered one case in its review of 99 randomly selected dose reconstructions for individual veterans in which the wrong disease was assumed (case #91); this error presumably could be corrected only if information sent to the veteran that summarizes the dose reconstruction and compensation decision specifically identifies the disease for which the claim was denied.

- Retroactive calculation of internal doses in claims that were denied before development of the FIIDOS code (Egbert et al., 1985) in cases in which internal dose was not estimated in the dose reconstruction (see case #89 and 99).
- Retroactive recalculation of inhalation doses in claims that were denied before an assumption of large particles (AMAD, 20 μm) was used routinely in exposure scenarios in which mostly large particles presumably were inhaled and an assumption of large particles, instead of respirable particles (AMAD, 1 μm), would increase the estimated dose to the organ or tissue of concern.
- Re-evaluation of cases in which a veteran did not file a claim for compensation but stated that he had a radiogenic disease specified in 38 *CFR* 3.311.[13]
- Retroactive re-evaluation of decisions to deny claims for compensation that were made before so-called screening doses (CIRRPC, 1988) discussed in Section III.E were used routinely in evaluating claims, especially when the screening dose to the organ or tissue of concern is lower than the dose criterion that was used previously in evaluating claims.

The issue of retroactive recalculations of dose and re-evaluations of decisions to deny compensation also will arise if methods of dose reconstruction are revised in response to the committee's evaluations presented in this report, because appropriate changes should result in increases in credible upper-bound estimates of dose in many cases.

It is not the committee's intent to criticize policies of the NTPR program and VA concerned with informing veterans whose claims for compensation were denied about changes in laws, regulations, or methods of reconstructing doses or estimating probability of causation that could affect their claims. The committee recognizes the effort that would be required to take the initiative to re-evaluate every denied claim whenever there is a change in some aspect of the program that could affect claims.

Nonetheless, veterans might view the NTPR program more favorably if, for example, individual veterans were informed when changes in methods of estimating doses are made that might result in increases in their previously assigned doses or when policies affecting evaluations of claims are changed and were reminded that they can request a revised dose reconstruction. For example, after recent changes in the methods of calculating beta dose to the skin and evaluating claims for compensation for skin cancer, individual veterans with a previously denied claim for skin cancer, and the community of atomic veterans as a whole, could have been informed that doses to the skin are now being calculated in a different way and, furthermore, that more claims for skin cancer are being granted on the basis of a re-evaluation (lowering) of doses that could cause skin cancer. A

---

[13]The committee encountered one such case in its review of dose reconstructions for individual veterans (case #55).

properly designed database would facilitate efforts to identify veterans with previously denied claims for particular diseases and to locate the veterans and inform them of changes in the dose reconstruction program and in the laws and regulations that could affect their claims.

VI.F IMPLICATIONS OF COMMITTEE'S FINDINGS

VI.F.1 The Central Issue

The committee's review of the program of dose reconstruction for atomic veterans was prompted by concerns about the adequacy of the methods used to estimate dose. Those concerns are important because estimated doses to atomic veterans are used in evaluating claims for compensation for nonpresumptive diseases. Although previous scientific reviews expressed concerns about some aspects of the methods of dose reconstruction (see Section V.C.2) and the atomic veterans themselves have questioned many assumptions used in dose reconstructions (see Section I.D), the central issue of concern to the veterans has been the apparently very small number of claims for compensation for nonpresumptive diseases that have been granted. When a legal and regulatory structure is established to provide compensation to atomic veterans who later experience radiation-related diseases on the basis, in part, of results of dose reconstructions but the odds of receiving compensation appear to be very low, it is understandable that the veterans would question their estimated doses and the value of dose reconstruction.

In accordance with the requirement to give the veterans the benefit of the doubt in evaluating claims for compensation for nonpresumptive diseases (see Section I.C.3.2), dose reconstructions should provide credible upper bounds of possible doses to participants, taking into account uncertainties in estimating dose that are an inherent part of any dose reconstruction. Indeed, it is the policy of the NTPR program to provide "high-sided" estimates of dose. However, as discussed at length in Chapter V, the committee has concluded that dose reconstructions for atomic veterans do not consistently provide credible upper bounds of possible doses. Rather, in many cases, credible upper bounds would be substantially higher than alleged "high-sided" dose estimates in existing dose reconstructions. The reasons for substantial underestimation of upper bounds of possible doses in many cases are related primarily to two recurring issues: assumptions about exposure scenarios that do not give the veterans the benefit of the doubt or do not conform to plausible conditions of exposure, taking into account information on the veterans' activities and reasonable assumptions about the radiation environment in which the activities took place; and inadequate accounting of uncertainty in data, models, and parameter values used to estimate external and internal doses.

Given the committee's findings on deficiencies in methods and assumptions used in dose reconstructions and the belief, based on these findings, that upper bounds of possible doses to atomic veterans are often underestimated substantially, the central question is: What are the implications of the committee's findings with respect to claims for compensation for nonpresumptive diseases? That is, if credible upper bounds of dose had been obtained in all dose reconstructions, what would have been the effect on the number of claims granted for nonpresumptive diseases?

VI.F.2 Number of Claims Granted for Nonpresumptive Diseases

Before addressing the question posed above, we should consider the issue of the number of claims for compensation for nonpresumptive diseases that have been granted. As of October 2001, about 14,000 claims had been filed for presumptive (38 *CFR* 3.309) and nonpresumptive (38 *CFR* 3.311) diseases, including about 4,400 claims from veterans who served in occupation forces in Japan (VA, 2001). On the basis of information provided by VA (1996), the atomic veterans believed that only about 50 claims had been awarded for nonpresumptive diseases.

If the estimate of about 50 claims granted for nonpresumptive diseases is correct, the odds of a successful claim indeed appear to be very low (about 1% or less).[14] However, in response to a written inquiry, the committee was informed that about 1,600 claims for compensation for nonpresumptive diseases had been granted, including about 350 claims by members of occupation forces in Japan (Flohr, 2001). If that estimate is correct, the odds of a successful claim would be much higher (perhaps 15-20%).

In an effort to resolve this important issue, which is central to the veterans' concerns about the value of dose reconstruction in their compensation program, the committee examined the outcomes of about 300 claims for compensation, some of which involved presumptive diseases, and a list of the outcomes of recent medical opinions by VA on claims for nonpresumptive diseases. The committee's investigation indicated that the proportion of successful claims for nonpresumptive diseases has been around 1% or less, excluding awards for skin cancer since 1998.[15] Therefore, statements that the number of successful claims for nonpresumptive diseases other than skin cancers since the regulations in 38 *CFR* 3.311 were promulgated is on the order of 50, as previously reported and as believed by the

---

[14]The rate of success of claims for nonpresumptive diseases cannot be estimated more precisely because VA does not maintain a database that gives the breakdown of successful and unsuccessful claims for presumptive and nonpresumptive diseases separately.

[15]Few, if any, claims for skin cancer were granted before 1998, but the number of claims granted since 1998 is such that the current rate of granting claims for all nonpresumptive diseases may be nearly 10%.

veterans, indeed appear to be reasonable. A report by GAO (2000) also indicated that the number of claims granted for nonpresumptive diseases is low.

VI.F.3 Implications of Findings for Past Compensation Decisions

As discussed in detail in Chapter V, the committee has concluded that there are many deficiencies in the methods of dose reconstruction being used in the NTPR program. However, the implications of the deficiencies with regard to evaluating claims for compensation for nonpresumptive diseases are difficult to assess. In many cases, it is unlikely that possible underestimation of upper bounds of doses to atomic veterans would have affected a decision about compensation, even though the degree of underestimation of upper bounds could be substantial.

As an example, consider the potential importance of deficiencies in the methods of estimating inhalation doses that were identified by the committee (see Section V.C). The committee's belief that underestimation of upper bounds of inhalation doses probably would not have affected a decision on compensation in many cases is based on two considerations. First, the lowest doses required to grant a claim for compensation can be high, depending on the cancer of concern and the participant's age at the time of exposure and time of diagnosis (see Section III.E). For example, if a veteran was a smoker and filed a claim for lung cancer, doses to the lung of greater than 25 rem normally were required on the basis of screening doses that have been used since the late 1980s (CIRRPC, 1988); doses of about 10 rem or higher were required for several other cancer types, including colon cancer (lung and colon cancers are common in veterans' claims). All screening doses were developed on the basis of an assumption that the risk of a specific cancer is a linear function of dose, without threshold. Second, difficulties in estimating inhalation doses notwithstanding, inhalation doses received by many participants almost certainly were low, judging by known conditions of exposure, and often were much lower than external doses that were monitored with film badges or field instruments. Similarly, the committee believes that external doses to most participants were sufficiently low that a re-evaluation of external doses based on the committee's findings about deficiencies in methods of estimating credible upper bounds probably would not result in estimates that exceed the screening doses used in evaluating claims.

Thus, for most veterans who filed claims for nonpresumptive diseases other than skin cancer, the committee believes that there is little chance that a reassessment of external and inhalation doses would affect a past compensation decision. That conclusion applies, for example, to participants for whom a credible upper bound of external dose is less than 1 rem and who were exposed only to descending fallout, resuspended neutron-activation products in soil, or deposited fallout that was resuspended by walking or other activities that did not involve vigorous disturbance of surface soil. In such cases, upper bounds of inhalation doses probably were less than 1 rem. Therefore, a credible upper bound of dose, taking into account uncertainty in estimates of external and internal doses, is unlikely to approach a screening dose that would qualify a veteran for compensation.

However, the committee also believes that there almost certainly have been cases in which a veteran's claim for compensation for a nonpresumptive disease was denied but would have been granted if a credible exposure scenario had been assumed and uncertainty in estimating dose had been taken into account properly. As an example, consider the committee's concern about neglect of the effects of a blast wave on inhalation exposures in areas of accumulated fallout at the NTS (see Section V.C.3.2, comments [6] and [7], and Appendix E). When exposure to accumulated fallout occurred, potential inhalation doses relative to external doses were substantially higher than in cases of exposure to descending or freshly deposited fallout, mainly because of the increased importance of plutonium relative to photon-emitting radionuclides, and a blast wave produced high concentrations of airborne radionuclides, compared with other causes of resuspension, over a substantial area. That type of exposure scenario occurred at several shots in later test series at the NTS. Furthermore, there are cancers for which the lowest dose required to grant a claim for compensation is not high. For example, claims for compensation for lung cancer in nonsmokers have been granted when the estimated dose to the lung from internal and external exposure combined is as low as 4 rem (Otchin, 2002; see also Section III.E). The committee's analysis of potential inhalation doses due to resuspension caused by a blast wave in areas at the NTS where fallout had accumulated throughout much of the period of atomic testing (see Appendix E) indicates that upper-bound estimates of lung doses of that magnitude are credible when uncertainties in data, models, and parameter values are taken into account.

A similar conclusion may apply in other scenarios at the NTS in which inhalation exposures resulted from resuspension of relatively large amounts of deposited fallout by vigorous disturbances of surface soil. An example is exposure during assaults or marches behind armored vehicles (see case #88). In this scenario, the resuspension factor normally assumed in dose reconstructions (see Table IV.C.2) probably is a reasonable upper bound, but consideration of the accumulation of deposited fallout throughout the period of atomic testing, fractionation of radionuclides, and uncertainty in inhalation dose coefficients could result in increases of more than a factor of 10 in credible upper bounds of inhalation dose compared with estimates obtained in dose reconstructions. Depending on the organ or tissue in which a veteran's cancer occurred, it is possible that credible upper bounds of the total dose from external and internal exposure could qualify the veteran for compensation in these cases.

The committee has not attempted to estimate the number of past claims for compensation for nonpresumptive diseases that could be affected by its findings on deficiencies in the methods of dose reconstruction; such an effort is beyond the scope of the committee's work. However, although the committee believes that most past claims would not be affected by a reassessment of doses on the basis of the findings described in this report, it believes that some claims that were not granted would have been.

VI.F.4 Implications of Findings for Future Compensation Decisions

The implications of the committee's findings on deficiencies in methods of dose reconstruction used in the NTPR program with regard to the number of claims for nonpresumptive diseases that might be granted in the future are similar to those described above. If methods of dose reconstruction were changed to be consistent with these findings, the committee expects that the outcome of most future claims would not be affected. That expectation is based on the presumption that the distribution of doses in future claims will be similar to the distribution in past claims and, therefore, that credible upper bounds of dose to most claimants would be too low for the VA to conclude that the veteran's disease was as least as likely as not caused by his radiation exposure and thus qualify the veteran for compensation.

Two additional factors that have not been important in the past may affect future claims for compensation. First, the list of presumptive diseases has been expanded to include 21 cancer types (see Section I.B.4). Of the cancer types for which radiation risks have been estimated from studies of the Japanese atomic-bomb survivors (Thompson et al., 1994), only cancers of the rectum, skin, uterus, prostate, and nervous system are now nonpresumptive; of these, skin and prostatic cancer appear frequently in veterans' claims, but the others apparently are rare. Estimates of the risk per unit dose for these cancers tend to be lower than estimates of risks for many cancers that are presumptive diseases, and this results in higher screening doses that might qualify a veteran for compensation.[16]

Second, VA is updating its methods of evaluating the probability of causation of radiation-related cancers (see Section III.E). Consequently, the lowest dose that would qualify a veteran for compensation will increase for many cancers. Such increases are a direct result of improvements in the data on cancer risks in humans and attendant decreases in uncertainty in cancer risk estimates (that is, decreases in credible upper bounds of cancer risks per unit dose).

Thus, if methods of dose reconstruction used by the NTPR program are changed to be consistent with the committee's findings, the effect on future claims for compensation for nonpresumptive diseases is not likely to be substantial. However, the importance of the committee's findings could increase when veterans file claims for compensation for presumptive diseases. In such cases, requirements for establishing a veteran's status as a participant are more demand-

---

[16]Expansion of the list of presumptive diseases means, of course, that it is much easier for many veterans whose claims for compensation for nonpresumptive diseases were denied in the past to be compensated now under the presumptive law.

ing, and if adequate proof of participation is not available, claims are evaluated under the nonpresumptive regulation and a dose reconstruction is required. The committee does not know how often claims for compensation for presumptive diseases have had to be evaluated under the nonpresumptive regulation, but the number of such cases could increase in the future, given the expansion of the list of presumptive diseases, thus increasing the importance of changes in methods of dose reconstruction.

VII Conclusions

In establishing the program of compensation for atomic veterans, Congress intended to create a program that was responsive, was based on sound science, and treated the veterans fairly and respectfully. The committee recognizes the challenge confronting the Defense Threat Reduction Agency and the Department of Veterans Affairs associated with reconstructing historical doses and making decisions about compensation to thousands of veterans who were exposed decades ago, using records and data that are incomplete and often difficult to piece together. And the committee recognizes that many improvements have been made in the NTPR program since its beginning. The committee has come to understand the frustrations of the veterans who willingly performed their duties under extraordinary circumstances and who are confronted with the burden of seeking compensation for diseases that they believe are related to the service they performed for their country.

The committee has undertaken its work fully cognizant of controversies associated with this important program since its inception. Therefore, it is to be expected that the committee, as discussed in Chapter II, has viewed its scope of work as somewhat broader than that specified in its charge. In addition to responding directly to the questions presented in the statement of task, the committee presents some conclusions related to other aspects of the dose reconstruction program that it hopes will respond to additional questions that have been raised about the atomic-veterans compensation program for many years. It should be understood that the committee's conclusions about the adequacy of the dose reconstruction program for atomic veterans and other findings represent consensus judgments that were developed on the basis of the preponderance of information available to the committee. Similar judgments are required in performing dose reconstructions. Since it is not possible to determine exactly what happened to the veterans, the committee's view is that the goal of dose reconstruction is to develop plausible assumptions that yield credible upper-bound estimates of dose, consistent with the requirement to give the veterans the benefit of the doubt.

The committee's conclusions are divided into two groups: those answering the four questions posed in the statement of task; and those related to the establishment of continuing review and oversight of the program. Finally, the committee offers some additional explanation about the implications of its findings.

VII.A RESPONSES TO QUESTIONS IN COMMITTEE'S CHARGE

Chapter II describes the committee's difficulty in understanding the intent of some of the questions that were presented in its charge and explains its interpretation of these questions. The questions in the statement of task, the committee's interpretation of the questions, and responses to the questions are presented below.

Question 1. Is the reconstruction of the sample(d) doses accurate? {Because dose reconstruction is inherently uncertain, the committee interprets this question to be whether uncertainty in the sampled doses has been appropriately considered and whether credible upper bounds of doses to atomic veterans have been obtained.}

According to the regulations and the objectives of the NTPR program, the goal is to report at least the 95th percentile upper bound of possible doses for each veteran. The committee has concluded, however, that upper-bound doses from external gamma, neutron, and beta exposure are often underestimated, sometimes considerably, particularly when doses are reconstructed as opposed to being based on film-badge data. A number of findings led the committee to that conclusion. Some of these are described below.

Methods used by the NTPR program to estimate average doses to participants in various military units from external exposure to photons (mainly gamma rays) and neutrons are generally valid. However, because the specific exposure conditions for any individual often are not well known and the available measurements used as input to calculation models are sparse and highly variable, the resulting estimates of total dose for many participants are highly uncertain.

Film-badge data, if available, are considered the dose data of record. The dose inferred from a film-badge reading is estimated by using "high-sided" assumptions. However, in some cases, even film-badge data are more uncertain than reflected by the corresponding upper-bound estimates.

Although it is difficult to define the degree of underestimation of credible upper bounds of reconstructed external gamma doses, the committee has concluded that a credible upper bound often could be 2-3 times the central estimate. In contrast, upper bounds reported by the NTPR program often were only 10-20% above the central estimates. The committee also has concluded that upper bounds of neutron doses are always underestimated, because of neglect of uncertainty in the biological effectiveness of neutrons relative to gamma rays.

Beta doses to skin are claimed to be "high-sided" because they are based on multiplying an upper-bound gamma dose by a presumed "high-sided" beta-to-gamma dose ratio. However, the upper-bound gamma dose based on a reconstruction is often too low, and the beta-to-gamma dose ratio is not evidently "high-sided" in all cases. In addition, it appears that estimates of beta dose to skin do not include the dose due to contamination of the skin or clothing.

The committee also has concluded that upper bounds of inhalation doses are underestimated in many cases. Estimation of internal dose—most important, the dose from inhalation—is an inherently more difficult problem than estimation of external dose because data that could be used to estimate intakes of radionuclides by the atomic veterans are not available. Given the lack of relevant data, the NTPR program relies on assumptions that are presumed to result in overestimates of concentrations of radionuclides in air, especially assumptions about resuspension factors.

The committee has identified many problems with the methods of estimating inhalation doses to atomic veterans. They center around three issues: the unknown reliability of methods of estimating airborne concentrations of radionuclides used by the NTPR program, including the assumption of no fractionation of radionuclides in fallout except for removal of noble gases; the lack of consideration of resuspension of previously deposited fallout by the blast wave produced in detonations at the NTS and the frequent neglect of aged fallout that accumulated during the period of atomic testing at the NTS; and the lack of consideration of uncertainty in inhalation dose coefficients for all radionuclides.

In spite of problems with the methods used in the NTPR program to estimate inhalation doses, inhalation doses assigned to many atomic veterans are probably "high-sided" and exceed the 95th percentile goal. However, there are important scenarios involving maneuver troops and close-in observers at the NTS in which credible upper bounds of inhalation doses would exceed the dose estimated by the NTPR program by large factors. Furthermore, in scenarios in which inhalation doses almost certainly were underestimated by large factors, credible upper bounds of organ equivalent doses could be important in some cases.

Thus, the committee has concluded that the methods that have been used to estimate inhalation doses to atomic veterans do not consistently provide credible upper bounds of possible doses and that this could be an important deficiency in some exposure scenarios.

The possibility of ingestion exposures apparently is not considered routinely in dose reconstructions for atomic veterans. However, except in rare situations, neglect of ingestion exposures does not have important consequences with regard to estimating credible upper bounds of total doses to the veterans.

The committee has concluded that veterans are not always given the benefit of doubt in developing exposure scenarios and assessing film-badge data. Veterans often were not contacted to verify their exposure scenario, even when such contact was feasible and could have been helpful. In some cases, there was inadequate follow-up with other participants who might have been able to clarify scenario assumptions.

In many cases, considerable judgment had to be used in developing exposure scenarios. Nonetheless, applicable regulations are quite clear that a veteran must be given the benefit of the doubt, which would lead to a higher dose, when there is a question regarding his exposure scenario. Although application of benefit of the doubt would not affect doses in all cases in our random sample, the committee found it to be a frequent problem.

Thus, the committee has concluded that upper bounds of total doses reported by the NTPR program have often been underestimated and therefore do not provide credible upper bounds (95th percentiles) of possible doses.

Question 2. Are the reconstructed doses accurately reported? {The committee interprets this question to be whether the doses that are calculated (regardless of their validity) are being reported accurately to the Department of Veterans Affairs.}

On the basis of its review of many case files, the committee has concluded that doses, as they have been calculated by the NTPR program, have been accurately reported to the VA, and to the veterans. However, the committee believes that uncertainty in assigned doses should be carefully explained and reported to the VA when they are used to evaluate claims for compensation and should be explained to the veterans.

Question 3. Are the assumptions made regarding radiation exposure based on the sampled doses credible? {The committee interprets this question to be whether the assumptions made to define the veterans' exposure scenarios and the methods and parameters used in dose reconstruction are reasonable and appropriate.}

This question is the most difficult of the four to answer. The committee has concluded that many assumptions regarding veterans' exposures during atmospheric nuclear-weapons tests are not reasonable and appropriate, given the objective of the NTPR program to estimate credible upper bounds of dose. A large number of separate assumptions are typically required to derive an estimate of dose for most veterans who were exposed. Many of the assumptions being used are indeed reasonable and based on current understanding of the science of historical dose reconstruction. Nevertheless, many key assumptions and methods being used are not appropriate and could lead to underestimation of upper bounds of doses to atomic veterans.

The committee's evaluation of this question was severely hampered by a lack of quality control over the conduct and documentation of dose reconstructions, which made it difficult to determine how doses were calculated in many cases. In some cases, documentation of a dose reconstruction is illegible. The lack of comprehensive quality control of dose reconstructions that have been performed diminishes the credibility of the work and has made it difficult for the committee to conduct its review.

Providing the veteran the benefit of the doubt when making assumptions about exposure scenarios and estimating dose is critical to implementation of applicable regulations. It is evident to the committee that there has not been a consistent application of the requirement to give the benefit of the doubt to atomic veterans. The inconsistency affected assumptions made about exposure scenarios and yielded upper-bound doses that clearly were too low in a number of cases.

The committee thus has concluded that in many of the cases reviewed, key assumptions about input values and exposure scenarios were not reasonable and appropriate and that this led to reported upper bounds of external and internal doses that fall short of the 95th percentile goal.

Question 4. Are the data from nuclear weapons tests used by DTRA as part of the reconstruction of sampled doses accurate? {The committee interprets this question to be whether the historical data and uncertainty in the data have been comprehensively compiled and are suitable for use in historical dose reconstruction.}

The committee believes that historical records provide sufficient data to permit doses to be reconstructed for atomic veterans. There is a large repository of information from which to draw data about exposures. In addition, the committee believes that the veterans themselves are a valuable source of information about their own exposures. Although some attempts have been made to contact veterans and seek their input about scenarios of exposure, this source of information seems to be underused.

The science of historical dose reconstruction has evolved over the last few decades. In some situations, historical doses have been estimated on the basis of less information than appears to be available for the atomic veterans. The necessary background information is available on which to base the atomic veterans' dose reconstructions.

All in all, the committee was impressed with the large amount of information that has been brought together. The committee has concluded that the radiological and historical information compiled by the NTPR program is suitable and sufficient for use in historical dose reconstruction for the atomic veterans.

VII.B RECOMMENDATIONS REGARDING A SYSTEM
FOR PERMANENT REVIEW OF THE
DOSE RECONSTRUCTION PROGRAM

The final charge in the statement of task asks the committee to provide recommendations, if appropriate, regarding a permanent system of review of the dose reconstruction program of DTRA.

Before discussing the issue of review, we should note that about 70% of all dose reconstructions have been performed in response to veterans' claims for compensation for nonpresumptive diseases. Furthermore, many of the diseases for which doses have been reconstructed are now included in the presumptive regulation. With the exception of reconstructions of doses to skin, including doses from exposure to beta particles, in response to skin-cancer claims, it is clear to the committee that in most cases, even revised upper-bound dose estimates would be too low to conclude that the veteran's disease was at least as likely as not caused by his radiation exposure and thus to justify awarding claims.

If the program of dose reconstruction continues, the committee believes that an external and independent system of review and oversight is needed. The degree of review and oversight should be commensurate with the anticipated scope of the compensation program in the future. Although the responsibility for a permanent system of review rests with DTRA and VA, the committee provides some guidelines below that may be helpful in its design and implementation.

- One approach to continuing review and oversight among possible alternatives is to create an advisory board. The board should consist of persons who can evaluate the many facets of the program, such as historical dose reconstruction, radiation risk and probability of causation, communication with the veterans and between VA and DTRA, quality assurance, and historical research related to service experience. The advisory board should include at least one representative of the atomic veterans.
- In addition to review and oversight of the dose reconstruction program of DTRA, review and oversight of the program as a whole, including the responsibilities of DTRA and VA in the administration of the atomic veterans' program, is desirable.
- The advisory board should meet frequently enough to understand the program fully, to conduct random audits of doses being reconstructed and decisions regarding claims, to review methods, and to recommend changes when needed.
- The advisory board should meet with atomic veterans regularly, listen to their concerns, and ensure that their concerns are addressed. The board should also help DTRA and VA in efforts to provide information to veterans that effectively communicates the program's mission and process and the science related to possible health effects of radiation exposures of atomic veterans.

VII.C EXPLANATION TO ATOMIC VETERANS REGARDING IMPLICATIONS OF COMMITTEE'S FINDINGS

The committee is sympathetic to many of the atomic veterans' concerns and frustrations. Furthermore, although the number is probably small, the committee believes that some veterans would have been compensated if more-credible upper bounds of dose had been estimated in dose reconstructions.

The committee also believes, as has been reported to and stated by the atomic veterans, that the total number of claims awarded under the nonpresumptive regulation since its promulgation is small (on the order of 50), excluding recent awards for skin cancer. That illustrates clearly that when a veteran files a claim for a disease under the nonpresumptive regulation, the probability is very low that an award will be granted. The committee is concerned that this small chance of success has not been clearly reported in the past.

The committee also believes, however, that it is important for the veterans to understand that there are legitimate reasons for the low number of successful claims for nonpresumptive diseases, and that these reasons are unrelated to deficiencies in the methods of dose reconstruction used in the NTPR program. On the basis mainly of data obtained from studies of the Japanese atomic-bomb survivors, it is evident that ionizing radiation is not a potent cause of cancer. That is indicated, for example, by the small number of excess cancers that have been observed in the atomic-bomb survivors, even though many in this population received doses much higher than the doses received by nearly all atomic veterans. That conclusion is also indicated by screening doses given in Table III.E.4 (see Section III.E) that are based on the current IREP method of calculating probability of causation of cancers. The screening doses, which correspond to a 99% confidence limit in an estimated probability of causation of 50% and are based on an assumption that cancer risk is a linear function of dose, without threshold, are 10 rem or greater for most cancers listed in the table; this indicates that high doses are required to give an appreciable probability of causation of cancer. For many cancers, the screening doses that have been used by VA to evaluate claims for compensation also are 10 rem or greater (see Tables III.E.3 and III.E.4). Furthermore, on the basis of what is known about conditions of exposure of atomic veterans, credible upper bounds of doses received by most veterans almost certainly would be so low that the probability that a cancer was due to radiation exposure in the atomic-testing program is small.

The committee notes that the established policy of using upper-bound estimates of dose (95th percentiles) with the more extreme lower-bound estimates of doses that correspond to a 50% probability of causation of various cancers is highly favorable to the veterans' interests. If credible upper bounds of dose are obtained in dose reconstructions, atomic veterans can be compensated for nonpresumptive diseases even when the true probability that radiation exposure caused the diseases is low.

None of that is to say that the veterans do not have legitimate complaints about their dose reconstructions; in many cases, the committee believes they do. Rather, the committee hopes that veterans will understand that their radiation exposure probably did not cause their cancers in most cases and that reasonable changes in methods of dose reconstruction in response to this report are not likely to greatly increase their chance of a successful claim for compensation when a dose reconstruction is required.

VIII Recommendations

The committee offers a number of recommendations that it believes would, if implemented, improve the dose reconstruction process and the atomic-veterans compensation program in general. Some have been mentioned previously; they are summarized here to provide a complete list.

1. There should be continuing external review and oversight of the atomic-veterans dose reconstruction and compensation programs. An independent advisory board could be established to implement this recommendation.

2. There should be a comprehensive re-evaluation of the methods being used to estimate doses and their uncertainties to establish more credible upper bounds of doses to atomic veterans.

3. A comprehensive manual of standard operating procedures should be developed and maintained.

4. A state-of-the-art program of quality assurance and quality control should be developed and implemented.

5. The principle of benefit of the doubt should be consistently applied in accordance with applicable regulations.

6. Interaction and communication with the atomic veterans should be improved. For example, veterans should be allowed to review the scenario assumptions used in their dose reconstructions before the dose assessments are sent to the Department of Veterans Affairs for claim adjudication.

7. More effective approaches should be established to communicate the meaning of information on radiation risk to the veterans. In addition to presenting general information on radiation risk, information should be communicated to

veterans who file claims regarding the significance of their doses in relation to their diseases.

8. The community of atomic veterans and their survivors should be advised when methods for calculating doses have changed so that they can ask for updated dose assessments.

References

Anspaugh, L. R., Shinn, J. H., Phelps, P. L., Kennedy, N. C. 1975. Resuspension and Redistribution of Plutonium in Soils. Health Physics 29:571-582.

Anspaugh, L. R., Church, B. W. 1986. Historical Estimates of External Gamma Exposure and Collective External Gamma Exposure from Testing at the Nevada Test Site. I. Test Series Through HARDTACK II, 1958. Health Physics 51:35-51.

Anspaugh, L. R., Ricker, Y. E., Black, S. C., Grossman, R. F., Wheeler, D. L., Church, B. W., Quinn, V. E. 1990. Historical Estimates of External γ Exposure and Collective External γ Exposure from Testing at the Nevada Test Site. II. Test Series after HARDTACK II, 1958, and Summary. Health Physics 59:525-532.

Apostoaei, A. I., Burns, R. E., Hoffman, F. O., Ijaz, T., Lewis, C. J., Nair, S. K., Widner, T. E. 1999a. Releases from Radioactive Lanthanum Processing at the X-10 Site in Oak Ridge, Tennessee (1944-1956)—An Assessment of Quantities Released, Off-Site Radiation Doses, and Potential Excess Risks of Thyroid Cancer. Nashville, TN: Tennessee Department of Health; Reports of the Oak Ridge Dose Reconstruction, Vol. 1.

Apostoaei, A. I., Blaylock, B. G., Caldwell, B., Flack, S., Gouge, J. H., Hoffman, F. O., Lewis, C. J., Nair, S. K., Reed, E. W., Thiessen, K. M., Thomas, B. A., Widner, T. E. 1999b. Radionuclides Released to the Clinch River from White Oak Creek on the Oak Ridge Reservation – An Assessment of Historical Quantities Released, Off-Site Radiation Doses, and Health Risks. Nashville, TN: Tennessee Department of Health; Reports of the Oak Ridge Dose Reconstruction, Vol. 4.

B2 Report. 1946. Radiological Situation on Target Ships. Appendix VII in Nuclear Radiation Effects in Test Able and Baker—Preliminary Report of. Available from Nuclear Test Perssonel Review Program, Defense Threat Reduction Agency; Fort Belvoir, VA.

Banks, J. E., Dick, J. L., Pinson, E. A. 1959. Contact Radiation Hazard Associated With Aircraft Contamination by Early Cloud Penetrations (U). Springfield, VA: National Technical Information Service; Air Force Special Weapons Center Report WT-1368.

Barrett, M., Goetz, J., Klemm, J., McRaney, W., Phillips, J. 1986. Low Level Dose Screen—CONUS Tests. McLean, VA: Science Applications International Corporation; Report DNA-TR-85-317.

Barrett, M., Goetz, J., Klemm, J., Ortlieb, E., Thomas, C. 1987. Analysis of Radiation Exposure for Military Participants, Exercises Desert Rock I, II, and III—Operation Buster-Jangle. McLean, VA: Science Applications, Inc.; Report DNA-TR-87-116.

Barss, N. M. 2000. Methods and Applications for Dose Assessment of Beta Particle Radiation. McLean, VA: Science Applications International Corporation; Report SAIC-001/2024.

Beck, H. L., Bennett, B. G. 2002. Historical Overview of Atmospheric Nuclear Weapons Testing and Estimates of Fallout in the Continental United States. Health Physics 82: 591-608.

Black, R. H. 1962. Some Factors Influencing the Beta-Dosage to Troops. Health Physics 8:131-141.

Bouville, A., Eckerman, K., Griffith, W., Hoffman, O., Leggett, R., Stubbs, J. 1994. Evaluating the Reliability of Biokinetic and Dosimetric Models and Parameters Used to Assess Individual Doses for Risk Assessment Purposes. Radiation Protection Dosimetry 53:211-215.

Brady, W. J., Nelson, A. G. 1985. Radiac Instruments and Film Badges used at Atmospheric Nuclear Tests. Alexandria, VA: JAYCOR; Report DNA-TR-84-338.

Brorby, G. P., diTommaso, D., Ting, D., Buddenbaum, J. E., Lee, K. H., Widner, T. E. 1994. Dose Assessment for Historical Contaminant Releases from Rocky Flats. Health Studies on Rocky Flats: Phase I: Historical Public Exposures. Denver, CO: Colorado Department of Public Health and Environment; Project Task 8 Report.

Bruce-Henderson, S., Gladeck, F., Hallwell, J., Martin, E., McMullan, F., Miller, F., Rogers, W., Rowland, R., Shelton, C. 1982. Operation REDWING: 1956. Washington, DC: Defense Nuclear Agency. DNA 6037F.

Caldwell, C. G., Kelly, D. B., Heath, C. W. 1980. Leukemia Among Participants in Military Maneuvers at a Nuclear Bomb Test. A Preliminary Report. JAMA 244:1575-1578.

Caldwell, C. G., Kelley, D., Zack, M., Falk, H., Heath, C. W., Jr. 1983. Mortality and Cancer Frequency Among Military Nuclear Test (Smoky) Participants, 1957 through 1979. JAMA 250:620-624.

Cederwall, R. T., Ricker, Y. E., Cederwall, P. L., Homan, D. N., Anspaugh, L. R. 1990. Ground-Based Air-Sampling Measurements Near the Nevada Test Site after Atmospheric Nuclear Tests. Health Physics 59:533-540.

CIRRPC (Committee on Interagency Radiation Research and Policy Coordination). 1988. Use of Probability of Causation by the Veterans Administration in the Adjudication of Claims of Injury Due to Exposure to Ionizing Radiation. Washington, DC: Office of Science and Technology Policy Report ORAU 88/F-4.

Decisioneering. 2001. Crystal Ball® 2000.2 user manual. Denver, CO: Decisioneering, Inc.

DHHS (U.S. Department of Health and Human Services). 2002. 42 CFR Part 82 – Methods for Radiation Dose Reconstruction under the Energy Employees Occupational Illness Compensation Program Act of 2000; Final Rule. Federal Register 67:22314-22336.

Dick, J. L., Baker, T. P., Jr. 1961. Monitoring and Decontamination Techniques for Plutonium Fallout on Large-Area Surfaces. Albuquerque, NM: Air Force Special Weapons Center; Report WT-1512.

DTRA (Defense Threat Reduction Agency). 1997. NTPR Standard Operating Procedures: Radiation Exposure Assessment. Alexandria, VA: Defense Threat Reduction Agency.

DOE (U.S. Department of Energy). 1988. Internal Dose Conversion Factors for Calculation of Dose to the Public. Washington, DC: U.S. Department of Energy; Report DOE/EH-0071.

DOE (U.S. Department of Energy). 2000. United States Nuclear Tests: July 1945 Through September 1992. Las Vegas, NV: U.S. Department of Energy Nevada Operations Office; Report DOE-NV 209 (Rev. 15).

Donaldson, L. R., Seymour, A. H., Nevissi, A. E. 1997. University of Washington's Radioecological Studies in the Marshall Islands, 1946-1977. Health Physics 73:214-222.

Dunning, D. E., Jr., Bernard, S. R., Walsh, P. J., Killough, G. G., Pleasant, J. C. 1979. Estimates of Internal Dose Equivalent to 22 Target Organs for Radionuclides Occurring in Routine Releases from Nuclear Fuel-Cycle Facilities, Vol. II. Springfield, VA: National Technical Information Service; Oak Ridge National Laboratory Report NUREG/CR-0150, Vol. 2, ORNL/NUREG/TM-190/V2.

Dunning, D. E., Jr., Schwarz, G. 1981. Variability of Human Thyroid Characteristics and Estimates of Dose from Ingested ^{131}I. Health Physics 40:661-675.

Durham, J. S. 1992. VARSKIN MOD2 and SADDE MOD2: Computer Codes for Assessing Skin Dose from Skin Contamination. Springfield, VA: National Technical Information Service; Battelle Pacific Northwest Laboratory Report NUREG/CR-5873, PNL-7913.

Eckerman, K. F., Wolbarst, A. B., Richardson, A. C. B. 1988. Limiting Values of Radionuclide Intake and Air Concentration and Dose Conversion Factors for Inhalation, Submersion, and Ingestion. Federal Guidance Report No. 11. Washington, DC: U.S. Environmental Protection Agency; Report EPA-520/88-020.

Eckerman, K. F., Leggett, R. W., Nelson, C. B., Puskin, J. S., Richardson, A. C. B. 1999. Cancer Risk Coefficients for Environmental Exposure to Radionuclides. Federal Guidance Report No. 13. Washington, DC: U.S. Environmental Protection Agency; Report 402-R-99-001.

Edwards, R., Goetz, J., Klemm, J. 1983. Analysis of Radiation Exposure, Task Force RAZOR, Exercise Desert Rock VI, Operation TEAPOT. McLean, VA: Science Applications, Inc.; Report DNA-TR-83-07.

Edwards, R., Goetz, J., Klemm, J. 1985. Analysis of Radiation Exposure for Maneuver Units, Exercise Desert Rock V, Operation Upshot-Knothole. McLean, VA: Science Applications International Corporation; Report DNA-TR-84-303.

Edwards, A. A. 1997. Relative Biological Effectiveness of Neutrons for Stochastic Effects, Documents of the NRPB 8, No. 2. Chilton, U.K.: National Radiological Protection Board.

Edwards, A. A. 1999. Neutron RBE Values and Their Relationship to Judgments in Radiological Protection. Journal of Radiological Protection 19:93-105.

Egbert, S. D., Kaul, D. C., Klemm, J., Phillips, J. C. 1985. FIIDOS – A Computer Code for the Computation of Fallout Inhalation and Ingestion Dose to Organs, Computer User's Guide. McLean, VA: Science Applications International Corporation; Report DNA-TR-84-375.

EPA (U.S. Environmental Protection Agency). 1997. Exposure Factors Handbook, Volume 1. General Factors. Washington, DC: U.S. Environmental Protection Agency; Report EPA/600/P-95/002Fa.

EPA (U.S. Environmental Protection Agency). 1999. Estimating Radiogenic Cancer Risks. Addendum: Uncertainty Analysis. Washington, DC: U.S. Environmental Protection Agency; Report EPA 402-R-99-003.

EPA (U.S. Environmental Protection Agency). 2001. Guidance for Preparing Standard Operating Procedures (SOPs). Washington, DC: U.S. Environmental Protection Agency; Report EPA QA/G-6.

Farris, W. T., Napier, B. A., Eslinger, P. W., Ikenberry, T. A., Shipler, D. B., Simpson, J. C. 1994a. Atmospheric Pathway Dosimetry Report, 1944-1992. Richland, WA: Battelle Pacific Northwest Laboratories; Report PNWD-2228 HEDR.

Farris, W. T., Napier, B. A., Simpson, J. C., Snyder, S. F., Shipler, D. B. 1994b. Columbia River Pathway Dosimetry Report, 1944-1992. Richland, WA: Battelle Pacific Northwest Laboratories; Report PNWD-2227 HEDR.

Finn, S. P., Simmons, G. L., Spencer, L. V. 1979. Calculation of Fission Product Gamma Ray and Beta Spectra at Selected Times After Fast Fission of U^{238} and U^{235} and Thermal Fission of U^{235}. McLean, VA: Science Applications, Inc.; Report SAI-78-782-LJ/F.

Flohr, B. 2001. Letter from U.S. Department of Veterans Affairs to J. E. Till, Chairman Committee to Review Dose Recoonstruction Program of DTRA (October 31).

Flor, W. J. 1992. Policy and Guidance for the Reporting of External Radiation Doses. NTPR Policy Note 02-92, Memorandum for Distribution.

Frank, G. 1982. Radiation Dose for Personnel of the 412th Engineer Construction Battalion, Exercise Desert Rock V, Operation Upshot-Knothole. McLean, VA: Science Application, Inc.; Memorandum to file.

Frank, G., Weitz, R., Goetz, J., Klemm, J., Schweizer, T. 1982. Analysis of Radiation Exposure, 2nd Marine Corps Provisional Atomic Exercise Brigade, Exercise Desert Rock V, Operation Upshot-Knothole. McLean, VA: Science Applications, Inc.; Report DNA-TR-82-03.

Freiling, E. C., Crocker, G. R., Adams, C. E. 1964. Nuclear-Debris Formation. In: Radioactive Fallout from Nuclear Weapons Tests, ed. by Klement, A. W., Jr. Germantown, MD: U.S. Atomic Energy Commission; pp. 1-43.

GAO (General Accounting Office). 1987. Nuclear Health and Safety: Radiation Exposures for Some Cloud-Sampling Personnel Need to Be Reexamined. Washington, DC: U.S. General Accounting Office.

GAO (General Accounting Office). 2000. Veteran's Benefits: Independent Review Could Improve Credibility of Radiation Exposure Estimates. Washington, DC: U.S. General Accounting Office.

Glasstone, S., Dolan, P. J. (Eds). 1977. The Effects of Nuclear Weapons, 3rd edition. Washington, DC: U.S. Government Printing Office.

Goetz, J. L., Kaul, D., Klemm, J., McGahan, J. T. 1979. Analysis of Radiation Exposure for Task Force Warrior-Shot SMOKY-Exercise Desert Rock VII-VIII, Operation PLUMBBOB. McLean, VA: Science Applications, Inc.; Report DNA 4747F.

Goetz, J. L., Kaul, D., Klemm, J., McGahon, J. T., McRaney, W. K. 1980. Analysis of Radiation Exposure for Task Force Big Bang, Shot Galileo. Exercise Desert Rock VII-VIII, Operation PLUMBBOB. McLean, VA: Science Applications, Inc.; Report DNA 4772F.

Goetz, J. L., Kaul, D., Klemm, J., McGahan, J., Weitz, R. 1981. Analysis of Radiation Exposure for Troop Observers, Exercise Desert Rock V, Operation UPSHOT-KNOTHOLE. McLean, VA: Science Applications, Inc.; Report DNA 5742F.

Goetz, J., Klemm, J., Thomas, C., Weitz, R. 1985. Neutron Exposure for DOD (Department of Defense) Nuclear Test Personnel. McLean, VA: Science Applications International Corporation; Report DNA-TR-84-405.

Goetz, J., Klemm, J., Phillips, J., Thomas, C. 1987. Analysis of Radiation Exposure – Service Personnel on Rongerik Atoll, Operation CASTLE – Shot BRAVO. McLean, VA: Science Applications International Corporation; Report DNA-TR-86-120.

Goetz, J., Klemm, J., McRaney, W., Barrett, M. 1991. Low Level Internal Dose Screen – Oceanic Tests. McLean, VA: Science Applications International Corporation; Report DNA-TR-88-260.

Grogan, H. A., McGavran, P. A., Meyer, K. R., Meyer, H. R., Mohler, J., Rood, A. S., Sinclair, W. K., Voilleque, P. G., Weber, J. M. 1999. Technical Summary Report of the Historical Public Exposures Studies for Rocky Flats, Phase II. Denver, CO: Colorado Department of Public Health and Environment; Radiological Assessments Corporation Report No. 14-CDPHE-RFP-1999-Final.

Hawthorne, H. A. (Ed.). 1979. Compilation of Local Fallout Data from Test Detonations 1945-1962 Extracted from DASA 1251. Volume I. Continental U.S. Tests. Santa Barbara, CA: General Electric Company; Report DNA 1251-1-FX.

Henderson, R. W., Smale, R. F. 1990. External Exposure Estimates for Individuals Near the Nevada Test Site. Health Physics 59:715-723.

Hicks, H. G. 1982. Calculation of the Concentration of Any Radionuclide Deposited on the Ground by Offsite Fallout from a Nuclear Detonation. Health Physics 42:585-600.

Hoffman, J. G. 1947. Nuclear Explosion 16 July 1945: Health Physics Report on Radioactive Contamination Throughout New Mexico Following the Nuclear Explosion; Part A—Physics and Part C—Transcripts of Radiation Monitor's [sic] Field Notes; Film Badge Data on Town Monitoring. Los Alamos, NM: Los Alamos Scientific Laboratory; Report LA-626.

IREP (Interactive RadioEpidemiological Program). 2000. Available at the NIOSH Web site at http://198.144.166.5/irep_niosh/.

ICRP (International Commission on Radiological Protection). 1975. Task Group Report on Reference Man. Oxford, U.K.: Pergamon Press.

ICRP (International Commission on Radiological Protection). 1977. Recommendations of the International Commission on Radiological Protection. ICRP Publication 26, Annals of the ICRP 1(3). Elmsford, NY: Pergamon Press.

ICRP (International Commission on Radiological Protection). 1979a. Limits for Intakes of Radionuclides by Workers. ICRP Publication 30, Part 1, Annals of the ICRP 2(3/4). Elmsford, NY: Pergamon Press.

ICRP (International Commission on Radiological Protection). 1979b. Limits for Intakes of Radionuclides by Workers. ICRP Publication 30, Supplement to Part 1, Annals of the ICRP 3(1-4). Elmsford, NY: Pergamon Press.

ICRP (International Commission on Radiological Protection). 1980a. Limits for Intakes of Radionuclides by Workers. ICRP Publication 30, Part 2, Annals of the ICRP 4(3/4). Elmsford, NY: Pergamon Press.

ICRP (International Commission on Radiological Protection). 1980b. Biological Effects of Inhaled Radionuclides. ICRP Publication 31, Annals of the ICRP 4(1/2). Elmsford, NY: Pergamon Press.

ICRP (International Commission on Radiological Protection). 1981a. Limits for Intakes of Radionuclides by Workers. ICRP Publication 30, Supplement to Part 2, Annals of the ICRP 5(1-6). Elmsford, NY: Pergamon Press.

ICRP (International Commission on Radiological Protection). 1981b. Limits for Intakes of Radionuclides by Workers. ICRP Publication 30, Part 3, Annals of the ICRP 6(2/3). Elmsford, NY: Pergamon Press.

ICRP (International Commission on Radiological Protection). 1982a. Limits for Intakes of Radionuclides by Workers. ICRP Publication 30, Supplement A to Part 3, Annals of the ICRP 7(1-3). Elmsford, NY: Pergamon Press.

ICRP (International Commission on Radiological Protection). 1982b. Limits for Intakes of Radionuclides by Workers. ICRP Publication 30, Supplement B to Part 3, Annals of the ICRP 8(1-3). Elmsford, NY: Pergamon Press.

ICRP (International Commission on Radiological Protection). 1989. Age-Dependent Doses to Members of the Public from Intake of Radionuclides: Part 1. ICRP Publication 56, Annals of the ICRP 17(2/3). Elmsford, NY: Pergamon Press.

ICRP (International Commission on Radiological Protection). 1991a. 1990 Recommendations of the International Commission on Radiological Protection. ICRP Publication 60, Annals of the ICRP 21(1-3). Elmsford, NY: Pergamon Press.

ICRP (International Commission on Radiological Protection). 1991b. The Biological Basis for Dose Limitation in the Skin. ICRP Publication 59, Annals of the ICRP 22 (2), p. 82. Elmsford, NY: Pergamon Press.

ICRP (International Commission on Radiological Protection). 1993. Age-Dependent Doses to Members of the Public from Intake of Radionuclides: Part 2, Ingestion Dose Coefficients. ICRP Publication 67, Annals of the ICRP 23(3/4). Elmsford, NY: Pergamon Press.

ICRP (International Commission on Radiological Protection). 1994a. Dose Coefficients for Intakes of Radionuclides by Workers. ICRP Publication 68, Annals of the ICRP 24(4). Elmsford, NY: Pergamon Press.

ICRP (International Commission on Radiological Protection). 1994b. Human Respiratory Tract Model for Radiological Protection. ICRP Publication 66, Annals of the ICRP 24(1-3). Elmsford, NY: Pergamon Press.

ICRP (International Commission on Radiological Protection). 1995. Age-Dependent Doses to Members of the Public from Intake of Radionuclides: Part 3, Ingestion Dose Coefficients. ICRP Publication 69, Annals of the ICRP 25(1). Elmsford, NY: Pergamon Press.

ICRP (International Commission on Radiological Protection). 1996a. Age-Dependent Doses to Members of the Public from Intake of Radionuclides: Part 4, Inhalation Dose Coefficients. ICRP Publication 71, Annals of the ICRP 25(3/4). Elmsford, NY: Pergamon Press.

ICRP (International Commission on Radiological Protection). 1996b. Age-Dependent Doses to Members of the Public from Intake of Radionuclides: Part 5, Compilation of Ingestion and Inhalation Dose Coefficients. ICRP Publication 72, Annals of the ICRP 26(1). Elmsford, NY: Pergamon Press.

ICRP (International Commission on Radiological Protection). 2002. The ICRP Database of Dose Coefficients: Workers and Members of the Public, Compact Disc Version 2.01. Stockholm, Sweden: International Commission on Radiological Protection.

ICRU (International Commission on Radiation Units and Measurements). 1986. The Quality Factor in Radiation Protection. ICRU Report 40. Bethesda, MD: International Commission on Radiation Units and Measurements.

IOM (Institute of Medicine). 1996. Mortality of Veteran Participants in the CROSSROADS Nuclear Test. Washington, DC: National Academy Press.

IOM (Institute of Medicine). 2000. The Five Series Study: Mortality of Military Participants in U.S. Nuclear Weapons Tests. Washington, DC: National Academy Press.

IOM/NRC (Institute of Medicine/National Research Council). 1995. A Review of the Dosimetry Data Available in the Nuclear Test Personnel Review Program. Appendix A in: The Five Series Study: Mortality of Military Participants in U.S. Nuclear Weapons Tests. Washington, DC: National Academy Press.

IOM/NRC (Institute of Medicine/National Research Council). 1999. Exposure of the American People to Iodine-131 from Nevada Nuclear-Bomb Tests. Washington, DC: National Academy Press.

IT and DRI (IT Corporation and Desert Research Institute). 1995. Evaluation of Soil Radioactivity Data from the Nevada Test Site. Las Vegas, NV: U.S Department of Energy Nevada Operations Office; Report DOE/NV-380.

Killough, G. G., Case, M. J., Meyer, K. R., Moore, R. E., Rope, S. K., Schmidt, D. W., Shleien, B., Sinclair, W. K., Voilleque, P. G., Till, J. E. 1998. Radiation Doses and Risk to Residents from FMPC Operations from 1951-1988. Neeses, SC: Radiological Assessments Corporation; RAC Report No. 1-CDC-Fernald-1998-Final, Vols. I and II.

Killough, G. G., Dunning, D. E., Jr., Bernard, S. R., Pleasant, J. C. 1978a. Estimates of Internal Dose Equivalent to 22 Target Organs for Radionuclides Occurring in Routine Releases from Nuclear Fuel-Cycle Facilities, Vol. 1. Springfield, VA: National Technical Information Service; Oak Ridge National Laboratory Report NUREG/CR-0150, ORNL/NUREG/TM-190.

Killough, G. G., Dunning, D. E., Jr., Pleasant, J. C. 1978b. INREM II: A Computer Implementation of Recent Models for Estimating the Dose Equivalent to Organs of Man From an Inhaled or Ingested Radionuclide. Oak Ridge, TN: Oak Ridge National Laboratory; Report NUREG/CR-0114, ORNL/NUREG/TM-84.

Kirchner, T. B., Whicker, F. W., Anspaugh, L. R., Ng, Y. C. 1996. Estimating Internal Dose Due to Ingestion of Radionuclides From Nevada Test Site Fallout. Health Physics 71:487-501.

Klemm, J. 1989. Application of NRC Film Badge Study. McLean, VA: Science Applications International Corporation; Memorandum, (December 15).

Kocher, D. C., Apostoaei, A. I., Hoffman, F. O. 2002. Radiation Effectiveness Factors (REFs) for Use in Calculating Probability of Causation of Radiogenic Cancers. Draft paper available at http://www.cdc.gov/niosh/ocas/ocasirep.html#review2.

Kocher, D. C., Eckerman, K. F. 1981. Electron Dose-Rate Conversion Factors for External Exposure of the Skin. Health Physics 40:467-75.

Kocher, D. C., Eckerman, K. F. 1987. Electron Dose-Rate Conversion Factors for External Exposure of the Skin From Uniformly Deposited Activity on the Body Surface. Health Physics 53:135-141.

Levanon, I., Pernick, A. 1988. The Inhalation Hazard of Radioactive Fallout. Health Physics 54:645-657.

Lorence, L. J., Jr., Morel, J. E., Valdez, G. D. 1989. User's Guide to CEPXS/ONEDANT: A One-Dimensional Coupled Electron-Photon Discrete Ordinates Code Package, Version 1.0. Springfield, VA: National Technical Information Service; Sandia National Laboratories Report SAND 89-1661.

Luna, R. E., Church, H. W., Shreve, J. D., Jr. 1969. A Model for Plutonium Hazard Assessments Based on Operation ROLLER-COASTER. Albuquerque, NM: Sandia Laboratories; Report SC-WD-69-154.

Malik, J. S. 1985. The Yields of the Hiroshima and Nagasaki Nuclear Explosions. LA-8819. Los Alamos, NM: Los Alamos National Laboratory; Report LA-8819.

Martin, E. J., Rowland, R. H. 1982. CASTLE Series, 1954. Washington, DC: Defense Nuclear Agency; Report DNA 6035F.

McArthur, R. D. 1991. Radionuclides in Surface Soil at the Nevada Test Site. Las Vegas, NV: Desert Research Institute; Water Resources Center Publication #45077, Report DOE/NV/10845-02.

McArthur, R. D., Kordas, J. F. 1983. Radionuclide Inventory and Distribution Program: The Galileo Area. Las Vegas, NV: Desert Research Institute; Water Resources Center Publication #45035, Report DOE/NV/10162-14.

McArthur, R. D., Kordas, J. F. 1985. Radionuclide Inventory and Distribution Program: Report #2. Areas 2 and 4. Las Vegas, NV: Desert Research Institute; Water Resources Center Publication #45-41, Report DOE/NV/10162-20.

McArthur, R. D., Mead, S. W. 1987. Radionuclide Inventory and Distribution Program: Report #3. Areas 3, 7, 8, 9, and 10. Las Vegas, NV: Desert Research Institute; Water Resources Center Publication #45056, Report DOE/NV/10384-15.

McArthur, R. D., Mead, S. W. 1988. Radionuclide Inventory and Distribution Program: Report #4. Areas 18 and 20. Las Vegas NV: Desert Research Institute; Water Resources Center Publication #45063, Report DOE/NV/10384-22.

McArthur, R. D., Mead, S. W. 1989. Radionuclide Inventory and Distribution Program: Report #5. Areas 5, 11, 12, 15, 17, 18, 19, 25, 26, and 30. Las Vegas, NV: Desert Research Institute; Water Resources Center Publication #45067, Report DOE/NV/10384-26.

McRaney, W., McGahan, J. 1980. Radiation Dose Reconstruction, U.S. Occupation Forces in Hiroshima and Nagasaki, Japan, 1945-1946. McLean, VA: Science Applications, Inc.; Report DNA 5512F.

Morgan, K. Z. 1946. Final Report of the Alpha, Beta Gamma Survey Team, August 6, 1946 (excerpt located in Case #57).

Muirhead, C. R., Cox, R., Stather, J. W., MacGibbon, B. H., Edwards, A. A., Haylock, R. G. E. 1993. Relative Biological Effectiveness. In: Board Statement on Diagnostic Medical Exposures to Ionising Radiation During Pregnancy and Estimates of Late Radiation Risks to the UK Population. Chilton, U.K.: National Radiological Protection Board; Documents of the NRPB 4(4):126-139.

NCI (National Cancer Institute). 1997. Estimating Exposures and Thyroid Doses Received by the American People from Iodine-131 in Fallout Following Nevada Atmospheric Nuclear Bomb Tests. Washington, DC: National Institutes of Health; Publication No. 97-4264.

NCI-CDC (National Cancer Institute-Centers for Disease Control and Prevention). 2002. Report of NCI-CDC Working Group to Revise the 1985 Radioepidemiological Tables. Bethesda, MD: National Cancer Institute; draft report (August 30).

NCRP (National Council on Radiation Protection and Measurements). 1971. Basic Radiation Protection Criteria. NCRP Report No. 39. Bethesda, MD: National Council on Radiation Protection and Measurements.

NCRP (National Council on Radiation Protection and Measurements). 1987a. Exposure of the Population in the United States and Canada from Natural Background Radiation. NCRP Report No. 94. Bethesda, MD: National Council on Radiation Protection and Measurements.

NCRP (National Council on Radiation Protection and Measurements). 1987b. Recommendations on Limits for Exposure to Ionizing Radiation. NCRP Report No. 91. Bethesda, MD: National Council on Radiation Protection and Measurements.

NCRP (National Council on Radiation Protection and Measurements). 1989. Limit for Exposure to "Hot Particles" on the Skin NCRP Report No. 108. Bethesda, MD: National Council on Radiation Protection and Measurements.

NCRP (National Council on Radiation Protection and Measurements). 1990. The Relative Biological Effectiveness of Radiations of Different Quality. NCRP Report No. 104. Bethesda, MD: National Council on Radiation Protection and Measurements.

NCRP (National Council on Radiation Protection and Measurements). 1993. Limitation of Exposure to Ionizing Radiation. NCRP Report No. 116. Bethesda, MD: National Council on Radiation Protection and Measurements.

NCRP (National Council on Radiation Protection and Measurements). 1997. Deposition, Retention and Dosimetry of Inhaled Radioactive Substances. NCRP Report No. 125. Bethesda, MD: National Council on Radiation Protection and Measurements.

NCRP (National Council on Radiation Protection and Measurements). 1998. Evaluating the Reliability of Biokinetic and Dosimetric Models and Parameters Used to Assess Individual Doses for Risk Assessment Purposes. NCRP Commentary No. 15. Bethesda, MD: National Council on Radiation Protection and Measurements.

NCRP (National Council on Radiation Protection and Measurements). 1999. Recommended Screening Limits for Contaminated Surface Soil and Review of Factors Relevant to Site-Specific Studies. NCRP Report No. 129. Bethesda, MD: National Council on Radiation Protection and Measurements.

Ng, Y. C., Anspaugh, L. R., Cederwall, R. T. 1990. ORERP Internal Dose Estimates for Individuals. Health Physics 59:693-713.

NRC (National Research Council). 1985a. Mortality of Nuclear Weapons Test Participants. Washington, DC: National Academy Press.

NRC (National Research Council). 1985b. Review of the Methods Used to Assign Radiation Doses to Service Personnel at Nuclear Weapons Tests. Washington, DC: National Academy Press.

NRC (National Research Council). 1985c. Radiation Dose Reconstruction for Epidemiological Uses. Washington, DC: National Academy Press.

NRC (National Research Council). 1989. Film Badge Dosimetry in Atmospheric Nuclear Tests. Washington, DC: National Academy Press.

NRC (National Research Council). 1990. Health Effects of Exposure to Low Levels of Ionizing Radiation (BEIR V). Washington, DC: National Academy Press.

Ortlieb, E. 1991. Radiation Dose Reconstruction for Personnel of Companies A and C of the 505th Military Police Battalion, Exercise Desert Rock VI, Operation TEAPOT. McLean, VA: Science Applications International Corporation; Memorandum to DNA-RARP/NTPR.

Ortlieb, E. 1995. Dose Tables for Camp Desert Rock Support Units, Operation UPSHOT-KNOT-HOLE. McLean, VA: Science Applications International Corporation; Memorandum to DNA-RAEM/NTPR.

Otchin, N. S. 2002. Summary of Radiation Medical Opinions 11/19/99-11/17/00 Including Results Using CIRRPC Screening Doses and Draft Interactive Radioepidemiological Program (IREP). Washington, DC: U.S. Department of Veterans Affairs; Personal Communication to I. Al-Nabulsi, National Academy of Sciences (August 23).

Phillips, J. 1983. Radiation Dose Estimates for Personnel of the 505th Military Police Battalion, Exercise Desert Rock VI, Operation TEAPOT. McLean, VA: Science Applications International Corporation; Memorandum to file.

Phillips, J., Klemm, J., Goetz, J. 1985. Internal Dose Assessment – Operation CROSSROADS. McLean, VA: Science Applications International Corporation; Report DNA-TR-84-119.

Robison, W. L., Bogen, K. T., Conrado, C. L. 1997a. An Updated Dose Assessment for Resettlement Options at Bikini Atoll—A U.S. Nuclear Test Site. Health Physics 73:100-114.

Robison, W. L., Noshkin, V. E., Conrado, C. L., Eagle, R. J., Brunk, J. L., Jokela, T. A., Mount, M. E., Phillips, W. A., Stoker, A. C., Stuart, M. L., Wong, K. M. 1997b. The Northern Marshall Islands Radiological Survey: Data and Dose Assessments. Health Physics 73:37-47.

Romanyukha, A. A., Desrosiers, M. F., Regulla, D. F. 2000. Current Issues on EPR Dose Reconstruction in Tooth Enamel. Applied Radiation and Isotopes 52:1265-1273.

Romanyukha, A. A., Mitch, M. G., Lin, Z., Nagy, V., Coursey, B. M. 2002. Mapping the Distribution of ^{90}Sr in Teeth with a Photostimulable Phosphor Imaging Detector. Radiation Research 157:341-349.

Rood, A. S., Grogan, H. A. 1999. Comprehensive Assessment of Exposure and Lifetime Cancer Incidence Risk from Plutonium Released from the Rocky Flats Plant, 1953-1989: Independent Analysis of Exposure, Dose, and Health Risk to Offsite Individuals. Denver, CO: Colorado Department of Public Health and Environment; Radiological Assessments Corporation Report No. 13-CDPHE-RFP-1999-FINAL.

Rood, A. S., Grogan, H. A., Till, J. E. 2002. A Model for a Comprehensive Assessment of Exposure and Lifetime Cancer Incidence Risk from Plutonium Released from the Rocky Flats Plant, 1953-1989. Health Physics 82:182-212.

Schaeffer, D. M. 1995. Further Guidance on the Handling of Operations REDWING and DOMINIC I Film Badges and Doses: NTPR Note 02-95; DNA Memorandum for the Record, 6/6/95. "Handling of Operations REDWING and DOMINIC I Film Badges and Doses, NTPR Note 01-95", DNA Memorandum for the Record.

Schaeffer, D. M. 1997. Letter from Nuclear Test Perssonel Review Program, Defense Threat Reduction Agency to L. Gervase, U.S. Department of Veterans Affairs (February 6) (included in Case #87).

Schaeffer, D. M. 2001a. Letter from Nuclear Test Perssonel Review Program, Defense Threat Reduction Agency to J. E. Till, Chairman Committee to Review Dose Reconstruction Program of DTRA (September 28).

Schaeffer D. M. 2001b. NTPR Dose Reconstruction. Briefing to the Committee. Fort Belvoir, VA: Defense Threat Reduction Agency (April 2).

Schaeffer, D. M. 2002a. Letter from Nuclear Test Perssonel Review Program, Defense Threat Reduction Agency to I. Al-Nabulsi, Study Director (October 10).

Schaeffer, D. M. 2002b. Letter from Nuclear Test Perssonel Review Program, Defense Threat Reduction Agency to I. Al-Nabulsi, Study Director (May 7).

Schaeffer, D. M. 2002c. Letter from Nuclear Test Perssonel Review Program, Defense Threat Reduction Agency to I. Al-Nabulsi, Study Director (August 30).

Schaeffer, D. M. 2002d. Letter from Nuclear Test Perssonel Review Program, Defense Threat Reduction Agency to I. Al-Nabulsi, Study Director (April 1).

Schaeffer, D. M. 2002e. Letter from Nuclear Test Perssonel Review Program, Defense Threat Reduction Agency to I. Al-Nabulsi, Study Director (July 16).

Schaeffer, D. M. 2002f. Plutonium Bioassay Testing of U.S. Atmospheric Nuclear Test Participants. Briefing to the Committee. Fort Belvoir, VA: Defense Threat Reduction Agency (February 21).

Schwendiman, L. C. 1958. Probability of Human Contact and Inhalation of Particles. Health Physics 1:352-355.

Sehmel, G. A. 1984. Deposition and Resuspension. In: Atmospheric Science and Power Production, ed. by Randerson, D. Washington, DC: U.S. Department of Energy; Report DOE/TIC-27601, pp. 533-583.

Simon, S. L., Lloyd, R. D., Till, J. E., Hawthorne, H. A., Gren, D. C., Rallison, M. L., Stevens, W. 1990. Development of a Method to Estimate Thyroid Dose From Fallout Radioiodine in a Cohort Study. Health Physics 59:669-691.

Simon, S. L., Till, J. E., Lloyd, R. D., Kerber, R. L., Thomas, D. C., Preston-Martin, S., Lyon, J. L., Stevens, W. 1995. The Utah Leukemia Case Control Study: Dosimetry Methodology and Results. Health Physics 68: 460-471.

Simon, S. L., Graham, J. C. 1997. Findings of the First Comprehensive Radiological Monitoring Program of the Republic of the Marshall Islands. Health Physics 73:66-85.

Sun, L. C., Clinton, J. H., Kaplan, E., Meinhold, C. B. 1997a. ^{137}Cs Exposure in the Marshallese Populations: An Assessment Based on Whole-Body Counting Measurements (1989-1994). Health Physics 73:86-99.

Sun, L. C., Meinhold, C. B., Moorthy, A. R., Kaplan, E., Baum, J. W. 1997b. Assessment of Plutonium Exposure in the Enewetak Population by Urinanalysis. Health Physics 73:127-132.

TGLD (Task Group on Lung Dynamics). 1966. Deposition and Retention Models for Internal Dosimetry of the Human Respiratory Tract. Health Physics 12:173-207.

Thomas, C. 1985. Revision to SAI Memorandum "Dose Estimates for Land-Based Personnel, Operation HARDTACK I," 18 January 1981. McLean, VA: Science Applications International Corporation; Internal memorandum to DNA-NTPR.

Thomas, C., Gm'inder, R., Stuart, J., Weitz, R., Goetz, R., Klemm, J. 1982. Analysis of Radiation Exposure for Naval Units of Operation GREENHOUSE. McLean, VA: Science Applications, Inc.; Report DNA-TR-82-15.

Thomas, C., Goetz, J., Klemm, J., Weitz, R. 1984. Analysis of Radiation Exposure for Naval Units of Operation CASTLE. McLean, VA: Science Applications International Corporation; Report DNA-TR-84-6.

Thomas, C., Stuart, J., Goetz, J., Klemm, J. 1983a. Analysis of Radiation Exposure for Naval Units of Operation SANDSTONE. McLean, VA: Science Applications International Corporation; Report DNA-TR-83-13.

Thomas, C., Stuart, J., Goetz, J., Klemm, J. 1983b. Analysis of Radiation Exposure for Naval Units of Operation IVY. McLean, VA: Science Applications, Inc.; Report DNA-TR-82-98.

Thompson, D. E., Mabuchi, K., Ron, E., Soda, M., Tokunaga, M., Ochikubo, S., Sugimoto, S., Ikeda, T., Terasaki, M., Izumi, S., Preston, D. L. 1994. Cancer Incidence in Atomic Bomb Survivors. Part II: Solid Tumors, 1958-1987. Radiation Research 137:S17-S67.

Till, J. E., Simon, S. L., Kerber, R., Lloyd, R. D., Stevens, W., Thomas, D. C., Lyon, J. L., Preston-Martin, S. 1995. The Utah Thyroid Cohort Study: Analysis of Dosimetry Results. Health Physics 68:472-483.

UNSCEAR (United Nations Scientific Committee on the Effects of Atomic Radiation). 2000. Exposures to the Public from Man-Made Sources of Radiation. Annex C In: Sources and Effects of Ionizing Radiation. Volume I: Sources. New York: United Nations.

U.S. Army. 1957. Exercise Desert Rock VII and VIII: Final Report of Operations. Presidio of San Francisco, CA: Sixth U.S. Army.

U.S. Senate Committee on Veterans' Affairs. 1998. Transcript of Hearing on Ionizing Radiation, Veteran's Health Care and Related Issues. U.S. Government Printing Office.

VA (U.S. Department of Veterans Affairs). 1996. Letter from K. Collier, Office of Compensation and Pension Service, to P. Broudy, Legislative Director, National Association of Atomic Veterans (April 23).

VA (U.S. Department of Veterans Affairs). 2001. Letter from Secretary Principi to the Honorable Patsy Mink, U.S. Senate.

VA (U.S. Department of Veterans Affairs). 2002. 38 CFR Part 3 — Diseases Specific to Radiation-Exposed Veterans. FEDERAL Register 67:3612-3616.

Weitz, R. 1995a. Dose Estimates for Residence Islands during Operation HARDTACK I. McLean, VA: Science Applications International Corporation; Internal memorandum to DNA-RAEM/NTPR.

Weitz, R. 1995b. Dose Assessment for Crew of USS BENNER (DDR 807) During and After Operation HARDTACK I (1958). McLean, VA: Science Applications International Corporation; Memorandum to DNA/DFRA-NTPR.

Weitz, R. 1997. Dose Assessment for Crew of USS BOLSTER (ARS 38) During and After Operation HARDTACK I (1958). McLean, VA: Science Applications International Corporation; Memorandum to DSWA-ESN.

Weitz, R., Stuart, J., Muller, E., Thomas, C., Knowles, H., Landay, A., Klemm, J., Goetz, J. 1982. Analysis of Radiation Exposure for Naval Units of Operation CROSSROADS. Volumes 1-3. McLean, VA: Science Applications, Inc.; Report DNA-TR-82-05.

Whicker, F. W., Kirchner, T. B., Anspaugh, L. R., Ng, Y. C. 1996. Ingestion of Nevada Test Site Fallout: Internal Dose Estimates. Health Physics 71:477-486.

Whicker, F. W., Schultz, V. 1982. Radioecology: Nuclear Energy and the Environment. Boca Raton, FL: CRC Press.

Wilson, D. W., Ng, Y. C., Robison, W. L. 1975. Evaluation of Plutonium at Enewetak Atoll. Health Physics 29:599-611.

Appendixes

Appendix A

Examples of Dose Reconstruction Memoranda from Sample Cases Reviewed by Committee

Case #22

DOSE SUMMARY REPORT

01/25/96

NAME/ EXPOSURE PERIOD	DOSE	SOURCE FB#	OPERATION	RANK	SERV #	SERV	UNIT(S)
531217 540228	0.000 R		CSTL	1LT	▓▓▓	A	8600TH ADMINISTRATIVE AREA UNIT, COMM SEC DET
		No exposure potential					
540301 540514	1.00 R		CSTL	1LT	▓▓▓	A	8600TH ADMINISTRATIVE AREA UNIT, COMM SEC DET
		May have been issued FB# 14689 on 29 Mar					
540515 540527	0.030 R		CSTL	1LT	▓▓▓	A	8600TH ADMINISTRATIVE AREA UNIT, COMM SEC DET
		Aboard USNS FRED C. AINSWORTH (T-AP 181)					

REF DATE 12/23/95	TOT	1.030 (ACTUAL)	NEUTRON DOSE: 0.000	
ORTLIEB	TOT	1.1 (REPORTED)	INTERNAL DOSE: COLON 0.8	
	TOT	1.1 (REPORTED - CALCULATED)	UPPER BOUND: 1.3	

MEMORANDUM

23 December 1995

To: DNA-DFRA/NTPR (CDR M. Ey)
From: SAIC (E. Ortlieb)
Subject: Radiation Dose Assessment Operation Castle (1954)

BACKGROUND (References 1, 2)
Operation Castle comprised six nuclear weapon tests conducted at the Pacific Proving Ground (PPG) between 1 March and 14 May 1954. The PPG, located in the Central Pacific Ocean area, consisted of the land areas, lagoons, and water areas within three miles of two Marshall Islands atolls, Enewetak and Bikini. Bikini Atoll is about 2200 nautical miles southwest of Hawaii, and Enewetak Atoll is about 195 nautical miles west of Bikini Atoll. The principal objective of Operation Castle was to test high-yield thermonuclear devices. All but the last were detonated at Bikini.

UNIT AND PERSONAL ACTIVITIES (References 1 through 4)
First Lieutenant was the Officer in Charge (OIC) of the Traffic Analysis Section of the Communication Security Detachment (Comm Sec Det), 8600 Administrative Area Unit, at Operation Castle. He arrived at Enewetak with the unit on 17 December 1953 by air transport from Travis AFB, California, and departed the PPG from Enewetak on board USNS FRED C. AINSWORTH (TAP 181) on 14 May 1954. The veteran was detached for return to his home station when AINSWORTH arrived in San Francisco on 27 May 1954.

While at PPG, unit personnel were stationed at various times on Enewetak, Bikini, and USS ESTES (AGC 12). Reference 4 includes a statement by the veteran indicating that he was on temporary duty (TDY) from Enewetak to Bikini until March 1954, "...or the big shot," and that at some time during the operation he was aboard a "communication" ship. These recollections are consistent with CASTLE documentation. As of 1 March 1954, 24 of the 34 unit personnel, including the veteran, were stationed at Enewetak Atoll (as per communication from CO, Comm Sec Det, to Rad Safe Officer, 18 March; in Reference 4); the 10 at Bikini had evacuated for Shot BRAVO aboard ESTES on the previous day. As Bikini was too radiologically contaminated by BRAVO to permit remanning, the detachment remained aboard until the unit departed the pro. Reference 3 states that the personnel of the unit were rotated frequently between Enewetak and ESTES but that a rank and MOS balance was retained. There is no direct evidence available of the veteran's rotation, but there was another lieutenant with his MOS.

An Employee Owned Company

EXTERNAL DOSE ASSESSMENT

Available Dosimetry: Film badge dosimetry records indicate that the veteran was to have been issued badge #14689, apparently on 29 March 1954 (two days after Shot ROMEO). If issued, there was no recorded return date or exposure reading. There are no other film badge records for the veteran; however, his personal dosimetry card shows assessed doses of 0.075 rem from Shot BRAVO and 0.500 rem from Shot ROMEO (periods not specified). These assignments indicate his presence on Enewetak for most, if not all, of March.

Radiation Environments: (Reference 2)
The veteran could have been exposed to fallout from the detonations indicated below:

Site	Shot	Date	Peak Intensities (mR/hr)	Time (H+ hours)
Enewetak	BRAVO	1 Mar 54	10	16
Enewetak	ROMEO	27 Mar 54	8.5	77.5
ESTES	ROMEO		12	42

While in ESTES, the veteran also would have been exposed to residual radiation from Shot BRAVO; and while in AINSWORTH, from both BRAVO and ROMEO. Neither Enewetak nor ESTES was exposed to fallout from subsequent shots prior to the veteran's departure. AINSWORTH departed PPG early on 14 May and did not receive any radioactive fallout from the last shot of the series (NECTAR, detonated at Enewetak).

Daily doses to generic personnel on Enewetak and ESTES are calculated in Reference 2. After the arrival of ROMEO fallout, the Enewetak daily doses are somewhat greater than those in ESTES. Given that the veteran's my to ESTES was for an unknown interval, apparently after ROMEO, his dose is high-sided by a reconstruction based on his continuous presence on Enewetak until departing in AINSWORTH.

Exposure to Initial Gamma and Neutron: Personnel at Enewetak Atoll were too distant from any CASTLE shots at Bikini Atoll to have received any measurable initial neutron or gamma dose (Reference 5). Similarly, there was no exposure to initial radiation while aboard ESTES or AINSWORTH.

The reconstructed doses from Reference 2 are zero for the period prior to Shot BRAVO (for both Enewetak aM Bikini); 0.288 rem at Enewetak. from 1 March to 26 March; 0.715 rem at Enewetak. from 27 March until 14 May; and 0.030 rem aboard AINSWORTH until detachment on 27 May 1954. Note that the reconstructed doses for Enewetak, are greater than the above CASTLE assessed doses associated with BRAVO and ROMEO (whatever their coverage dates may have been).

INTERNAL DOSE ASSESSMENT

The (50-year) committed dose equivalent to the colon (lower large intestine) is calculated using the methodologies presented in References 6 and 7. A nominal breathing rate of 1.2 m^3/hr, a characteristic resuspension factor of $10^{-5}m^{-1}$. and particle size of 20 μm activity median aerodynamic diameter are applied for the calculations. The bulk of resultant dose was from inhalation of descending fallout from Shots BRAVO (0.07 rem) and ROMEO (0.59 rem). Additional accumulations resulted from inhalation of resuspended fallout particles on Enewetak from BRAVO (0.03 rem) and ROMEO (0.06 rem), and in AINSWORTH (0.00 rem). The total committed dose equivalent to the colon is 0.8 rem.

TOTAL DOSE SUMMARY

External Dose:

Neutron:	Inclusive Dates at PPG	Dose (rem)	Method
	17 Dec 53 – 14 May 54	0.000	Reconstruction

Gamma:	Inclusive Dates	Dose (rem)	Method
	17 Dec 53 – 28 Feb 54	0.000	Reconstruction
	1 Mar 54 – 14 May 54	1.00	Reconstruction
	15–27 May 54	0.030	Reconstruction
	Total	1.1	(Upper bound of 1.3)

Internal Dose: The total committed dose equivalent to the colon is 0.8 rem.

References

1. "CASTLE Series—1954," DNA 6035F, Defense Nuclear Agency, 1 April 1982.

2. "Analysis of Radiation Exposure for Naval Personnel at Operation CASTLE," DNA-TR-84-6, Defense Nuclear Agency, 28 February 1984.

3. "Unit Final Report for Operation CASTLE," Communication Security Detachment, 8600 AAU, APO 187, 8 May 1954.

4. Documents located in the veteran's NTPR files.

5. "Neutron Exposure for DOD Nuclear Test Personnel," DNA-TR-84-405, Defense Nuclear Agency, 15 August 1985.

6. "FIIDOS—A Computer Code for the Computation of Fallout Inhalation and Ingestion to Organs," DNA-TR-84-405, Defense Nuclear Agency, 15 August 1985.

7. "Low Level Internal Dose Screen—Oceanic Tests, Nuclear Test Personnel Review," DNA-TR-88-260, Defense Nuclear Agency, October 1991.

Case #60

DOSE SUMMARY REPORT

05/14/99

NAME/ EXPOSURE PERIOD	DOSE	SOURCE	FB#	OPERATION	RANK	SERV #	SERV	UNIT(S)
510202 510407	0.00	Recon		GH	CPL	▓	AF	TU 3.4.1; 3200 DRONE SQDN
			No Exposure Potential					
510408 510618	3.76	Recon	Enewetak	GH	CPL	▓	AF	TU 3.4.1; 3200 DRONE SQDN

REF DATE 04/30/99	TOT	3.760	(ACTUAL)	NEUTRON DOSE: 0.00
BARSS	TOT	3.8	(REPORTED)	INTERNAL DOSE: Spine <0.1
	TOT	3.8	(REPORTED - CALCULATED)	

EXTERNAL UPPER BOUND: 4.4

DATA CHECKED
13 Nov 98, UB rptd

DTRA-NSSN
1767

RADIATION DOSE ASSESSMENT

Operation GREENHOUSE (1951)

Background Information

Operation GREENHOUSE was a series of four nuclear weapons tests conducted at Enewetak Atoll, Marshall Islands, from 8 April to 25 May 1951. Enewetak Atoll is about 2370 nautical miles (nmi) southwest of Hawaii. All four shots were tower bursts. The shots were detonated on the northeast islands of Enewetak Atoll, at a minimal of 8 nmi from the southeast islands, Enewetak and Parry, on which most land-based support personnel were located. Additional personnel were based at Kwajalein Atoll, about 370 nmi southeast of Enewetak. Fallout from all shots except GEORGE impacted the southeast islands of Enewetak Atoll. No fallout from GREENHOUSE shots was detected at Kwajalein. The shot times, dates, locations, and yields in kilotons (kT) are identified in Table 1. (References It 2)

Table 1. Nuclear shot and geographic data, Operation GREENHOUSE.

Shot	Local Time/Date (1951)	Enewetak Atoll Locations	Yield (kT)
DOG	0634/ 8 Apr	Runit Island	81
EASY	0627/ 21 Apr	Enjebi Island	47
GEORGE	0930/ 9 May	Eleleron Island	225
ITEM	0617/ 25 May	Enjebi Island	45.5

Unit and Personal Activities

During GREENHOUSE, the veteran was a Corporal and an Apprentice Auto Mechanic assigned to the 3200th Drone Squadron, based at Eglin Air Force Base, Florida. From 27 February through 18 June 1951, he was on temporary duty with Task Unit (TU) 3.4.1 (Headquarters and Headquarters Squadron) at Enewetak Atoll. TU 3.4.1 operated the airbase on Enewetak Island, including the base operations facilities and maintenance and supply support for aircraft. The task unit also had a detachment on Kwajalein Atoll to provide maintenance and supply support to aircraft landing at this location. The veteran could have been alternately assigned to duty on Enewetak Island and Kwajalein Atoll during his GREENHOUSE tour, but no records were located that placed him at Kwajalein. As an auto mechanic, he would not have been involved in drone decontamination or other maintenance activities that were the responsibility of TU 3.4.2 (Experimental Aircraft). (References 1,3,4)

There are no indications that the veteran was involved in any activity that may have resulted in radiation exposures related to GREENHOUSE other than those documented herein.

External Dose Assessment

Dosimetry: No film badge record was found for the veteran (References 3, 5).

Exposure Scenarios: The veteran was potentially exposed to the following sources of radiation:

- Initial gamma and neutron radiation.
- Residual fallout at Enewetak Island.

Reconstructed Dose

Initial Radiation: Because of his location at Enewetak Island and its distance from the detonations, the veteran received no doses from initial gamma and neutron radiation (References 6-8).

Residual Fallout at Enewetak Island: The veteran was exposed to radioactive fallout during GREENHOUSE while billeted and working on Enewetak Island. Table 2 identifies the peak intensity measured in roentgens per hour (R/hr) and the time of this peak intensity following each shot.

Table 2. Peak intensities on Enewetak Island after GREENHOUSE shots.

Shot	Peak Intensity (R/hr)	Time of Peak after Shot (hr)
DOG	0.04	6.4
EASY	0.001	24
GEORGE	0.0	No fallout
ITEM	0.118	13.7

From Table 2, it is apparent that most of the dose accrued by personnel at Enewetak Island was due to fallout from ITEM. Doses received during the veteran's dates of duty on Enewetak are summed over time using the radiation decay that followed each peak intensity (References 1, 9-12). The veteran is assumed to have been exposed outdoors for 60 percent of his time while on Enewetak and exposed indoors with 50-percent shielding for 40 percent of his time. A reconstruction based on the above parameters results in a 3.76 rem dose during the period of time the veteran was at Enewetak (References 9-12). His gamma radiation dose is summarized below.

External Dose Summary

Neutron: 0.00 rem.

Gamma	Inclusive Dates (1951)	Dose (rem)	Method
	2 Feb – 7 Apr	0.00	No exposure potential
	8 Apr - 18 Jun	3.76	Reconstruction (Enewetak Island)
	Total	3.8	(upper bound 4.4)*

*Based on uncertainty factors for reconstructed doses (References 9, 11).

Internal Dose Assessment

The methodology of References 13 and 14, together with the FIIDOS code (Reference 15), provides a means of calculating the internal dose to a given organ or tissue, based on existing intensity measurements, film badge readings, or dose reconstructions. For internal exposures, airborne radioactive particles are assumed to be fully respirable with an Activity Median Aerodynamic Diameter of 1 micrometer (Reference 16). Based on radiochemistry data for Shots DOG, EASY, and ITEM, and the aforementioned intensity measurements, the veteran's (50-year) committed dose equivalent to spinal nerve tissue from the inhalation of descending nuclear debris in the fallout on Enewetak Island is less than 0.01 rem. For the inhalation of resuspended fallout, a resuspension factor of $10^{-5} m^{-1}$ is applied to the assumed 60 percent of the time spent outdoors, resulting in less than 0.01 rem to spinal nervous tissue.

Internal Dose Summary: The veteran's total committed dose equivalent to spinal nervous tissue is less than 0.1 rem.

References

1. "Operation GREENHOUSE, 1951" DNA 6034F, Defense Nuclear Agency, 15 June 1983.

2. "United States Nuclear Tests, July 1945 through September 1992," U.S. Department of Energy, DOE/NV-209 (Rev. 14), December 1994.

3. Veteran's Nuclear Test Personnel Review (NTPR) file: service record, monitoring reports, and medical records.

4. "Operation GREENHOUSE, Communications Technical Report, 1951," WT-45, Joint Task Force 3, 1951.

5. Microfiche records of film badge readings.

6. "Neutron Exposure for DoD Nuclear Test Personnel," DNA-TR-84-405, Defense Nuclear Agency, 15 August 1985.

7. "DoD Nuclear Test Personnel Not Exposed to Neutron Radiation," SAIC Memorandum to DNA-DFRA/NTPR, 7 September 1995.

8. The Effects of Nuclear Weapons, third edition, compiled and edited by Samuel Glasstone and Philip J. Dolan, United States Department of Defense and Department of Energy, 1977.

9. "Analysis of Radiation Exposure for Naval Personnel at Operation GREENHOUSE," DNA-TR-82-15, Defense Nuclear Agency, 30 July 1982.

10. "Daily Dose Tables for GREENHOUSE Residence Islands," SAIC Memorandum to DNA-RAEM/NTPR, 8 March 1994.

11. "Analysis of Radiation Exposure for Personnel on the Residence Islands of Enewetak Atoll after Operation GREENHOUSE, 1951-1952," DNA-TR-85-390, Defense Nuclear Agency, 20 April 1987.

12. "Calculated Daily Doses for Personnel on the Residence Islands of Enewetak Atoll, April-June 1951," SAIC memorandum, 22 June 1987.

13. "Low Level Internal Dose Screen - OCEANIC Tests," DNA-TR-88-260, Defense Nuclear Agency, October 1991.

14. "Low-Level Internal Dose Screen - CONUS Tests," DNA-TR-85-317, Defense Nuclear Agency, 22 December 1986.

15. "FIIDOS - A Computer Code for the Computation of Fallout Inhalation and Ingestion Dose to Organs, Computer User's Guide," DNA-TR-84-375, Defense Nuclear Agency, 12 December 1985.

16. "Limits for Intakes of Radionuclides by Workers," Publication No. 30, International Commission on Radiological Protection, Pergamon Press, New York, July 1978.

Appendix B

Sampled Case Files of Dose Reconstructions Reviewed by Committee

Case Number	SERVICE	SERIES CODE	CLAIM FILED? (YEAR IF YES)	STATUS OF CLAIM
1	Army	GH, B-J	NO	
2	Army	GH, B-J, IVY	YES (1980)	Not rated - no radiogenic disease
3	Air Force	CSTL	NO	
4	Army	GH	YES (1992)	NP-Denied
5	Air Force	GH	NO	
6	Navy	CSTL	NO	NP-Denied
7	Army	U-K	YES - 2000	NP-Denied
8	Air Force	CSTL	YES - 1990	
9	Navy	CSTL	YES (2) 1978 and 2000	NP-Granted
10	Navy	RW	YES - 1981	P-Granted
11	Air Force	CSTL, TP, RW	YES - 1994	NP-Denied
12	Air Force	GH	YES - 1999	NP-Denied
13	Navy	XRDS	YES - 1996	P-Granted
14	Army	TRIN	NO	
15	Army	TRIN	NO	
16	Air Force	HT I	NO	
17	Navy	XRDS	YES - 1993	NP-Denied
18	Navy	IVY, CSTL	YES - 1990	NP-Denied
19	Army	CSTL	YES - 1994	NP-Denied
20	Army	B-J	YES - 1990 & 1997 personal; 1991 VA	NP-Denied
21	Marine	U-K	YES - 1990	NP-Denied
22	Army	CSTL	YES - 1995	NP-Denied
23	Navy	B-J	YES - 1985 VARO& 19 92 VA	Not rated-did not respond
24	Navy	HT I	YES - 1997	NP-Denied
25	Navy	HT I	YES	NP-Denied

EXTERNAL GAMMA DOSE (rem)	UPPER BOUND DOSE (rem)	ORGAN DOSE (rem)	TYPE OF DISEASE AS BASIS OF CLAIM	DATE OF LAST DOSE RECON-STRUCTION
1.4	1.6			Feb-2000
1.2		No calculation	Skin cancer	Feb-1983
2.6	4.1	CEDE 0.2		Jul-1997
3.5	4.1	No calculation.	Skin cancer	Jan-1993
3.8	4.4	Thyroid 2.3		Mar-1999
1.6	2.5	EDE 0.2		May-1999
1.3	1.8	Prostate 0	Prostate cancer	Sep-2000
2.3		Lung 0.5620; Bone <0.15	Lung cancer	Jul-1990
3.2	5.2	Eye 8.7; Thyroid 0.4; Forearm 55.5; Face 30.3; Head 30.3; Neck 30.3; Upper arm 3 0. 1; Back 8.5	Skin cancer, Thyroid cancer, and cataract	Dec-2000
3.3	4.6		Blood problem and bone deterioration	Mar-1996
5.6	7.9		Heart disease	Dec-1994
3	3.7	Lower leg 77.5	Basal cell carcinoma	Oct-1999
1.3	3.2	Bladder & Prostate <0.15	Bladder and Prostate	Feb-1996
10	20			Oct-1985
2.1	3.8	CEDE <0.15		Sep-1996
1.6	1.8	CEDE 0.2		Aug-1998
2.3	4.5	No calculation	Skin cancer	Mar-1994
5.7		Lung 0.58; Bone <0.15	Lung and Skin cancers	Sep-1990
4.7	7.3		Cataract	Jun-1995
0.1	0.1	Prostate <0.1	Lung disease, Prostate cancer, Basal call carcinoma, Kidney problem	Oct-1997
4.7	6.8	Colon & Bone <0.15	Colon cancer	Aug-1990
1.1	1.3	Colon 0.8	Intestinal cance	Dec-1995
0.2		Colon <0.15	Colon cancer	Jan-1994
0.5	0.5	Brain <0.1	Brain cancer	Jan-1998
1.3	1.5	Forehead 7.1; Lung 0.9	Basal cell carcinoma and lung cancer	Oct-2000

Case Number	SERVICE	SERIES CODE	CLAIM FILED? (YEAR IF YES)	STATUS OF CLAIM
26	Navy	XRDS	YES	P-Granted
27	Marine	U-K	YES -1992	NP-Denied
28	Navy	XRDS	YES- 1996	NP-Denied
29	Army	U-K	YES-1985	NP-Denied
30	Army	U-K	YES	NP-Denied
31	Air Force	POST HT I	YES	P-Denied
32	Army	CSTL	YES - 1994 & 1995	NP-Denied
33	Army	H, N	YES	NP-Denied
34	Navy	XRDS	YES - 1995	NP-Denied
35	Navy	RW	YES - 1994	NP-Denied
36	Air Force	HT I	YES - 1980	NP-Denied
37	Army	TP, PB	YES - 2000	P-Granted
38	Air Force	CSTL	YES - 1991	NP-Denied Previously granted for skin cancer based on UV radiation
39	Navy	CSTL	YES 2000	NP-Denied
40	Army	U-K	NO	
41	Navy	XRDS	YES - 1987	NP-Denied
42	Navy	RW	YES - 1996	NP-Denied
43	Army	GH	YES - 1993	NP-Denied
44	Navy	RW	NO	
45	Navy	XRDS	YES - 1997	NP-grant
46	Army	DOM I	YES - 1999	Not rated - Veteran died
47	Army	IVY, CSTL	YES - 1987	NP-Denied
48	Navy	XRDS	YES	NP-Denied
49	Navy	N, XRDS	YES - 1999	NP-Denied
50	Air Force	TP	NO - NA 1993; Congress 1997	Not a participant

EXTERNAL GAMMA DOSE (rem)	UPPER BOUND DOSE (rem)	ORGAN DOSE (rem)	TYPE OF DISEASE AS BASIS OF CLAIM	DATE OF LAST DOSE RECON-STRUCTION
1.8		Bone 0.32		Jan-1989.
3.7	4.5	Lung <0.15	Lung cancer	Dec-1992
0.7	1.1	No calculation	Hodgkin's diseases; Skin cancer; Prostate cancer	Aug-1996
2.4	3.9	No calculation	Skin cancer	Oct-1985
3.1	3.4			Dec-1999
0.1	0.1	All <0.1		Nov-1998
1.2	1.4	Colon 0.4 Liver & SI <0.15	Small Intestine cancer	Sep-1995
0	0.01	Any 0.0	?	Jun-1997
0.8	2	No calculation	Stomach problem	Aug-1995
0.8	1.1	No calculation	Skin cancer	May-1994
0.1		Colon & Bone <0.15	Colon cancer	May-1990
1.2	2.1	Red marrow (mult. Myeloma) <0.01	Multiple Myeloma	Sep-2000
1.1		Colon 0.29; Bone I <0.15; Lung 0.53	sigmoid colonic cancer, skin cancer	Nov-1991
6.8	7.3	Eye 25.5; Prostate <0.1	Cataract; Prostate cancer; moderately differentiated Adenocarcinome	Sep-2000
0.1	0.2	Face 0.3; Prostate 0	Skin Cancer mentioned	Jul-1999
1.4	1.8	Bladder <0.15	Bladder	Sep-1993
0.5	0.6	No calculation	Lung disease	Aug-1996
3.7	4.4	Colon 0.32	Colon cancer	Apr-1994
0.9	1.1	CEDE 0.2		May-1998
2.4	2.8	No calculation		May-1997
0.07	0.1	No calculation	Lung/not cancer	Apr-1996
2.8		Lung 1.1; Bone <0.15	Lung and Liver cancers	Mar-1989
1.2			Ulcer	Jun-1993
1.6	2.3	Prostate 0	Prostate cancer	Feb-2000
0.1			Eye problem	Jul-1997

Case Number	SERVICE	SERIES CODE	CLAIM FILED? (YEAR IF YES)	STATUS OF CLAIM
51	Army	U-K	NO	
52	Air Force	HT I	YES	NP-Denied
53	Navy	XRDS	YES - 1997	NP-Denied
54	Navy	RW	NO	
55	Army	TP, RW	NO	
56	Army	RW	NO	
57	Navy	XRDS	NO	
58	Air Force	RW	YES - 1994	NP-Denied
59	Navy	XRDS	YES - 1987	Not rated - veteran died
60	Air Force	GH	YES - 1981 Personal	NP-Denied
61	Army	U-K	YES	NP-Denied
62	Air Force	U-K	YES - 1995	NP-Denied
63	Air Force	HT I	YES - 1996	NP-Denied
64	Navy	XRDS	NO	
65	Navy	IVY, WW	YES - 1996	NP-Denied
66	Navy	XRDS	YES	NP-Denied
67	Navy	PB	YES-1995	P-grant
68	Navy	CSTL	YES-1990	P-grant
69	Navy	HT I	NO	
70	Navy	DOM I	YES	Pending
71	Navy	GH	YES - 1997	NP-Denied
72	Navy	HT I	NO	
73	Navy	XRDS	YES (2) 1987, 1988, and 1993	P granted, NP denied
74	Air Force	U-K	NO	
75	Army	U-K	YES	NP-Denied
76	Navy	XRDS	YES - 1990	NP-Denied
77	Army	TP	NO	
78	Army	GH, IVY	YES - 1989	NP-Denied
79	Army	PB		Not Participant

EXTERNAL GAMMA DOSE (rem)	UPPER BOUND DOSE (rem)	ORGAN DOSE (rem)	TYPE OF DISEASE AS BASIS OF CLAIM	DATE OF LAST DOSE RECON-STRUCTION
2.3	3.1	CEDE <0.01		Feb-2000
1.6	1.8	No calculation		Apr-1997
1.1	1.8	Colon <0.01; Any <0.15	Colon cancer	Jan-1997
4.4	5.9		Incurable skin disease mentioned	Jul-1995
3.6	4.5		Skin cancer mentioned	Dec-1994
4.7	6.4			Mar-1995
0.55	1.2	Colon <0.15		Aug-1993
4.2	5.6	Lung 2.3	Lung cancer	Mar-1995
0.81		Colon & Bone <0.15	Colon cancer	Nov-1988
3.8	4.4	Spine <0.1		Apr-1999
0	0	No calculation	personal remark Lung cancer?	Sep-1999
0.001	0.001	No calculation	Lung?	Jul-1995
1.6	1.8	Prostate <0.15	Prostate cancer	Nov-1996
0.7	1.6	Skin 1.6; Eye 1.6; All 0.0		Jan-1999
0.1	0.1	CEDE <0.15	Skin cancer	Feb-1997
0.7	1.8	Arms 1.8	Skin cancer	Jan-1999
0.1	0.1	Lung & Pancreas <0.1	Lung cancer	Apr-1998
5.081		Kidney & Bone <0.15	Kidney cancer	Jul-1990
1	1.1	CEDE <0.1		Mar-1999
0		Any 0.0	Skin; Liver	Oct-1998
1	1.4	No calculation	macular degeneration	Nov-1997
0.87	1.1			Jan-1993
2.1	2.7	Lung 1.6; Bone 0.37; Bladder, Pancreas, and Colon <0.15	Lung and Bladder cancers	Jan-1989
0.3	0.8			Sep-1999
0.04		No calculation	Skin cancer-	Oct-1985
0.977		Lung 0.46; Bone <0.15	Lung cancer	Apr-1991
6				Jun-1988
3.114		lung 0.79; bone <0.15	Lung cancer	May-1989
1.3	1.7			Aug-1999

Case Number	SERVICE	SERIES CODE	CLAIM FILED? (YEAR IF YES)	STATUS OF CLAIM
80	Army	U-K	YES - 1999 & 2000	P-Granted
81	Army	U-K	YES - 1992	NP-Denied
82	Army	RW, HT I	YES	P-Granted
83	Army	B-J	YES - 2000	Pending
84	Army	U-K	NO	
85	Army	HT I	YES	NP-Denied
86	Army	U-K	YES - 1995	NP-Denied
87	Army	U-K	YES -1996	NP-Denied
88	Army	TP	YES 1980	NP-Denied
89	Army	U-K	YES-1980	NP-Denied
90	Army	TP, RW	NO	
91	Army	U-K	YES	NP-Denied
92	Air Force	DOM I	NO	
93	Army	B-J	YES - 1995	NP-Denied
94	Army	HT I	YES-1988?	NP-Denied
95	Air Force	RW	NO	
96	Air Force	GH	YES- 1983 & 1998	NP-Denied
97	Air Force	HT I	NO	
98	Army	GH, B-J, IVY	YES-1988?	P-Granted
99	Army	U-K	YES- 1983	NP-Denied

Hiroshima and Nagasaki Cases

A	Army	H	YES- 1986	
B	Army	H		
C	Army	Non-participant	YES- 1993 and 1997	
D	Navy	H	YES-1997	
E	Navy	N	YES-1987	

EXTERNAL GAMMA DOSE (rem)	UPPER BOUND DOSE (rem)	ORGAN DOSE (rem)	TYPE OF DISEASE AS BASIS OF CLAIM	DATE OF LAST DOSE RECONSTRUCTION
0.1	0.1	Red marrow and All organs <0.001	multiple Melanoma	Sep-2000
3.1	3.4	Colon <0.15	Colon cancers	Jun-1992
7.2	8.5		Kidney & Bone cancers	May-1996
0	0.1	Kidney & All organs 0.0	Kidney	Oct-2000
2.4	3.9	CEDE <0.01	Skin; Thyroid; and hearing problems	May-1999
2.1				May-1990
1.35	2.4		Skin cancer	May-1996
1.7	3	CEDE <0.15	Skin rash, discolored spots	Jan-1997
0.95		Thyroid 0.4; Bone 0.002; Whole Body 0.003	Skin and Nerve disability	Aug-1983
2.7		All 0.0	Rectal cancer	Jul-1983
4.4	5.9	Heart <0.1		Jul-1998
0.1	0.1	Bladder <0.1	Lung cancer	Jan-1998
1.8	1.9			Feb-2001
0.6	1.1	No calculation	Melanoma	Dec-1995
0.7		Bladder & Bone <0.15	Bladder cancer	Feb-1990
0	0			Jul-1992
3.7	4.3	Arm 35; Prostate & Bladder <0.15	Prostate cancer and Skin (melanoma)	Aug-1998
1.9	1.9	Eye 2.4; CEDE <0.1		Apr-1999
3.9		Bone & Esophagus <0.15	Esophagus cancer	Sep-1988
3.8		No calculation	Kidney cancer	Mar-1983
0.1		Bone <0.15	Skin cancer	Apr-87
0.1	0.1			Feb-96
0.0		Skin (chest) 0	Breast cancer	Feb-99
3.9	4.3	Prostate and lymph nodes <0.1	Prostate and lymphoma cancers	Jan-98
0.1		Colon 0.4	Colon cancer	June-88

Case Number	SERVICE	SERIES CODE	CLAIM FILED? (YEAR IF YES)	STATUS OF CLAIM
F	Navy	H-N		
G	Marine corps	N	YES-1999	
H	Army	H	YES-2000	
I	Navy	N	YES-1989 And 1998	
J	Marine corps	N	YES-1985	
K	Army	Non-participant	YES- 1995	
L	Army	Non-participant	YES-1987	

P-presumptive disease (38 *CFR* 3.309)
NP-nonpresumptive disease (38 *CFR* 3.311)
CEDE-committed effective dose equivalent

Series Code	Series Name
GH	GREENHOUSE
B-J	BUSTER-JANGLE
IVY	IVY
CSTL	CASTLE
U-K	UPSHOT-KNOTHOLE
RW	REDWING
TP	TEAPOT
XRDS	CROSSROADS
TRIN	TRINITY
HTI	HARDTACK I
POST HT I	POST HARDTACK I
H	HIROSHIMA
N	NAGASAKI
PB	PLUMBBOB
DOM I	DOMINIC I
WW	WIGWAM

EXTERNAL GAMMA DOSE (rem)	UPPER BOUND DOSE (rem)	ORGAN DOSE (rem)	TYPE OF DISEASE AS BASIS OF CLAIM	DATE OF LAST DOSE RECON- STRUCTION
0.0	<1.0			Aug-93
0.0	<0.01	Skin (ear) <0.01	Basal cell carcinoma	July-00
<0.001	<0.001			May-01
1.0	1.0		Disability	June-98
0.7	0.7	Whole body 0.088	Death/lung cancer	Dec-85
0.0			Death/liver cancer	Feb-99
0.0		Bone <0.018	Disability/ multiple myeloma	Apr-87

Appendix C

Names of Invited Speakers and Interactions with Veterans

NAMES OF INVITED SPEAKERS

D. Michael Schaeffer, Senior Manager, Defense Threat Reduction Agency

Steve Powell, JAYCOR

W. Jeffrey Klemm, SAIC

Julie Fischer, Congressional staff

Cindy Bascetta, General Accounting Office staff

Neil Otchin, Program Chief for Clinical Matters, Office of Public Health and Environmental Hazards, Veterans Health Administration, Department of Veterans Affairs

Bradley Flohr, Chief, Judical/Advisory Review, Compensation and Pension Service, Department of Veterans Affairs

Pat Broudy, veteran's widow, official of the National Association of Atomic Veterans

Richard Conant, Former Commander, National Association of Atomic Veterans

Andy Nelson, retired Navy Captain, former manager of JAYCOR-NTPR effort and Chief of the U.S. Navy NTPR team

Barry Pass, Division Head, Dalhousie University

Alex Romanyukha, Research Associate, Department of Radiology, Uniformed Service University of the Health Sciences

INTERACTIONS WITH VETERANS

Last Name	First Name	Permission to review personal records	Personal letter/ documentation/email
Acciardo	Gilbert		x
Avans	James	x	
Bradley	James		x
Brenner	Robert		
Bushey	Frank		x
Caffarello	Thomas		x
Caldwell III	Boley	x	
Ceonzo	Joe		
Clapp	Fred		x
Clark	Fred		x
daughterrad[a]			x
Duffy	William		x

Dvorak	Theodore		x
Fancieullo	Frank		x
Fish	William	x	
Furbee	Walter	x	
Gilson	Richard	x	x
Howard	Glen	x	
Hughes	Thomas		x
Jones	Jennifer		x
Kinney	Martin		x
Kolb	Harold		x
Lloyd	David		x
Locke	John		x
Lynch	Michael		x
McDonald	James		x
Nelson	Jack		x
Peden	James Robert		x
Pettet	Howard		x
Pierson	Howard	x	
Reynolds	Bernard	x	
Richard	Claude		x
Schwenk	Keith		x
Scott	James Warren	x	
Seidler	Rodney		
Sterling	Delinda		x
Stockwell	R		x
Stone	Gerald		x
Stoyle	Richard		x
Stradley	Herb	x	
Strawbry[a]			x
Templin	Arthur	x	x
Thomas, Jr.	James		x
Tokar	John	x	
Tutas	Paul		x
Wagner	Lawrence	x	
Wolfeld	Sidney		x

[a] E-mail names

Appendix D

Responses to Committee's Questions

Defense Threat Reduction Agency
8725 John J Kingman Road MS 6201
Ft Belvoir, VA 22060-6201

SEP 2 8 2001

John Till, Ph.D.
Chairman, Committee to Review
DTRA Dose Reconstruction Program
National Academy of Sciences
Board on Radiation Effects Research
2101 Constitution Avenue, NW
Washington, DC 20418

Dear Dr. Till:

This is in reply to your letter dated September 6, 2001. The following responses are provided with respect to the questions posed by the National Research Council Committee to Review the Dose Reconstruction Program of the Defense Threat Reduction Agency and the Board on Radiation Effects Research. Per our conversation with NAS' Dr. Al-Nabulsi on 7 September 2001, we indicated it would be appropriate for DTRA to respond to questions 5, 6, 8, 9, 10, 11, 12, 13, 14, and 16 and provide comments on the flow chart pertaining to DTRA responsibilities. We recommended questions 1 through 4 should be referred to Department of Veterans Affairs (VA), Office of the Undersecretary for Benefits and questions 7, 8, 10, and 15 to VA, Office of the Undersecretary for Health.

Question 5. When a claimant is referred to JAYCOR, is the dose of record then sent to VA and DTRA or only to DTRA?

Reply: When a dose is provided for a claim referred to JAYCOR, the dose information is given to DTRA for review and approval. DTRA then sends the dose information to both the VA and the claimant. In cases where the claimant's address is not known, DTRA requests the VA to forward a copy to the claimant.

Question 6. Under what circumstances is the file referred to SAIC for a dose and in what circumstances is this not done?

Reply: Claimant files not referred to SAIC are under the following circumstances:
1997 non-participant cases (these cases are not verified participants);
1998 presumptive cases (these cases do not require dose information); and

1999   most Hiroshima/Nagasaki cases (these cases receive a generic
        reconstructed total dose, upper bound less than 1 rem).

All other cases are referred to SAIC.

Question 8. Explain the use of the "dose screen." At what point in the process
is the CIRRPC screen applied? Which of the confidence limits are used, 90%,
95%, or 99%? Are different screens applied at different stages in the process?

Reply: The CIRRPC dose screen is not, and has never been used by DTRA's
Nuclear Test Personnel Review (NTPR) program.

Question 9. Provide a history of the dosimetry process. How has it changed
over the years since its beginning? Also, provide history on the use of the PC
tables.

Reply: The history of the dosimetry process and its significant changes are
enumerated below:

1978   External gamma dose from film badge dosimetry applied by service
        NTPR teams
1979
1980   Neutron dose reconstruction added to film badge doses and bounding
        estimates for major troop units performed, i.e. generic doses applied to
        military units performing common activities
1981
1982   Internal dose reconstruction (50-year) committed dose equivalent to
        target organ introduced
1983
1984   Statistical application of military unit film badge readings in lieu of
        missing film badge readings introduced
1985   Standardized NTPR guidance for dose assessment published in the
        Federal Register (32 CFR 218)
        FIIDOS computer code applied to internal dose methods
1986   Generic internal dose screen developed - no impact on individual organ
        doses from intake of fallout for VA claims
1987   Defense Nuclear Agency consolidates service NTPR teams into a
        centralized DNA program
1988   Dose reconstruction applied to periods of incomplete film badge
        coverage
1989   Upper-bound dose from individual film badge readings applied
        Doses extended to VA-defined operational test periods as published in

the Federal Register under implementing guidelines for Public Law
100-321

1990   Deep dose equivalent factor applied as defined in 1989 NAS Film
        Badge Dosimetry publication
        Readings from verified damaged film superseded by dose
        reconstruction, primarily applicable to Operations REDWING and
        DOMINIC film badges

1991

1992   Deep dose equivalent factor application rescinded
        Total upper-bound dose, coupling uncertainties for film badge and
        reconstructed doses, introduced for application to some VA claims

1993

1994

1995   Single contract for Jaycor and SAIC team introduced by DTRA
        Coordination of records research and reporting of dose information
        improved
        Reduction of data inconsistencies and improved internal (contractor)
        QA introduced as a result of collecting data for NAS mortality study

1996   Upper-bound dose included for all VA claims
        Doses accrued past the VA-defined operational period assigned vs.
        truncated at end of period included in reconstructions

1997   Standard operating procedures (SOP) for radiation exposure assessment
        introduced
        Case by case limited-scope guidance from previous decade rationalized
        in SOP
        Mean and upper-bound (95%) dose reporting emphasized
        NAS Five Series Study Dose Methodology Report details applied to
        specific test series

1997   Skin dose reconstruction added in response to VA review of scientific
        literature

1998   Limited one-time plutonium urine bioassay conducted - no impact on
        internal doses

2000

2001

PC tables are not, and has never been used by DTRA's Nuclear Test Personnel
Review (NTPR) program.

Question 10. Who has the authority to communicate with the veteran and/or
the veteran's family?

Reply: The DTRA Program Manager has authorized the Jaycor/SAIC Program
Management team to designate hotline operators and analysts to communicate

with veterans and family members. I am not aware of any restriction on any DTRA employees communicating with veterans or family members.

Question 11. Are there any written procedures guiding the conduct of dose reconstructions other than those already provided to the committee? If so, please send us a copy of them. We assume the Standard Operating Procedure drafted in March 1997 is the basic procedure for assigning doses.

Reply: The SOP and 32CFR Part 218 standard serve as the Program's written dose reconstruction guidelines and procedures.

Question 12. Are there any written procedures guiding the quality assurance measures applied during the dose reconstructions other than those already provided to the committee? If so, please identify them. Currently the primary document describing the dose reconstruction process is "NTPR Standard Operating Procedures: Radiation Exposure Assessment" dated 8 March 1997; is this the current guidance for quality control as well?

Reply: There are no additional quality assurance written procedures other than those provided to the committee. The SOP indicates what constitutes a quality dose reconstruction and directs review for conformity with the SOP's procedures, and appropriateness and responsiveness to the correspondence or request received by the NTPR program. The DTRA Program Manager conducts the final review/approval.

Question 13. Is the master file still in use in Las Vegas and updated by JAYCOR/DTRA? Does the JAYCOR/SAIC database have different data from the master file?

Reply: The Reynolds Electrical and Engineering Company (REECo), in Las Vegas, Nevada, was a prime support contractor to the Department of Energy (DOE), formerly the Atomic Energy Commission, throughout most of the U.S. nuclear testing program. As a result, the company maintained an archive of dose records, historical source documents, and original film badges for DOE and DoD participants in both atmospheric and underground nuclear testing. In 1978, DTRA (then DNA) began funding REECo (now Bechtel Nevada) for a dosimetry project to establish a database of all U.S. DoD-affiliated atmospheric nuclear testing records. NTPR dosimetry research and dose reconstruction results were added to the master file as they became available, primarily through routine data exchanges. As a result of DTRA's database upgrade project, which resulted in the Nuclear Test Review Information System (NuTRIS), DTRA assumed full responsibility for maintaining dosimetry data for all U.S. DoD-affiliated U.S. atmospheric nuclear test participants. DOE continues to provide DTRA dosimetry documents and extant film badge analysis support. Also, when DOE receives a direct request for dosimetry information from a DoD-affiliated individual, Bechtel Nevada (formerly REECo) contacts DTRA to request available dose data from the NTPR database. This case-by-case exchange of data ensures consistency when doses for DoD-affiliated individuals are reported by Bechtel for DOE.

Question 14. Is file A separate from the JAYCOR/DNA database?

Reply: File "A" was a satellite database of DTRA's retired legacy system database. It contained anecdotal and unverified information collected from callers and correspondents. The File "A" database has been integrated into DTRA's newer, upgraded NuTRIS database system.

Question 16. What is the policy for contacting veterans during the dose assignment process? Does SAIC often check with the veteran or references he may have provided by phone to verify information? Please explain the frequency of this occurrence during the past.

Reply: Contact with the veterans (and/or their representatives or references) is an integral part of the dose reconstruction process. There is some form of pertinent participation statement regardless of the method by which they contact the program (personal inquiry, VA claim, FOIA, Congressional inquiry, etc.). The participation statements are contained within the responses to an NTPR program questionnaire, a statement in support of claim, or simply within the text of a letter the veteran wrote, or transcribed from conversation notes from when he called to request information through the NTPR hotline regarding his dose.

An analyst can, at his/her discretion, contact the veteran or review any references they may have been provided by telephone (if contact information is available) in the absence of veteran-supplied statements, if the information in the statements leaves large uncertainties in the assessment which might be reduced by follow-up, or if there are inconsistencies in the statements that need to be resolved. The most effective and expedient way to obtain this information is by direct contact between the analyst and the veteran. Even after the assessment is complete and the veteran is informed of the resulting dose(s), he is free to contact the program if he feels that the events described in the assessment do not properly reflect the scope or circumstances of participation. If new and pertinent event information is proffered and after review is deemed to be within the credible range of possible activities, documented by record sources, in which the veteran could have participated, then the analysis is revisited and, if necessary, a new assessment is performed.

While written statements and correspondence have always been a part of the dose reconstruction process, contact by telephone has, until recent years, been relatively rare, due in large part to the inherent public relations risk when contacting a veteran. However, the increased complexity of the reconstructions, due to both the incorporation of additional historical information as well as new tasks into the scope of the program (internal doses, skin doses, etc.) has resulted in a need for increased levels of detail in the veteran's participation statements. While direct contact is not always necessary, the veterans can often provide invaluable information, sometime even beyond what was expected, that has a direct impact on their dose assessment.

Enclosed is a copy of the flow chart with our corrections as they pertain to the DTRA responsibilities in the process. I recommend that you contact VA, Office of the Undersecretary for Benefits, to obtain specific information for the VA parts of the process.

Regarding the input numbers for the DTRA part of the process, DTRA receives typically about 600 VA claim-related inquiries a year. The outcomes of these claims are: about 150 are found to be non-participants, about 90 are presumptive, participation only responses, and about 360 are responses for which participation and dose information is provided to VA.

I hope these comments will be of assistance to your committee and board. Please feel free to contact me at 703-325-2407 or by e-mail at dennis.schaeffer@dtra.mil if there are additional questions.

Sincerely,

D. M. Schaeffer
Program Manager
Nuclear Test Personnel Review
Technology Development
Directorate

Copy to:
NAS BRER (Attn: Dr. Evan Douple)
NAS BRER (Attn: Dr. Isaf Al-Nabulsi)

Defense Threat Reduction Agency
8725 John J. Kingman Road MSC 6201
Ft Belvoir, VA 22060-6201

3 NOV 2001

Department of Veterans Affairs
Central Office
Compensation and Pension Service
ATTN: Mr. Bradley Flohr
810 Vermont Avenue, NW
Washington, DC 20420

Dear Mr. Flohr:

The National Academy of Sciences has requested the Defense Threat Reduction Agency (DTRA) to provide the VA with identifying information for 99 case files the Academy is sampling as part of its dose reconstruction review. The Academy would like to investigate actions VA may have taken in processing claims on the 99 individuals.

A review of our files indicates that the Nuclear Test Personnel Review Program has received VA claim inquiries for only 62 of the 99 sample cases. Attached is a spreadsheet containing pertinent identifying information for these 62 cases; a second spreadsheet contains identifying data for the 37 cases for which we have no record of a VA contact. It is possible that for these 37 cases, some individuals may not have submitted claims to the VA, and others may have involved VA action but without referral to DTRA. The Academy has requested that you indicate only key number and adjudication outcome in your response.

Sincerely,

D. M. Schaeffer
Nuclear Test Personnel Review
Technology Development Directorate

Enclosures
as stated

cc:
I. Al-Nabulsi (w/o encls)
E. Douple (w/o encls)

DEPARTMENT OF VETERANS AFFAIRS
Veterans Benefits Administration
Washington DC 20420

October 31, 2001

John E. Till, Ph.D
Chairman, Committee to Review
 the Dose Reconstruction Program of DTRA
National Academy of Sciences
Board on Radiation Effects Research
2101 Constitution Ave., NW
Washington, DC 20418
Attn: Ms. Isaf AI-Nabulsi, Ph.D.
 Study Director

Dear Dr. Till:

I am responding to your request of September 7, 2001, seeking answers to certain questions from the NRC committee to review the dose reconstruction program of the Defense Threat Reduction Agency. Please excuse the delay in responding to your request.

I have been asked to respond to questions 1, 2, 3, 4, 7, and 10.

Question #1: Clarify and resolve the number of claims awarded on the basis of non-presumptive, atmospheric testing and Hiroshima and Nagasaki, "radiogenic" diseases, especially the issue of the number 50.

Reply: According to the information contained in our special issues database, non-presumptive service connection has been awarded in 1,260 claims based on atmospheric nuclear testing, and in 357 claims on the basis of the occupation of Hiroshima or Nagasaki. As we discussed during your information gathering session earlier this year, the number "50", which has been used, unfortunately, to represent the number of claims for which service connection has been granted under the provisions of 38 CFR § 3.311(b), has no basis in fact.

Question #2: Can you provide us a list of names/ID numbers for veterans who filed a claim under "non-presumptive" law?

Reply: The information you are requesting is protected by the Privacy Act (5 USC § 552a (b)) and 38 USC § 5701. We may release names of veterans and addresses, but not identifying claim numbers, to DoD, but not to NRC.

Question #3: When did the VA's claims database get started and what information is available in the database?

Reply: VA began tracking special issue cases, of which radiation-related cases is a category, in 1982. In November 1998, the initial tracking system was migrated into the Veterans Issue Tracking Adjudication Log (VITAL). VITAL tracks information input into it on a number of special issue claims, e.g., Agent Orange, post-traumatic stress disorder, and radiation. Information can be extracted concerning the number of granted and denied claims in different categories, such as grants and denials of claims based on atmospheric nuclear testing on a non-presumptive basis and grants and denials of claims on the basis of the presumptive provisions of Pub. L. 100-321.

Question #4: Clarify the VA's policy on handling claims retroactively that would have been different due to changes in policy (such as the change in the list of presumptive cancers).

Reply: When a cancer is added to the list of presumptive diseases contained at 38 CFR § 3.309(d), benefits which had previously been denied may be granted upon request of a claimant, or on VA's own review, but no earlier than the date of the legislation. This is not a matter of VA policy, but a matter of a legislative change with a resulting amendment to the regulation.

Question #7: Is a compensation decision for a "radiogenic" disease ever made without a SAIC reconstruction?

Reply: There are two categories of claims where determinations of entitlement to benefits based on a radiation-related illness are made without an SAIC reconstruction. (1) Claims based on the presumptive provisions of Pub. L. 100-321 and (2) claims based on occupational or therapeutic exposure to radiation while in the active military service.

Question #10: Who has the authority to communicate with the veteran and/or the veteran's family?

Reply: Under the general delegation of authority provided by 38 U.S.C. § 512(a), any employee of the Department of Veterans Affairs, in the performance of their duty, has the authority to communicate with the veteran and/or the veteran's family.

I hope this satisfactorily responds to your questions. Should you require further information, please do not hesitate to contact me.

Bradley Flohr, Chief
Judicial/Advisory Review

Defense Threat Reduction Agency
8725 John J. Kingman Road MSC 6201
Ft Belvoir, VA 22000-6201

April 1, 2002

Dr. Isaf Al-Nabulsi
Board on Radiation Effects Research
National Academy of Sciences
2001 Wisconsin Avenue, NW
Washington, DC 20007

Dear Dr. Al-Nabulsi:

In response to your 26 February 2002 letter, we provide an explanation of the calculations in the computer printouts for cases 344921 and 201699. We are including the documented method for performing the calculations, its basis, and policy memorandum implementing the method's use (see enclosed).

The documented method is contained in the attached computer code printout. The codes are what produce the computer printouts contained in your February 2002 letter. The basis for the codes is the National Academy of Sciences (NAS) report of "Film Badge Dosimetry in Atmospheric Nuclear Tests" (1989) (see excerpts enclosed). We implemented the performance of combined uncertainty analysis on multiple film badges in July 1992 (see policy memorandum enclosed). This memorandum implements the combined uncertainty analysis as prescribed in the 1989 NAS film badge study, except for the application of the deep-dose equivalence conversion and the use of 95% confidence intervals. We use the uncorrected film badge readings (factor of 1.0) for determining the total dose vice the deep-dose factor. We do this to preserve a one-to one correlation to the film badge record for the veteran to see evidence that original records are being used in the dose reconstruction and avoid the perception that we are lowering recorded doses. We use a 90% confidence interval vice 95% so that we have a consistent basis for combining the overall film badge reading uncertainty with reconstructed dose uncertainties, compiled and published in historical records at the 90% level.

The code computes the individual entries by selecting the applicable test series by inclusive operation dates. It then selects the bias factors (B) and uncertainty factors (K) for the applicable test series (see computer code printout as annotated). The Band K factors originate from pages 136, 143, and 152 of the 1989 NAS report (excerpts enclosed). They correspond to Operations CASTLE,

TEAPOT, and REDWING in file 201699's printout and CASTLE in file 344921's printout (both enclosed as marked). The B and K factors are applied in the code as prescribed in pages 63 through 66 of the 1989 NAS report. From this application, the "lower, mean, and upper" entries are produced for each film badge entry in each of the file printouts. The code next applies the methodologies and considerations of pages 67 through 72 of the 1989 NAS report to combine the uncertainties of the individual film badges. The last line marked TOTALS results from these combined uncertainties.

Please contact me if you have questions or need additional information. I can be reached at (703) 325-2407 or at dennis.schaeffer@dtra.mil.

Sincerely,

D. Michael Schaeffer
Program Manager
Nuclear Test Personnel Review Program
Technology Development Directorate

Enclosures as stated

Defense Threat Reduction Agency
8725 John J. Kingman Road MSC 6201
Ft Belvoir, VA 22060-6201

May 7, 2002

Dr. Isaf Al-Nabulsi
Board on Radiation Effects Research
National Academy of Sciences
2001 Wisconsin Avenue, NW
Washington, DC 20007

Dear Dr. Al-Nabulsi

In response to your 17 April 2002 letter, we provide responses to your eight
questions, and narrative explanations and public law citations for the DTRA
flow charts attached to your letter. Enclosed are the supporting materials for
the answers to your questions.

**Question 1 - Specific Statements regarding REDWING Film Badge
Damage:** Regarding the statements about damaged film badges at REDWING
in files 321458,489573, 199993, we cite two passages from the 1989 NAS film
badge report and one reference entitled, "Operation REDWING, Radiological
Safety," WT-1366 [EX] (excerpts enclosed). The specific passage in WT-1366
referring to the proportion of damaged badges is (see page 38 excerpt from
WT): "As the operation progressed, it was found that badges worn in excess of
four weeks were badly watermarked, showed evidence of severe light leaks,
and were generally quite difficult to read." The NAS report states (see page 50
from NAS): "During Operation REDWING, operation or series badges were
initially issued for 4 to 6 week intervals. When unprotected, those badges used
for longer periods showed frequent evidence of light leaks and water damage."
The NAS report (see page 149 from NAS) further states: "After light damage
and water damage were detected in a few badges after the initial deployment of
six weeks, subsequent film packets were dipped in ceresin wax before sealing."
The latter statement demonstrates that film badges issued in the first badge
exchange were not sealed in this fashion, as the problem had yet to be
discovered. Thus, most second period badges worn for long intervals also show
damage, as has been verified visual examinations of archived REDWING films
by Bechtel Nevada, Inc. (formerly REECo). Badges issued in a third exchange
period tend not to show damage. This may reflect a combination of successful
sealing, a shorter wear period, and masking of damage by high optical densities
registered by rem-level exposure to Shot TEWA fallout on the residence
islands.

Question 2 - Calculating and Reporting Upper Bound Doses: SAIC started calculating upper bound doses in 1978. The first instance of this was for Task Force WARRIOR as can be found in DNA Report 4747F. The first upper bound dose reported to VA was in 1979, starting with a reconstruction for John E. Knights and likely other reconstructions processed by NTPR Service Teams.

It is DTRA's policy to report upper bound doses to all veterans who have been confirmed as participants.

We do not know when the upper bounds were first used as a basis for probability of causation calculations and their application to decisions regarding compensation awards. The answer to that question should be sought from the VA as DTRA plays no role in VA's application of probability of causation calculations.

Upper bound doses reported to the VA have always been reported to the veterans who have been confirmed as participants.

The reporting of upper bound doses have gone through some evolutions. From the start of the NTPR program and as later codified in 1985 in the Federal Register (32 CFR part 218), upper bound doses, if available were reported to the VA and claims-filing veterans. Through this time frame (1978-1988), available upper bound doses were drawn from DNA-published dose reconstruction reports for major participating military units, In 1989, after the NAS film badge report, upper bound doses on film badge became available and then were reported. From 1989 to 1992, upper bounds were reported on the individual components making up the veteran's total dose, i.e. on the individual film badge results and on the specific reconstruction activities. An overall uncertainty (for example combining the individual components of film badges per the 1989 NAS publication and then combining that result with individual reconstruction uncertainties in quadrature) was not calculated and reported prior to 1992. From 1992 to present, individual components of uncertainty are no longer reported, but are combined and reported as an overall upper bound uncertainty for the summed components of a veteran's dose (see policy note enclosed).

Question 3 - Copies of Citations Listed in File 489573: The citations for the file are attached or, for the NAS film badge reference, given in the reply to question 1 above.

Question 4 - Selection of Resuspension Factors and Policy Guidance: The written documentation on the selection of resuspension factors that apply to Pacific and NTS, and ship and land is contained in three references: "Low-

Level Internal Dose Screen - CONUS Tests," DNA Report, DNA- TR-85-317 (Table 5); "Internal Dose Assessment, Operation CROSSROADS," DNA Report, DNA- TR-84-119; and "Low-Level Internal Dose Screen - Oceanic Tests," DNA Report, DNA- TR-88-260. The references explain how resuspension factors are selected and applied. If novel and unique exposure situations arise, the guidance from the above three references are adapted to fit these situations) subject to DTRA technical review. The documented policy is in the Federal Register, 32 CFR part 218 and the Standard Operating Procedure (provided earlier to you) implements the Federal Register policy through further guidance.

Question 5 - Use of Screening Factors for Evaluating Inhalation Doses: DTRA/SAIC does not and has not used internal dose screening factors to evaluate inhalation doses to individuals' organs or tissues. The FIIDOS report, DNA-TR-84-75, provides the principal methodology for preparing an individual's dose reconstruction. The screen reports provide an underlying scientific methodological basis which augments the principal FIIDOS internal dose methodology.

The dose screens were prepared as generic internal dose estimates for groups of participants belonging to common military units, but never implemented because the focus of Public Law 98-542 and its implementing regulations issued by VA and DTRA (38 CFR part 3.311 and 32 CFR part 218) were on preparing individualized internal organ dose reconstructions vice generic internal dose reconstructions. The FIIDOS methodology was developed and placed in use for individualized dose reconstructions.

If the screens had been used, they would have provided a means of specifying which military units had a potential for intake that would result in a non-zero internal dose estimate from those that had no potential for an intake that would indicate no internal dose. It further subdivided non-zero dose units into those that had nominal less than 0.150 rem dose (chosen as 1% of the then-occupational organ dose limit) and those that were above that dose. Bone was used as the reference organ as it was the one that provided the highest valued internal dose for most of the units' participation activities. For a few of the activities, for example, the internal dose to the lung is higher than bone.

The examples, which you cited as a source for confusion on the use of the screens, show both the internal bone dose as given by the screen and the specific organ dose as computed by FIIDOS. Note that FIIDOS computes larger organ doses than the bone dose for many of the cited examples. Why both doses were reported in individual dose reconstructions is unclear. In the mid 1990s, reporting the bone dose from the screen was dropped after VA indicated that it was not used to evaluate claims.

Question 6 - Use of 20 Micron vs 1 Micron AMAD Particles: The standard practice for internal dose reconstructions is to provide the veteran a single, high-sided dose in the absence of bioassay data. This practice is accomplished through the choice of 1 micron or 20 micron AMAD particle sizes for inhalation. For most organs or tissues, one micron AMAD particle sizes are used when the particle size is unknown as is the guidance in ICRP Report 30.

We have recognized from the 1980's that a 20 micron AMAD particle size maximizes the dose to certain organs, especially in the gastrointestinal tract. As given in the FIIDOS report, the larger of the particle sizes are cleared from the naso-pharyngeal region into the GI tract rather than into the pulmonary region. For a few organs, the high-siding of the particle size depends on the radionuclide inventory available for the given situation and is determined using FIIDOS.

Question 7 - Target Ship Dose Uncertainties: Although DNA Report DNA-TR-82-05 estimates a target ship intensity uncertainty factor of 1.5, this factor was reduced to 1.2 in 1986 for most applications involving topside intensities. There is no quantified derivation available for the 1.5 uncertainty factor in that DNA report. The motivation for the change to a 1.2 factor can be explained through Figure 2-7 of the DNA report (copy enclosed). The depicted topside intensities are reasonably characterized as falling within a factor of 1.5 of the trend line. If a single day's intensity were relied on, an uncertainty factor of 1.5 could be justified. However, the standard error of the mean, closer to a factor of 1.2, is usually more appropriate.

The scatter in the data most likely results from inconsistencies in characterizing the topside average intensities from scattered radiation survey measurements. Thus, for any situation in which reboarding personnel were not confined to a limited area on deck, the standard error of the mean provides the better measure of uncertainty. If you note the cases cited in the question, involving submarine reboardings, the radiological data depicted in Figures A-35 and A-36 (see enclosed) of the above referenced DNA report have less scatter than in the example of Figure 2-7.

Question 8 - Support of Individual Dose Reconstructions: SAIC is funded under a level-of-effort subcontract through Jaycor Corporation, the prime contractor to DTRA for the NTPR veterans support program. The subcontract contains a maximal annual ceiling for dose reconstruction support. Payment is not based on the completion of each dose reconstruction. There is no fixed dollar limit for the completion of each reconstruction. The complexity of the dose reconstruction does not factor into payments for dose reconstructions. There is no imposed time limit for the completion of dose reconstructions. The origin of the dose reconstruction request does not factor into its completion time or payment. There is no limitation on the amount of effort to be expended on an individual assessment The contractual performance standards that apply to the Jaycor/SAlC team are to complete actions on veterans' cases within 90 days for 70% of inquiries, to have no more than 5% of cases pending longer than 180 days, and to have no pending transactions exceeding 365 days. If veterans inquiry demands exceed the capacity of the maximal annual support ceiling, DTRA can increase funding resources to expand the level of effort.

Narrative Explanations: DTRA will provide narratives as follows for the flow charts it originally submitted to you. NAS should ask the VA to provide narratives on the charts they provided.

Chart pertaining to VA claim inquiries requiring participation and dose information: This chart shows the organizations handling a Department of Veterans Affairs (VA) claim initiated by a veteran through a VA Regional Office. The chart depicts the detailed actions and steps through which a veterans claim inquiry transits after the VA Regional Office submits it to Defense Threat Reduction Agency (DTRA). DTRA has a veterans support effort (steps 4 through 9), a teamed contract with Jaycor Corporation and SAIC, to conduct historical research on veterans participation activities and to determine radiation doses related to those activities. DTRA reviews and approves the results of the contract effort before submitting them to the VA Regional Office. The requirement for providing participation and dose information supporting a veteran's claim originates from Public Law 98-542 as implemented by VA regulation under 38 CPR part 3.311 of the Code of Federal Regulations (CPR). The VA Regional Office specifies the above cited CFR in its inquiry to DTRA for processing the claim per this chart.

Chart pertaining to VA claim inquiries requiring participation only: This chart shows the organizations handling a Department of Veterans Affairs (VA) claim initiated by a veteran through a VA Regional Office. The chart depicts the detailed actions and steps through which a veterans claim inquiry transits after the VA Regional Office submits it to Defense Threat Reduction Agency (DTRA). DTRA has a veterans support effort (step 4), a teamed contract with Jaycor Corporation and SAIC, to conduct historical research on veterans participation activities and to determine radiation doses related to those activities. For this chart, only participation information is required and does not involve the SAIC team partner to complete the action. DTRA reviews and approves the results of the contract effort before submitting them to the VA Regional Office. The requirement for providing only participation information supporting a veteran's claim originates from Public Law 100-321 as implemented by VA regulation under 38CFR part 3.309 of the Code of Federal Regulations. The VA Regional Office specifies the above cited CFR in its inquiry to DTRA for processing the claim per this chart.

I trust that the above answers provide a complete and comprehensive treatment of your questions. Please contact me if you have questions or need additional information. I can be reached at (703) 325-2407 or at dennis.schaeffer@dtra.mil.

Sincerely,

D. Michael Schaeffer
Program Manager
Nuclear Test Personnel Review Program
Technology Development Directorate

Enclosures
as stated

Defense Threat Reduction Agency
8725 John J. Kingman Road
MSC 6201
Ft Belvoir, VA 22060-6201

July 16, 2002

Dr. Isaf Al-Nabulsi
Board on Radiation Effects Research
National Academy of Sciences
500 5th Street, NW
Washington, DC 20001

Dear Dr. Al-Nabulsi

In response to your 18 June 2002 letter, we provide replies to your seven questions. In a recent conversation with you to clarify certain questions, you indicated that the Dose Reconstruction Committee preferred to receive an informal reply to the questions at the upcoming August 2002 Committee meeting. We intend to support this meeting and reply to the questions in person. Additionally, we believe it would help to provide written answers in advance, especially to question 6. Our reply to that question is detailed and lengthy and important that you receive the full reply in advance of the meeting. We feel that a verbal reply in the absence of a formal response could be misconstrued and possibly result in continued and open-ended questions. We ask for your indulgence in this regard.

Question 1- Second Period REDWING Film Badge Damage: Verified visual examinations of archived REDWING films performed by Bechtel Nevada, Inc. (formerly REECo) in support of individualized dose reconstruction provides the evidence of damaged film badges.

Question 2 -Fixed Rate Compensation for Dose Reconstruction: SAIC never received fixed rate compensation for the performance of dose reconstructions.

Question 3 - Update on Bioassay Program: The enclosed spreadsheet of actual vs. predicted bioassay measurements for 100 eligible NTPR Program veterans was provided on 18 March 2002 to NAS at the request of the Committee for its review. We will update the Committee at its August 2002 meeting on other follow-on activities.

Question 4 -Date Pair/Upper Bound Meaning: Date pair request means reconstructing a dose for periods for which film badge readings do not exist or for which a film badge reading is not valid. Upper bound request means computing a 95% percentile total external dose which was not calculated previously and also computing a internal dose to a target organ, skin dose, or eye dose for a radiogenic disease specified in a VA claim.

Question 5 - Use of 20 Micron Particle Sizes: To the best of our knowledge, we instituted the practice of using 20 micron particle sizes to maximize internal organ doses sometime in the late 1980s.

Question 6 - Top Level Quality Assurance Guidance: Quality Assurance has always been a key element in our management and direction of the NTPR program. Specifically, the DTRA solicitation for NTPR Program Support, Statement of Work (June of 2000), contained a program management requirement for quality assurance monitoring in the program areas of database management, dose assessment and veteran assistance. Additionally, the solicitation designated quality assurance process as one of the contract evaluation factors for award.

In response to the solicitation, Jaycor/SAIC submitted a technical proposal that specified quality assurance measures in the program task areas of database management, radiation exposure assessment, and veteran assistance. Our approach is consistent with many of the key elements of both ASME NQA-1 and ANSI/ASQC E4-1994.

Key elements of NTPR quality assurance include:
- Designated senior management responsibilities
- Designated responsibilities and roles by program task area
- Integrated and interdisciplinary work processes established IAW technical and administrative standards
- Approved work processes incorporated into standard operating procedures
- Comprehensive case tracking / monitoring system, records management system, and records back up system

The key feature of NTPR Program Support's quality assurance is the use of standard operating procedures (SOPs) in all program areas, to include dose reconstruction. As part of an earlier response to NAS, we provided a copy of the SOPs. As noted in the EPA Quality Assurance guidelines (EPA QA/G-6, March 2001): *"The development and use of SOPs are an integral part of a successful quality system as it provides individuals with the information to perform the job properly, and facilitates consistency in the quality and integrity of a product or end result.*" The NTPR SOPs were developed by Jaycor/SAIC as standard procedures and guidelines for company proprietary work in support of the NTPR program and were reviewed and approved by us to ensure consistency with federal guidelines and standards applicable to the program.

The fully integrated Jaycor/SAIC NTPR team, under our direction, uses an interdisciplinary approach to ensure the work in all task areas meets Government-established performance standards. The following summary illustrates the comprehensive quality assurance actions implemented in the program's three key task areas.

Database Management

Application Development
- Requirements review by working groups, team leader, program manager
- Application testing by selected users and team leaders
- Major modifications review by team leaders

LAN Administration
- Standard procedures review by team leader
- Enhancements review by team leader, corporate LAN engineers, program manager

Database Administration / Analysis
- Data entry review by Quality Assurance Specialist and team leader
- Data structure modifications review by programmers, team leader, program manager
- Dose data accuracy and database integrity review by data analyst and database administrator

Documentation / Training
- Training materials review by selected users and team leader
- Training procedures review by team leader and program manager

Radiation Dose Assessment

Dose Triage
- Correspondence case review by health physicist

Radiation Exposure Data
- Film badge review by health physicist
- Film badge versus reconstructed dose analysis/review by dose analysts
- Test series operational detail knowledge by dose analysts

Technical Documentation
- Assessments and technical documents review by senior scientist and senior historian/editor

Radiation Environment
- Scenario development review by senior scientist
- Scenario refinement and enhancement by senior scientist
- Methodology application consistency and accuracy IAW dose SOP review by senior scientist

Veteran Assistance

Research documentation:
- Correspondence actions review by deputy team leader, team, leader, health physicist (if dose info provided), admin supervisor, operations director (selected actions), program manager

Outreach Admin:
- Outreach data collection review by team leader
- Word processing review by team leader
- Administrative action review by team leader

St. Louis operation:
- Records retrieval and review cross-checks by team leader and researchers

Special Projects
- Data abstraction review by deputy team leader and team leader
- Library maintenance review by team leader
- Special taskings review by operations director, program manager, health physicist (if dose related)

Question 7 - Generic Dose Assignment vs. Dose Assessment Statistic:
We are unable to sort out the information in our NTPR database according to these parameters to provide you a reply. We would have to hand sort NTPR files in order to make a meaningful reply. We would prefer to discuss alternatives with the Committee for obtaining the needed data.

I trust that the above answers provide the groundwork for the reply to your questions at the August Committee meeting. Please contact me if you have questions or need additional information. I can be reached at (703) 325-2407 or at dennis.schaeffer@dtra.mil.

Sincerely,

D. Michael Schaeffer
Program Manager
Nuclear Test Personnel Review Program
Technology Development Directorate

Enclosure
as stated

Defense Threat Reduction Agency
8725 John J. Kingman Road MSC 6201
Ft Belvoir, VA 22060-6201

August 30, 2002

Dr. Isaf Al-Nabulsi
Board on Radiation Effects Research
National Academy of Sciences
500 5th Street, NW
Washington, DC 20001

Dear Dr. Al-Nabulsi:

In response to your 13 August 2002 letter, we provide responses to your six questions, concerning report SAIC - 0012024 "Methods and Applications for Dose Assessments of Beta Particle Radiation".

Question 1: P3, second bullet. Please elaborate on how adding the gamma dose to the beta dose leads to an upper bound of the skin dose. What method is used to provide a central estimate of skin dose?

Response: We do not provide a central estimate for the skin dose. For the purposes of beta dose calculations, the nominally uniform external whole body gamma dose (including its upper bound) is considered to be applicable at 70 μm. A more central estimate of skin dose can be given by using the mean whole body gamma dose (i.e., in lieu of the terms annotated by $D\gamma$/ub/fallout and $D\gamma$/ub/total in Equation 7 on page 17). In addition, because the assumed retention of fallout on surfaces tends to high-side the beta component of the skin dose, we do not have a means of providing a central tendency value for the beta skin dose component.

Question 2: PI5, table 10. What is the meaning of the numbers in parentheses in the column identified as "βdose[4] (CEPXS)"?

Response: The dose conversion factor in parentheses was based on an air density of 1.189 mg/cc, whereas the other value was based on an air density of 1.293 mg/cc.

Question 3: Please help us understand the apparent contradiction between the statement on pg.13 "Figure 5 depicts…and illustrates reasonable justification to use U-235 based generic spectra for all shots during the testing era," and the statement on pg.17 , "the determination of what constitutes the appropriate beta/gamma (β/γ) ratio must include evaluations such as the type of weapon used. . . ."

Response: Fission byproduct radionuclides predominated both fission and thermonuclear-type detonations. Figure 5 indicates that in the worst known deviation from a pure fission product inventory based on pure U-235 (i.e., CASTLE Bravo being thermonuclear) the difference in β/γ ratios is modest. Given that the majority of shots are much closer to a pure fission product inventory than were the CASTLE shots, the solid-line relationship in Figure-5 is typically representative. However, the deviation for CASTLE shots (similar to BRAVO) is enough to warrant applying the dashed-line relationship in Figure-5 for these shots. The statement made on page 17 regarding type of weapon is applied only in this context.

Question 4: Page 28, last paragraph. What is the basis for using "l/r" estimation for an infinite plane source geometry factor?

Response: There is none. The l/r estimate was not meant to indicate a mathematical functional relationship between gamma intensity and height from a plane source, but rather to indicate that there is an inverse proportional relationship with height. Note that the sentence could be clarified to indicate that an linear interpolation of the (β/γ) ratio between a height of 1 and 20 centimeters (to obtain an estimate of the ratio at 10 cm from Table-7) would result in an expected beta dose about 9 times greater than that at a height of 120 cm (rather than the stated factor of 12).

Question 5: P44. It is implied in the second complete paragraph that an uncertainty factor of 2 in the beta dose component is reasonable. Should this be interpreted as the entire uncertainty of the beta dose or the uncertainty in the beta/gamma ratio?

Response: Neither should be implied. This statement was an interpretation relative to the ratio of the beta dose to gamma dose cited in Reference 1 and does not pertain to current skin dose methodology. The ratio cited was 8 to 16 (i.e., a factor-of-2 range for this ratio). Note that while this range for the ratio may be representative (and generally corresponds to SAIC values at 1 meter from a surface), it does not address dependence on distance from surface and radionuclide inventory (e.g., with time for mixed-fission products).

Question 6: What procedures were used for assessing beta dose prior to publication of SAIC-001/2024?

Response: Prior to 1998, skin doses were not performed in the NTPR program, except on a case-by-case basis. Prior to the publication of SAIC-001/ 2024, beta doses were computed by applying the references therein directly - principally (and as cited in the radiation dose assessments) the user's manual for the CEPXS radiation transport code and the SAI report that specified the beta and gamma energy spectra as a function of time after a detonation.

I trust that the above answers provide a complete and comprehensive treatment of your questions. Please contact me if you have questions or need additional information. I can be reached at (703) 325-2407 or at dennis.schaeffer@dtra.mil.

Sincerely,

D. Michael Schaeffer
Program Manager
Nuclear Test Personnel Review Program
Technology Development Directorate

Defense Threat Reduction Agency
8725 John J. Kingman Road MSC 6201
Ft Belvoir, VA 22060-6201

Department of Veterans Affairs
Central Office   SEP _ 5 2002
Compensation and Pension Service
ATTN: Mr. Bradley Flohr
810 Vermont Avenue, NW
Washington, DC 20420

Dear Mr. Flohr:

The National Academy of Sciences has requested the Defense Threat Reduction Agency (DTRA) to provide the VA with identifying information for 200 case files the Academy is sampling as part of its dose reconstruction review. The Academy would like to investigate actions VA may have taken in processing claims on the 200 individuals.

A review of our files indicates that the Nuclear Test Personnel Review (NTPR) Program reported radiation exposure information for 141 of the 200 sample cases. The remaining 59 cases did not require the NTPR Program to furnish dose information to the VA.

Enclosed is a spreadsheet containing pertinent identifying information for all 200 cases. The first worksheet of this spreadsheet includes the dose information reported by the NTPR Program to the VA for 141 cases. The second worksheet provides pertinent information for the remaining 59 cases. As can be seen from this enclosure, some VA claim data, i.e., claim number and/or Regional Office information, is available from the NTPR case files for some of these 59 cases. This claim data was obtained from various sources, e.g., service medical records, reviewed by the NTPR Program during the course of researching individual inquiries. The available data is being provided to assist you with your response to the NAS.

Sincerely,

D. M. Schaeffer
Nuclear Test Personnel Review
Technology Development Directorate

Enclosures
as stated

cc:
✓ I. Al-Nabulsi (w/o encls)
  E. Douple (w/o encls)

Defense Threat Reduction Agency
8725 John J. Kingman Road MSC 6201
Ft Belvoir, VA 22060-6201

Dr. Isaf Al-Nabulsi   OCT 1 0 2002
Board on Radiation Effects Research
National Academy of Sciences
500 5ᵗʰ Street, NW
Washington, DC 20001

Dear Dr. Al-Nabulsi:

    In response to your early September request, we are providing an update to our dose
reconstruction summary which extends to 30 September 2002 (copy attached). Please contact
me if you have questions or need additional information. I can be reached at (703) 325-2407 or
at dennis.schaeffer@dtra.mil.

                                        Sincerely,

                                        D. Michael Schaeffer
                                        Program Manager
                                        Nuclear Test Personnel Review Program
                                        Technology Development Directorate

RECEIVED

OCT 1 6 2002

BOARD ON RADIATION
EFFECTS RESEARCH

Defense Threat Reduction Agency
8725 John J. Kingman Road MSC 6201
Ft Belvoir, VA 22060-6201

December 03, 2002

Dr. Isaf Al-Nabulsi
Board on Radiation Effects Research
National Academy of Sciences
500 5th Street, NW
Washington, DC 20001

Dear Dr. Al-Nabulsi:

In response to your November 5, 2002 letter, we will clarify the dose reconstruction process when unit dose reconstructions form the basis for an individual's dose assessment. Current DTRA policy requires that SAIC provide all doses prepared in support of VA claims.

At one point in the NTPR Program before the current policy, there were cases where Jaycor applied available SAIC-prepared generic external dose reconstructions if the participation scenario for that reconstruction was applicable to the veteran's participation scenario and if SAIC had previously determined the internal dose to a particular organ. As noted in your letter, the assigned doses then went to DTRA for review, approval, and release to the VA.

Please contact me if you have questions or need additional information. I can be reached at (703) 325-2407 or at dennis.schaeffer@dtra.mil.

Sincerely,

D. Michael Schaeffer
Program Manager
Nuclear Test Personnel Review Program
Technology Development Directorate

DEPARTMENT OF VETERANS AFFAIRS
Veterans Health Administration
Washington DC 20420

In Reply Refer To:

September 17, 2001

13

Isaf AI-Nabulsi, Ph.D.
Study Director
National Academies
2101 Constitution Avenue, N.W.
Washington, DC 20418

Dear Dr. AI-Nabulsi:

This is in reply to your letter dated September 7, 2001. As we discussed, I am responding to question numbers 8, 10, and 14 as they pertain to responsibilities of our office.

Question 8. Explain the use of the "dose screen". At what point in the process is the CIRRPC screen applied? Which of the confidence limits are used, 90%, 95%, or 99%? Are different screens applied at different stages in the process?

Reply: Screening doses are used when a compensation claim requires a medical opinion on the likelihood that radiation exposure in service was responsible and when the CIRRPC report provides screening doses relevant to the veteran's disease. The CIRRPC screening doses are applied after the case has been sent to the Office of Public Health and Environmental Hazards for a medical opinion. The screening doses based on the 99% confidence limits are used. The same screening doses based on 99% confidence limits are used for medical opinions provided at all stages of the adjudication process.

Question 10. Who has the authority to communicate with the veteran and/or the veteran's family?

Reply: I am not aware of any restriction on any VA employees communicating with veterans or family members.

Question 14. Are the CIRRPC screening doses used in actual compensation decisions? If not, what PC labels are used?

Reply: The CIRRPC screening doses when applicable are used in formulating a medical opinion but other factors are considered as well.

I hope these comments will be of assistance. Please feel free to contact me at 202-273-8452 or by e-mail at neil.otchin@hq.med.va.gov if there are additional questions.

Sincerely,

Neil S. Otchin, M.D.
Program Chief for Clinical Matters
Office of Public Health and Environmental Hazards

Appendix E

Analysis of Potential Inhalation Doses Due to Blast-Wave Effects at Operation PLUMBBOB, Shot HOOD, and Implications for Dose Reconstructions for Atomic Veterans

E.1 INTRODUCTION

This appendix presents an example analysis to investigate potential inhalation doses to participants in forward areas after detonation of Operation PLUMB-BOB, Shot HOOD. This analysis is intended to illustrate the potential importance of resuspension caused by the blast wave produced in aboveground detonations at the NTS. The effect of a blast wave on resuspension of fallout that was previously deposited on the ground has been ignored in all dose reconstructions for atomic veterans, but the committee believes that the effect is potentially important and should be taken into account (see Section V.C.3.2, comment [7]).

The results of the example analysis have important implications for dose reconstructions for other exposure scenarios involving inhalation of resuspended fallout, and these implications also are discussed in this appendix.

E.2 RADIATION ENVIRONMENT IN FORWARD AREAS AT SHOT HOOD

Shot HOOD was detonated on July 5, 1957, and was one of the later shots during the period of aboveground testing at the NTS (Hawthorne, 1979). Shot HOOD provides a good example of the potential importance of resuspension caused by a blast wave on inhalation doses to participants because this shot had the largest yield of any aboveground test at the NTS (Hawthorne, 1979) and produced the most violent blast wave, an extensive cloud of dust was produced in the detonation (Maag et al., 1983; USMC, 1957), there was considerable fallout from previous shots in the area of the detonation (see Section IV.C.2.1.1, Table IV.C.1, and Section V.C.3.2, comment [6]), and participant groups undertook activities in the forward area soon after detonation (Maag et al., 1983; Frank et al., 1981; USMC, 1957).

The locations of all shots at the NTS through Operation PLUMBBOB—excluding shots in Operation RANGER, which did not result in significant fallout on the NTS (Hawthorne, 1979)—are shown in Figures V.C.2 through V.C.6 (see Section V.C.3.2). As discussed in Section V.C.3.2, comment [6], the area near Shot HOOD experienced fallout from several previous shots, and there probably was additional fallout from other shots not mentioned. On the basis of locations of PLUMBBOB shots shown in Figure V.C.6 and fallout patterns reported by Hawthorne (1979), Shot SMOKY, which occurred after Shot HOOD, apparently deposited fallout in the same area. In addition, later safety shots in Operation HARDTACK-II—including OTERO, VESTA, JUNO, and GANYMEDE—deposited fallout in the area of Shot HOOD (Hawthorne, 1979).

Concentrations of radionuclides in surface soil in Area 9, where Shot HOOD occurred, were measured during the 1980s (McArthur and Mead, 1987; McArthur, 1991). Estimated distributions of the photon-emitting radionuclides ^{241}Am, ^{60}Co, ^{137}Cs, and ^{152}Eu in the vicinity of Shot HOOD are shown in Figures E.1 through E.4; additional data on ^{154,155}Eu are not shown. The high radionuclide concentrations to the south and southwest of Shot HOOD presumably are due mainly to fallout from HARDTACK-II safety shots that occurred after Shot HOOD (Hawthorne, 1979). Distributions of ^{90}Sr and plutonium in surface soil were derived from the distributions of ^{137}Cs and ^{241}Am, respectively, and measured ^{90}Sr-to-^{137}Cs and ^{239,240}Pu-to-^{241}Am activity ratios in soil samples obtained from various locations in the area. Distributions of ^{238}Pu also were estimated from measured ^{238}Pu-to-^{241}Am ratios in soil samples.

The information summarized above can be used to estimate concentrations of important radionuclides that were present in the area of Shot HOOD at the time of detonation. On the basis of measured concentrations of longer-lived radionuclides in surface soil after the period of atomic testing shown in Figures E.1 through E.4, measured activity ratios obtained from soil samples (McArthur, 1991), and the ICRP's current dose coefficients for inhalation of respirable particles (AMAD, 1 μm) given in Table V.C.2 (see Section V.C.3.1), ^{239,240}Pu probably posed the most important inhalation hazard at the time of Shot HOOD, and the presence of ^{60}Co, ^{90}Sr, ^{137}Cs, ^{152,154,155}Eu, ^{238}Pu, and ^{241}Am probably increased potential inhalation doses by less than a factor of 2.[1]

---

[1]A more rigorous analysis would need to consider the presence of additional radionuclides with half-lives sufficiently short that they were no longer present in detectable amounts when measurements were made during the middle 1980s. Additional fission products that could be important in aged fallout in the area of Shot HOOD include ^{89}Sr, ^{95}Zr, ^{106}Ru, and ^{144}Ce (see Table V.C.2).

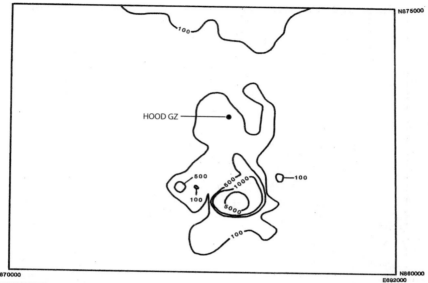

FIGURE E.1   Distribution of concentrations of ^{241}Am in surface soil (nCi m^{-2}) in vicinity of ground zero (GZ) of Operation PLUMBBOB, Shot HOOD based on measurements in middle 1980s (McArthur and Mead, 1987). Area shown is about 6.7 by 4.6 km.

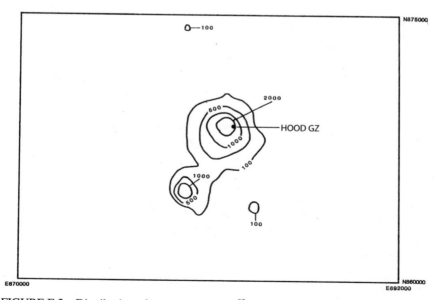

FIGURE E.2   Distribution of concentrations of ^{60}Co in surface soil (nCi m^{-2}) in vicinity of ground zero (GZ) of Operation PLUMBBOB, Shot HOOD based on measurements in middle 1980s (McArthur and Mead, 1987). Area shown is about 6.7 by 4.6 km.

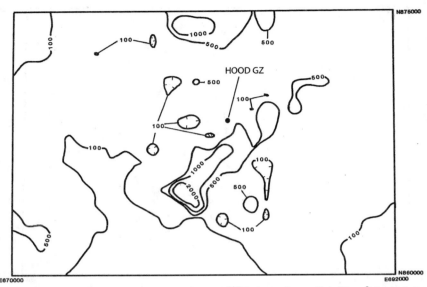

FIGURE E.3 Distribution of concentrations of ^{137}Cs in surface soil (nCi m^{-2}) in vicinity of ground zero (GZ) of Operation PLUMBBOB, Shot HOOD based on measurements in middle 1980s (McArthur and Mead, 1987). Area shown is about 6.7 by 4.6 km.

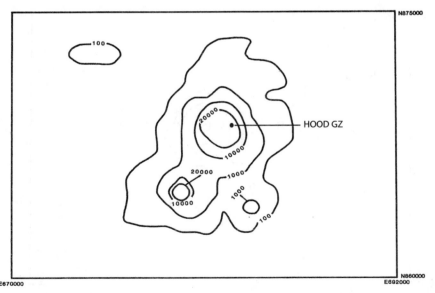

FIGURE E.4 Distribution of concentrations of ^{152}Eu in surface soil (nCi m^{-2}) in vicinity of ground zero (GZ) of Operation PLUMBBOB, Shot HOOD based on measurements in middle 1980s (McArthur and Mead, 1987). Area shown is about 6.7 by 4.6 km.

E.3 ANALYSIS OF POTENTIAL INHALATION DOSES

This section presents an analysis of potential inhalation doses due to resuspension of radionuclides in surface soil caused by the blast wave at Shot HOOD. On the basis of data discussed above, the analysis assumes that only plutonium (^{239,240}Pu) was present and ignores potential doses from inhalation of other radionuclides. This analysis is presented for illustrative purposes only, and it should not be interpreted as providing definitive estimates of inhalation doses to any atomic veterans who participated in activities in forward areas shortly after detonation of Shot HOOD.

In this analysis, potential inhalation doses due to effects of the blast wave at Shot HOOD are estimated by taking into account subjective estimates of uncertainty in all input parameters. Assumed uncertainties in parameters that are used to estimate airborne concentrations of plutonium are intended to represent a range of plausible conditions at different distances from ground zero and at various times after the blast wave occurred; the assumed uncertainties are not intended to represent plausible conditions of exposure that would result in the highest estimates of dose at a particular location and time. When uncertainty in input parameters is taken into account, the result of the analysis is a subjective probability (uncertainty) distribution of potential inhalation doses. Again, inhalation doses are assumed to be due only to the presence of plutonium, and estimated concentrations of plutonium in surface soil are based on survey data on ^{241}Am and measured ^{239,240}Pu-to-^{241}Am ratios in soil samples. The analysis is based on assumptions summarized in Table E.1 and described as follows.

- Concentrations of ^{241}Am in surface soil in the forward area at Shot HOOD after the period of atomic testing ended are described by a lognormal probability distribution with an 80% confidence interval of 20-200 nCi m^{-2}. This uncertainty takes into account that exposures to resuspended fallout occurred at various locations in the forward area as the participants carried out their assigned activities. The lower confidence limit of 20 nCi m^{-2} is an estimate in regions of Area 9 away from any shot locations (McArthur, 1991), and it represents a general background level of local fallout in Area 9 from all shots at the NTS. The upper confidence limit of 200 nCi m^{-2} is based on data in the area of Shot HOOD shown in Figure E.1. Much higher concentrations to the south and southwest of ground zero for Shot HOOD are excluded because, as discussed in the previous section, these concentrations presumably are due primarily to fallout from later safety shots in Operation HARDTACK-II (Hawthorne, 1979).
- The fraction of the activity of ^{241}Am in surface soil in the forward area at Shot HOOD defined above that was present at the time of detonation is described by a uniform probability distribution over the range of 0.5-1.0. That is, at least half the contamination in the forward area at the time of Shot HOOD is assumed to be due to fallout from prior shots. Because areas of high contamination from later safety shots are excluded from the analysis, it appears unlikely that a smaller

TABLE E.1   Summary of Assumed Probability (Uncertainty) Distributions of Input Parameters Used in Example Analysis of Inhalation Doses Caused by Blast-Wave Effects at Shot HOOD

Parameter	Assumed probability distribution
Concentrations of ^{241}Am in surface soil at end of period of atomic testing	Lognormal distribution with 80% confidence interval of 20-200 nCi m^{-2}
Fraction of ^{241}Am in surface soil at end of period of atomic testing that was present at time of Shot HOOD	Uniform distribution over range of 0.5-1.0
Concentrations of ^{239,240}Pu in surface soil relative to ^{241}Am[a]	Lognormal distribution with 90% confidence interval of 4–18
Resuspension factor associated with blast wave	Lognormal distribution with 90% confidence interval of 10^{-4}–10^{-2} m^{-1}
Fraction of resuspended activity attached to respirable particles (AMAD, 1 μm)	Uniform distribution over range of 0.2-1.0
Breathing rate	Lognormal distribution with 90% confidence interval of 0.6–1.7 m^3 h^{-1}
Organ-specific dose coefficients for inhalation of plutonium attached to respirable particles	To account for uncertainties in dosimetric and biokinetic models used to calculate dose coefficients, lognormal distributions with 90% confidence interval of 0.1–10 times current ICRP recommendations for adult workers
	To account for uncertainty in biological effectiveness of alpha particles relative to photons and electrons, replacement of standard value of 20 used in ICRP dose coefficients with lognormal distribution with 95% confidence interval of 3.2-100
Time after exposure when cancer occurred in exposed person	Reduces dose of concern to all organs and tissues other than the lung by factor of 2

[a]Other radionuclides in fallout from prior shots that were present at time of Shot HOOD and could contribute to inhalation doses are neglected.

fraction of the assumed contamination at the time of detonation is due to fallout from prior shots, and it is possible that nearly all the contamination is due to prior fallout.

• Concentrations of ^{239,240}Pu relative to ^{241}Am in surface soil in the forward area at Shot HOOD are described by a lognormal probability distribution with a 90% confidence interval of 4-18. That distribution is based on data obtained from soil samples in several areas on the NTS where extensive fallout occurred (McArthur, 1991; IT and DRI, 1995). However, measured ^{239,240}Pu-to-^{241}Am ratios in soil samples taken near the location of Shot HOOD (McArthur and Mead, 1987) are not used in this analysis. Only three soil samples were taken in the vicinity of Shot HOOD, and two of the samples, in which the concentrations of plutonium relative to ^{241}Am of about 24 and 32 are much higher than the ratios in any other areas on the NTS, were taken in the vicinity of later safety shots. Ratios of plutonium to ^{241}Am in safety shots are expected to be much higher than ratios in normal nuclear detonations and therefore are not relevant to estimating plutonium in deposited fallout at the time of Shot HOOD.

• Concentrations of plutonium in air caused by the blast wave at Shot HOOD relative to concentrations in surface soil are described by a resuspension factor that has a lognormal probability distribution with a 90% confidence interval of 10^{-4}-10^{-2} m^{-1}. The upper confidence limit is based on an assumption that all activity in surface soil was resuspended and that the height of the resulting dust cloud was about 100 m. The lower confidence limit is intended to take into account several factors, including that some of the activity in surface soil may not have been resuspended by the blast wave, that some of the resuspended material may have been in the form of large particles that fell to Earth quickly, that the height of the dust cloud may have been greater than 100 m, and that the effect of the blast wave should have diminished with increasing distance from ground zero. Resuspension factors substantially less than 10^{-4} m^{-1} do not appear to be credible, however, given that values as high as 10^{-4} m^{-1} have been observed under conditions of stress considerably less vigorous than a blast wave (Sehmel, 1984).

• The fraction of resuspended material that was in the form of small, respirable particles (AMAD, 1 μm) and remained in the air at times after detonation when exposures in the forward area occurred is described by a uniform probability distribution over the range of 0.2-1.0. That factor is intended to take into account that not all resuspended materials may have been in the form of respirable particles and that dilution of the dust cloud due to surface winds may have occurred at some locations in forward areas during the first several hours after detonation when participants were exposed.

• The breathing rate of a participant is described by a lognormal probability distribution with a 90% confidence interval of 0.6-1.7 m^3 h^{-1}. The lower confidence limit is the same as the value assumed in Section V.C.3.3 for participants engaged in light activity. The increase in the upper confidence limit compared with the assumption in Section V.C.3.3 is intended to take into account that heat exhaustion occurred in some participants (Maag et al., 1983) and that the excitement of being in the presence of dust and fires on a warm summer day could have resulted in an increase in breathing rate compared with that during normal light activity.

• Organ-specific dose coefficients for inhalation of plutonium attached to respirable particles (AMAD, 1 μm) by adult workers currently recommended by ICRP (2002) are assumed (see Section V.C.3.1, comment [1], and Table V.C.2), and the dose coefficient for the lung is assumed to represent the dose to regions of the lung where most lung cancers caused by radiation exposure occur (see Section V.C.3.1, comment [7]). Inhalation of large particles is assumed to be relatively unimportant after a blast wave, because of the vigorous nature of the stress causing resuspension and the lower dose coefficients for inhalation of large particles compared with respirable particles (see Section V.C.3.1, comments [1] and [2], and Tables V.C.2 and V.C.4).

    • Uncertainty in dose coefficients for inhalation of plutonium due to uncertainties in the dosimetric and biokinetic models used to calculate them is described by a lognormal probability distribution with a 90% confidence interval of 0.1–10 times the ICRP recommended values (see Section V.C.3.2, comment [1], and Table V.C.6).

    • The uncertainty in dose coefficients for inhalation of plutonium due to the uncertainty in the biological effectiveness of alpha particles relative to photons and electrons is described by a lognormal probability distribution with a 95% confidence interval of 3.2-100 (see Section V.C.3.2, comment [2]). That probability distribution replaces the standard assumption of 20 used in calculating dose coefficients for plutonium and other alpha-emitting radionuclides (ICRP, 1991).

    • On the basis of an assumption that a cancer was diagnosed in an exposed veteran at 35 years after exposure at Shot HOOD, the dose to all organs or tissues other than the lung calculated with the ICRP-recommended dose coefficients is reduced by a factor of 2. That reduction accounts for the difference between 50-year committed doses, as embodied in the ICRP dose coefficients, and the dose that could have caused the cancer, which is assumed to be the dose received during the first 25 years after exposure in this example (see Section V.C.3.1, comment [6]).

The inhalation dose to an organ or tissue of concern per hour of exposure is the product of the assumed plutonium concentration in air, the breathing rate, and the organ-specific dose coefficient (see Section IV.C.2, Equation IV.C-1). By multiplying the assumed probability distributions of the different parameters described above with Latin Hypercube sampling techniques and the Crystal Ball® 2000 software (Decisioneering, 2001), central estimates (50th percentiles) and upper confidence limits (95th percentiles) of estimated inhalation doses per hour of exposure given in Table E.2 are obtained. Again, the estimated doses are based on an analysis only for plutonium (^{239,240}Pu), and the estimates would increase somewhat if other radionuclides that were present in surface soil at the time of Shot HOOD were included.

E.4 DISCUSSION OF EXAMPLE ANALYSIS AND IMPORTANCE OF INHALATION DOSES

    Assumptions about probability (uncertainty) distributions of parameter values used to obtain the results in Table E.2 clearly are subjective, and other

TABLE E.2   Calculated Probability Distributions of Doses Due to Inhalation of Plutonium in Example Analysis of Blast-Wave Effects at Shot HOOD

Organ or Tissue	Probability Distribution of Equivalent Dose (rem h^{-1})[a]	
	50th percentile	95th percentile
Lung	0.06	3.7
Lymphatic tissues	0.3	19
Lower large intestine	0.0001	0.007
Red bone marrow	0.003	0.2
Bone surfaces	0.06	4.0
Liver	0.01	0.8
Bladder wall	0.0001	0.007

[a]Fiftieth percentile represents central estimate of uncertain dose, and 95th percentile represents upper confidence limit used in dose reconstructions for atomic veterans.

choices are plausible. It is partly for that reason that the results are intended to be illustrative rather than definitive. For example, the resuspension factor that describes the effect of a blast wave is highly uncertain. If the lower confidence limit of the resuspension factor were increased from 10^{-4} to 10^{-3} m^{-1}, and all other assumptions were unchanged, central estimates (50th percentiles) of the doses in Table E.2 would increase by a factor of about 3 and the 95th percentiles would increase by a factor of about 1.7. Those comparisons also indicate, however, that substantial changes in assumed probability distributions of individual parameters would be required to give substantial changes in the results, given that at least some of the parameters, such as inhalation dose coefficients, would have large uncertainty in any credible analysis.

The committee believes that two seemingly contradictory conclusions can be drawn from the example analysis summarized in Table E.2. First, because central estimates (50th percentiles) of inhalation doses are less than 0.1 rem h^{-1} in all organs or tissues except lymphatic tissues, high doses should occur only under conditions of unusually high concentrations of longer-lived radionuclides in surface soil and very high resuspension factors. For example, on the basis of data summarized by McArthur (1991), central estimates of inhalation doses due to resuspension caused by a blast wave in areas away from ground zeros of previous shots at the NTS would be about the same as doses in Table E.2 at locations, such as Area 4, where the background of local fallout is the highest. Thus, high inhalation doses are likely to have occurred only in cases of exposure in small areas near locations of previous shots where radionuclide concentrations are unusually high. Similarly, in scenarios in which resuspension was caused by light activity, such as walking, rather than a blast wave, a central estimate of the resuspension factor should be at least a factor of 1,000 less than the central estimate of 10^{-3} m^{-1} for a blast wave assumed in this analysis (see Section V.C.3.1, comment [8]), and central estimates of inhalation doses even at locations where radionuclide concentrations are the highest, such as locations in Area 4 (McArthur and Kordas, 1985), should be less than central estimates obtained in this analysis.

Second, although central estimates of inhalation doses obtained in this analysis are not high, credible upper bounds are considerably higher. If a credible upper bound of an uncertain dose is represented by the 95th percentile of a probability distribution, as assumed in the NTPR program, estimated upper bounds of doses given in Table E.2 exceed the central estimates by a factor of about 70. Estimated upper-bound doses to the lung, lymphatic tissues, bone surfaces, and liver are comparable to or greater than 1 rem h^{-1}, and doses of this magnitude could be important in evaluating claims for compensation for radiation-related diseases when veterans are to be given the benefit of the doubt in estimating dose. Such doses clearly are important, for example, in cases of lung cancer in nonsmokers and liver cancer, for which claims for compensation normally have been granted when upper-bound estimates of dose to participants exceeded about 4 rem and 1 rem, respectively (see Section III.E).

A blast wave occurred at many aboveground shots at the NTS, and participant groups engaged in various activities in forward areas after many of these shots (Barrett et al., 1986). Given the magnitude of credible upper bounds of inhalation doses obtained in the example analysis at Shot HOOD, it is evident to the committee that blast-wave effects are potentially important in dose reconstructions for the thousands of participants who engaged in activities in forward areas soon after shots at the NTS. The example analysis is but one of several instances in which substantial inhalation doses could have resulted from a blast wave.

E.5 IMPLICATIONS OF EXAMPLE ANALYSIS FOR OTHER EXPOSURE SCENARIOS

The upper-bound estimates of inhalation dose in Table E.2 also indicate that credible upper bounds could be important in other exposure scenarios in which resuspension was caused by vigorous disturbance of surface soil. For example, as discussed in Section IV.C.2.1.3 and summarized in Table IV.C.2, resuspension factors as high as 10^{-3} m^{-1} are assumed in scenarios involving assaults or marches behind armored vehicles at the NTS, and credible upper-bound estimates of inhalation doses in these scenarios could be within a factor of 10 of the estimated upper bounds in Table E.2.

However, the upper-bound estimates of inhalation dose in Table E.2 indicate that credible upper bounds are unlikely to be important in exposure scenarios in which resuspension was caused by walking or other activities that did not involve vigorous disturbance of surface soil. In those types of scenarios, credible upper-bound estimates of inhalation dose probably are at least a factor of 100 less than upper bounds given in Table E.2. Such scenarios often occurred, for example, during normal activities of most participants on residence islands in the Pacific or when participants at the NTS engaged in various light activities in forward areas before a detonation, rather than immediately afterward when blast-wave effects were important. Pre-shot activities in forward areas occurred at several tests at the NTS.

E.6 DISCUSSION OF DOSE RECONSTRUCTIONS FOR PARTICIPANT GROUPS AT SHOT HOOD

As noted in Section V.C.3.2, comment [7], the effects of a blast wave on resuspension of radionuclides were ignored in all dose reconstructions for participant groups at the NTS. Furthermore, as noted in Section V.C.3.2, comment [6], the presence of previously deposited fallout in the forward area at Shot HOOD was not accounted for in dose reconstructions for participant groups at that test.

The approach to assessing inhalation doses at Shot HOOD is indicated by a dose reconstruction for one of the participant groups (Frank et al., 1981). In that analysis, the possibility that participants received an inhalation dose due to deposited fallout that was resuspended by the blast wave was dismissed with the statement that "what dust was lofted by the shock wave had either settled or blown out of the shot area, away from the troops, before the HOOD radiation field was entered" (Frank et al., 1981). In addition, as noted in Section V.C.3.2, comment [6], dose reconstructions for all participant groups in forward areas at Shot HOOD apparently assumed that no fallout from prior shots was present (Barrett et al., 1986).

The statement by Frank et al. (1981) given above is directly contradicted by a report of activities of participant groups at Shot HOOD (Maag et al., 1983). With reference to Figure E.5, which shows routes of troop movements after the detonation, this report states that "because dust was obscuring visibility in Loading Zone Two, the helicopters delayed their departure from Yucca Pass one hour [after detonation]"; an after-action report indicates that the elapsed time was about 85 min (USMC, 1957). The helicopters picked up troops who marched from the trenches toward ground zero and back starting within 15 min after detonation (Maag et al., 1983). Because Loading Zone Two is farther from ground zero than the trenches and the line of march of troops who went toward ground zero, the only reasonable assumption is that the troops encountered high dust concentrations during the maneuver, which lasted about 2 hours (Maag et al., 1983). The report by Maag et al. (1983) also notes that "because heavy dust obscured ground points, the . . . aerial survey team could not perform its survey until about six hours after the detonation." Thus, the heavy dust cloud caused by the blast wave evidently persisted for several hours in the area near ground zero, and it is virtually certain that this cloud was encountered by troops who marched toward ground zero soon after the detonation. It also is plausible that significant remnants of the dust cloud, especially the smaller, respirable particles with very low deposition velocities (Sehmel, 1984; Luna et al., 1969), were present when

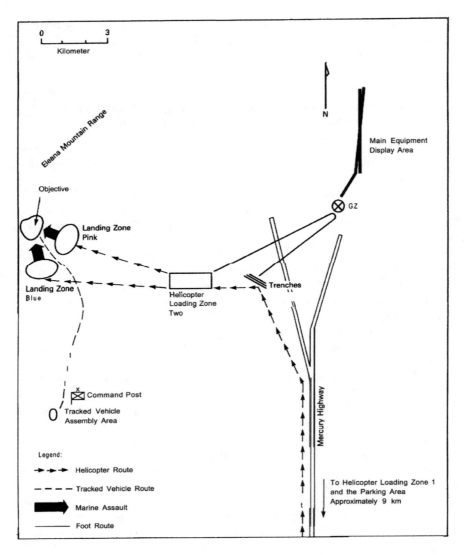

FIGURE E.5 Diagram of troop movements in forward areas after detonation of Operation PLUMBBOB, Shot HOOD at location marked "GZ" (Maag et al., 1983). Movement of troops to Main Equipment Display Area after detonation is not shown.

troops were trucked to the Main Equipment Display Area; trucking began within 6 hours of detonation and continued for about 3 hours (Maag et al., 1983).

The committee, of course, does not know the exact conditions (airborne concentrations of radionuclides) that were encountered by maneuver troops in forward areas at Shot HOOD, nor do the analysts who performed dose reconstructions. And it is documented that troops who were in forward trenches at the time of detonation were issued gas masks in anticipation of high concentrations of dust and were instructed to wear them at least until the blast wave passed (Maag et al., 1983). However, the committee believes that it is implausible to assume that troops continued to wear gas masks during the entire time spent in forward areas after the initial blast wave passed, even though dust concentrations remained high. The high temperatures on a July day and the presence of many brush fires in the area, which added to the heat and obscured visibility, made it difficult to carry out maneuvers while wearing respiratory protection. Thus, without regard for whether a participant in the trenches maneuvered toward ground zero, marched west toward Loading Zone Two to await helicopter airlift, or waited in the trench area for transport to the vehicle assembly area (Maag et al., 1983), it is highly plausible that inhalation of dust resuspended by the blast wave occurred throughout the period from shortly after detonation until the tour of the equipment display area ended several hours later. Although airborne concentrations of radionuclides undoubtedly decreased over that time, the concentrations of respirable particles probably did not decrease by large amounts, because of their low deposition velocities and the low wind speeds. Furthermore, this dust cloud undoubtedly contained plutonium and other longer-lived radionuclides of importance in estimating inhalation dose.

Dose reconstructions are supposed to be based on plausible assumptions that give the veterans the benefit of the doubt (see Section I.C.3.2). The approach taken in dose reconstructions for participant groups at Shot HOOD (Barrett et al., 1986; Frank et al., 1981) of completely ignoring blast-wave effects and inhalation of resuspended dust containing plutonium and fission products that undoubtedly was present in the exposure environment in the forward areas is not plausible, and it certainly does not give the affected veterans the benefit of the doubt.

REFERENCES IN APPENDIX E

Barrett, M., Goetz, J., Klemm, J., McRaney, W., Phillips, J. 1986. Low Level Dose Screen – CONUS Tests. McLean, VA: Science Applications International Corporation; Report DNA-TR-85-317.

Decisioneering. 2001. Crystal Ball® 2000.2 User Manual. Denver, CO: Decisioneering, Inc.

Frank, G., Goetz, J., Klemm, J., Thomas, C., Weitz, R. 1981. Analysis of Radiation Exposure, 4th Marine Corps Provisional Atomic Exercise Brigade, Exercise Desert Rock VII, Operation PLUMBBOB. McLean, VA: Science Applications, Inc.; Report DNA 5774F.

Hawthorne, H. A. (Ed). 1979. Compilation of Local Fallout Data from Test Detonations 1945-1962 Extracted from DASA 1251. Volume I. Continental U.S. Tests. Santa Barbara, CA: General Electric Company; Report DNA 1251-1-FX.

ICRP (International Commission on Radiological Protection). 1991. 1990 Recommendations of the International Commission on Radiological Protection. ICRP Publication 60, Annals of the ICRP 21(1-3). Elmsford, NY: Pergamon Press.

ICRP (International Commission on Radiological Protection). 2002. The ICRP Database of Dose Coefficients: Workers and Members of the Public, Compact Disc Version 2.01. Stockholm, Sweden: International Commission on Radiological Protection.

IT and DRI (IT Corporation and Desert Research Institute). 1995. Evaluation of Soil Radioactivity Data from the Nevada Test Site. Las Vegas, NV: U.S. Department of Energy Nevada Operations Office; Report DOE/NV-380.

Luna, R. E., Church, H. W., Shreve, J. D., Jr. 1969. A Model for Plutonium Hazard Assessment Based on Operation ROLLER-COASTER. Albuquerque, NM: Sandia Laboratories; Report SC-WD-69-154.

Maag, C., Wilkinson, M., Striegel, J., Collins, B. 1983. Shot HOOD – A Test of the PLUMBBOB Series. McLean, VA: JRB Associates, Inc.; Report DNA 6002F.

McArthur, R. D. 1991. Radionuclides in Surface Soil at the Nevada Test Site. Las Vegas, NV: Desert Research Institute; Water Resources Center Publication #45077, Report DOE/NV/10845-02.

McArthur, R. D., Kordas, J. F. 1985. Radionuclide Inventory and Distribution Program: Report #2. Areas 2 and 4. Las Vegas, NV: Desert Research Institute; Water Resources Center Publication #45-41, Report DOE/NV/10162-20.

McArthur, R. D., Mead, S. W. 1987. Radionuclide Inventory and Distribution Program: Report #3. Areas 3, 7, 8, 9, and 10. Las Vegas, NV: Desert Research Institute; Water Resources Center Publication #45056, Report DOE/NV/10384-15.

Sehmel, G. A. 1984. Deposition and Resuspension. In: Atmospheric Science and Power Production, ed. by D. Randerson. Washington, DC: U.S. Department of Energy; Report DOE/TIC-27601, pp. 533–583.

USMC (U.S. Marine Corps). 1957. Provisional Atomic Exercise Brigade, Report of Exercise Desert Rock VII, Marine Corps. Camp Pendleton, CA: Headquarters, 4th Marine Corps.

Appendix F

Unit Dose Reconstruction for Task Force WARRIOR at Operation PLUMBBOB, Shot SMOKY

F.1 INTRODUCTION

This appendix presents an evaluation of a unit dose reconstruction for a participant group called Task Force WARRIOR at Operation PLUMBBOB, Shot SMOKY at the NTS (Goetz et al., 1979). The committee's evaluation focuses on development of a plausible exposure scenario for members of the task force who received unusually high external doses. In dose reconstructions for atomic veterans, assumptions about exposure scenarios are of paramount importance for obtaining credible upper-bound estimates of dose for use in evaluating claims for compensation for radiation-related diseases (see Section I.C.3.2 and introduction to Chapter IV).

F.2 ACTIVITIES OF TASK FORCE WARRIOR AT SHOT SMOKY

The discussions in this appendix concern only particular activities of Task Force WARRIOR in the first few hours after detonation of Shot SMOKY, specifically attempts by two units of the task force, the 2nd and 3rd Platoons of Company C, to reach the objectives of a planned maneuver shown in Figure F.1. Activities of interest are summarized in the following paragraphs (Jensen, 1957; Harris et al., 1981). The exposure-rate contour lines shown in Figure F.1 are discussed in the next section.

Shot SMOKY was detonated on August 31, 1957, at 0530 hours in Area 8, shot tower location $T2_C$ (see Figure V.C.6 in Section V.C.3.2). Some units observed the detonation at a location about 12 km south-southwest of ground zero (GZ) indicated in Figure F.1, and they moved to the nearby loading area shortly

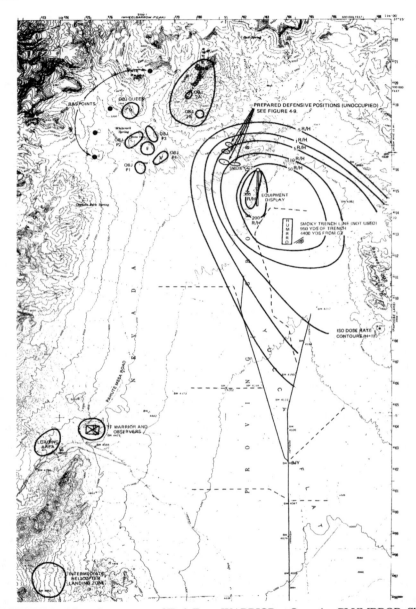

FIGURE F.1   Map of maneuvers of Task Force WARRIOR at Operation PLUMBBOB, Shot SMOKY (Harris et al., 1981) (approximate scale: 1 cm = 1.14 km). Exposure-rate contours at H + 1 hour in R h^{-1} ("R/H") are as given by Hawthorne (1979). However, assumption that contour lines were closed to northwest of ground zero was not based on measurement and was not confirmed by later survey data (REECO, undated); see Appendix F.3.

thereafter. Beginning shortly after 0700 hours, those units were transported by helicopter to landing sites about 5 km west-northwest of ground zero. By 0740 hours, the 2nd and 3rd Platoons had landed and seized Objectives P3 and P4.

At 0830 hours, the 2nd and 3rd Platoons were ordered to attack and seize Objectives 2A and 2B, which were on Quartzite Ridge about 4.5 and 3.5 km northwest of ground zero, respectively. The two platoons immediately moved to accomplish those missions. The next report of activity was at 0915 hours, or 45 min after the assault began. At that time, the commander of Task Force WAR-RIOR reported to the battle group commander that "the 2nd and 3rd Platoons, attacking to seize Objectives 2A and 2B, had advanced to the points permitted by [radiation safety] personnel and had been halted prior to seizure of Objective 2" (Jensen, 1957). At 0935 hours, a request was made for delivery of water to the west slope of Quartzite Ridge to supply the troops in that area. The last report of activity was at 0945 hours, when the exercise was terminated; the request for water was not honored for this reason.

Termination of the exercise 45 min after the assault on Objectives 2A and 2B began was unexpected because planning for the exercise called for resupply of the objective area by helicopters "for a period of not less than two days" (Jensen, 1957). The request for delivery of water noted above presumably was part of the plan of normal resupply.

F.3 RADIOLOGICAL CONDITIONS IN MANEUVER AREA

At the time Shot SMOKY was detonated, surface winds were calm, and winds above about 6,000 ft (1,800 m) were from the north and northwest (Hawthorne, 1979). Thus, as anticipated during planning for the test and post-shot maneuvers, the pattern of fallout near ground zero was mainly to the south and southeast. The exposure-rate contours at H + 1 hour shown in Figure F.1 (Hawthorne, 1979) were based on survey data at H + 8 hours and at 1, 3, and 5 days after detonation (REECO, undated). However, it is important to emphasize that survey data were not taken in the vicinity of Objectives 2A and 2B on Quartzite Ridge, because of rough terrain in the area, and the assumption in Figure F.1 that exposure-rate contours were closed to the northwest of ground zero (that is, that there was no significant fallout beyond the contour lines in this direction) was not confirmed by the later surveys (REECO, undated). Thus, survey readings do not directly address the question of whether there was significant fallout at locations of the planned assaults by the 2nd and 3rd Platoons.

Reports of activities of the 2nd and 3rd Platoons during their assaults on the two objectives on Quartzite Ridge, as described in the previous section, provide compelling evidence that these units encountered unexpected radiological conditions. The assault was halted before they seized their objectives (and much sooner than planned), and this action was taken by radiation-safety personnel who accompanied the platoons and monitored radiological conditions during the assault.

There is no apparent reason that the exercise would have been terminated if the expected low levels of radiation had been encountered.

Another interesting aspect of the early phases of the exercise was the apparent presence of a dense dust cloud produced by the blast wave of Shot SMOKY. At 15–18 minutes after detonation, the first helicopters left an assembly area about 25 km south of ground zero and headed toward landing sites west-north-west of ground zero. The task of the pathfinder team in the initial landings was to determine that radiological conditions would permit deployment of troops in the task force who awaited transport at the loading area. A report of the initial helicopter flights (Jensen, 1957) includes the following statement:

> The pathfinder serial was forced to deviate from its proposed flight path because of a dense smoke and dust cloud which lay between it and the objective area. Taking advantage of a west wind which was beginning to move the cloud back in the direction of ground zero, the flight flew around the cloud and landed in an eastern approach, on appointed landing sites under conditions of visibility that did not exceed 800 yards . . . Visibility did not permit the establishment of the designated release point for a period of 30 minutes after the pathfinder landings.

Because the pathfinder team landed at 0617 hours, dust evidently was a problem for more than an hour after detonation, and some dust could have persisted after the 2nd and 3rd Platoons landed at about 0715 hours.

In addition, an aerial survey team that began making measurements one-half hour after detonation observed large dust clouds in the area of ground zero "which persisted for several hours" (Harris et al., 1981). The presence of an extensive dust cloud is significant because the area affected by the blast wave from Shot SMOKY was contaminated by fallout from previous shots, including PLUMBBOB Shots BOLTZMANN, DIABLO, and SHASTA (see Table IV.C.1 in Section IV.C.2.1.1 and Figure V.C.6 in Section V.C.3.2).

F.4 DOSE RECONSTRUCTION FOR TASK FORCE WARRIOR

This section describes aspects of the unit dose reconstruction for Task Force WARRIOR (Goetz et al., 1979) that apply to the 2nd and 3rd Platoons that assaulted Objectives 2A and 2B on Quartzite Ridge. The chronology of events up to when the exercise was halted by authority of radiation-safety personnel who accompanied the platoons is as described in Appendix F.2. The dose reconstruction then included information described in the following paragraphs.

Section 2.6 of the report documenting the dose reconstruction (Goetz et al., 1979), which describes operations on the day of Shot SMOKY, includes the following statement:

> The extent of fallout patterns . . . indicates that [the radiation safety] criterion should not have been a factor in halting the advance, if the path of the assault was as planned. Because the planned path of direct assault would have encoun-

tered some very steep slopes, the assault may have deviated to the south and east. This excursion could have led the 2nd Platoon toward the SMOKY fallout field where residual radiation levels were sufficient to cause [radiation-safety personnel] to halt the attack.

The fallout patterns referred to in that statement include the exposure-rate contours at H + 1 hour in Figure F.1. It is again important to emphasize that the assumption that the contour lines were closed to the northwest of ground zero and did not extend as far as Quartzite Ridge was not based on measurement shortly after detonation and was not confirmed by later surveys (REECO, undated). The 2nd Platoon assaulted Objective 2B from Objective P4 in the landing area.

Section 6 of the report by Goetz et al. (1979) discusses film-badge dosimetry for Task Force WARRIOR. Film-badge records for members of the task force were available for use in dose reconstruction. The report notes that many participants were issued two film badges, the first covering pre-shot operations up to August 27 and the second covering the period from August 27 to September 2, including operations on the day of Shot SMOKY.

The analysts noted that the film-badge readings cluster into two groups (Goetz et al., 1979). In the first cluster, which included about 95% of the film badges, doses for the period covering pre-shot operations (390 ± 150 mrem) were about twice the doses for the period covering operations on shot day (185 ± 60 mrem), indicating that most of the dose was due to exposure to residual fallout from the prior PLUMBBOB shots noted in Appendix F.2. In the second cluster, which included 20 film badges, doses for the period covering operations on shot day (1140 ± 150 mrem) were nearly 3 times as high as doses for the period covering pre-shot operations (405 ± 130 mrem). The agreement of pre-shot doses in the two clusters was noted, and the report also discusses how the film-badge readings during the first period are in reasonable agreement with estimates of dose that were based on the exposure-rate contours for the previous shots that deposited fallout in the area (Hawthorne, 1979), knowledge of the locations and times of activities of members of the task force, and assumptions about decreases in dose rates from deposited fallout over time.

Goetz et al. (1979) discuss the cluster of 20 unusually high film-badge readings during the period of operations on shot day. Those doses ranged from 800 to 1,400 mrem, and there was evidence that the spectrum of radiation was similar in all 20 badges. On the basis of that information, the analysts inferred that the 20 badges were worn by members of a group that stayed together during the operation, and their doses were assumed to apply to members of the 2nd Platoon, which was supposed to assault Objective 2B. On the basis of the presumption that the high doses were received on the day of Shot SMOKY, the report states (Goetz et al., 1979) that:

[I]t follows that the group would have been closer to SMOKY [ground zero] than the task force as a whole. Nowhere else on 31 August could the 800 to 1,400 mrem dose level have been achieved unless the group was subjected to the radiation field of SMOKY itself. It could be postulated from the film badge evidence . . . that [a group] could have proceeded due east from the objective area toward Smoky Hill and the Phase I positions instead of assaulting Quartzite Ridge to the northeast. Whether such an excursion was by oversight or design is immaterial. In either case, it is undocumented. The excursion would explain, however, why the assault was halted due to [radiation safety] considerations, presumably at the 500 [mR h^{-1}] level. An examination of the SMOKY residual contamination contours would support this hypothesis—the group would have halted short of Smoky Hill near the Phase I defensive positions, having proceeded less than two miles in about 45 min. . . . They could have remained in the vicinity of the 500 [mR h^{-1}] line until exercise termination at 0945, inspecting the post-shot damage to the defensive positions . . . If they had adhered to the 500 [mR h^{-1}] limit, had not encountered any hot spots, and had departed promptly at exercise termination, their total dose from this excursion would have been about 300 mrem. They may have encountered hot spots, however. It is also possible that they ventured toward the close-in positions where intensities were greater, or stayed long enough to view all the positions. Given the uncertainties of this excursion, doses on the order of 1,000 mrem cannot be ruled out.

The Phase I defensive positions noted in these statements are the positions (trenches) within about 1,850 m northwest of ground zero shown in Figure F.1. The presumption that the excursion was halted when the exposure rate exceeded 500 mR h^{-1} may be questioned because a later report states that the limit for maneuver troops probably was 10 mR h^{-1} and that an exposure rate of 100 mR h^{-1} defined full radiological exclusion areas (Harris et al., 1981). However, such discrepancies are not important when doses are measured by film badges.

Finally, the analysts developed an upper-bound estimate of external dose to individual members of Task Force WARRIOR of 1,530 mrem based on film-badge readings during pre-shot and post-shot phases of the operation. In discussing the upper-bound estimate, the report notes that "if the group of 20 [film badges] are excluded due to evident incompatibility with the troop movements as known, the film badge equivalent of the upper exposure limit compares favorably with the highest combined film badge readings for any individual" (Goetz et al., 1979). The term "film badge equivalent" refers to a reconstructed dose based on exposure-rate contours estimated from survey data, as described previously.

F.5 DISCUSSION OF EXPOSURE SCENARIO ASSUMED IN DOSE RECONSTRUCTION

In reconstructing external doses received by the two platoons that assaulted objectives on Quartzite Ridge, the committee believes that it is reasonable to assume that the 20 film badges discussed in the previous section that had unusually high readings for the period including the day of Shot SMOKY were worn by members of the 2nd Platoon, as assumed in the dose reconstruction (Goetz et al., 1979).

The essence of the analysts' explanation of the unusually high film-badge readings, as given in the previous section, is that the 2nd Platoon disobeyed orders, either deliberately or inadvertently, and went toward ground zero instead of the planned objective. The reason given is that the planned path of assault on Objective 2B would have encountered very steep slopes and that a substantial deviation from the planned path of assault is the only way that unexpectedly high radiation levels could have been encountered, according to an assumption that the exposure-rate contours shown in Figure F.1 adequately describe the radiation environment to the northwest of ground zero.

The committee does not believe that the analysts' explanation of the unusually high readings on the 20 film badges is credible. There is no documentation that either of the two platoons encountered inaccessible terrain in assaulting the objectives. Furthermore, Figure F.1 indicates that the steepest terrain lay beyond the locations of the objectives. It also does not seem reasonable that the objectives would have been placed at locations that were inaccessible by direct assault, especially inasmuch as two plans for the exercise were rehearsed in advance (Jensen, 1957), as acknowledged in the dose reconstruction (Goetz et al., 1979). The committee also notes that a deviation from the planned path of assault by the 2nd Platoon to an extent sufficient to approach ground zero, instead of Objective 2B, would have required a change in direction of about 90°. Such a large deviation was not required to avoid steep terrain.

It is, of course, possible that the 2nd Platoon deliberately disobeyed orders and marched toward the defensive positions (trenches) much closer to ground zero and therefore encountered higher radiation levels. However, such an action would have entailed substantial risk of discovery with no evident benefit, given that maneuver units were to be resupplied regularly (Jensen, 1957) and that their absence from the area of the objectives would have been noticed. The committee does not believe that it is reasonable to assume that military units deliberately disobeyed orders unless there is documented evidence of such actions.

F.6 ALTERNATIVE EXPOSURE SCENARIO FOR TASK FORCE WARRIOR

The committee believes that there is a plausible explanation for the unusually high film-badge readings that apparently occurred for members of at least one of the two platoons during the aborted assault on Objectives 2A and 2B on Quartzite Ridge. This explanation does not require assumptions that a platoon did not follow the plan of assault, and it is based on observations at a later shot in the area of Shot SMOKY.

There is no doubt that at least one of the two platoons encountered unexpected radiological conditions during the period after the assault on Objectives

2A and 2B began. With reference to Figure F.1, Shot SMOKY was detonated in an area of north Yucca Flat that is ringed to the north and west by mesas, as indicated by the dense contour lines to the northwest of the landing areas. The elevation at ground zero is about 1,400 m, and the elevation of the mesas reaches about 2,300 m. The detonation occurred about one-half hour before sunrise. The upper part of the cloud of debris that rose above the elevation of the mesas was carried to the south and southeast by the winds aloft (Hawthorne, 1979), as indicated by the fallout pattern in Figure F.1. However, as the sun rose, it is plausible that warming of the east- and south-facing slopes of the mesas caused an updraft and, therefore, a northwestly wind at low elevations in the area of the maneuvers. As a result, the stem and lower portions of the cloud that had not risen above the mesas and portions of the remaining dust cloud caused by the blast wave could have been transported in the direction of the platoons on Quartzite Ridge and resulted in fallout along the planned path of the assaults.

The scenario described above does not contradict a statement in Appendix F.3 that a dust cloud in the vicinity of the landing areas to the north-northwest of ground zero was soon transported back toward ground zero by a west wind, because the eastward movement of the dust cloud, which could have been caused by reflection of the blast wave from the slopes of the mesas to the west, occurred 3 hours before the assault by the 2nd and 3rd Platoons was halted (Jensen, 1957). Thus, the two occurrences probably are unrelated.

The committee believes that two pieces of evidence support an assumption that the platoons assaulting Quartzite Ridge encountered unexpected fallout that was not lofted above the mesas but was transported to the northwest of ground zero at low elevations. First, such an occurrence after underground Shot BANE-BERRY in December 1970 is documented. As shown in Figure F.2, photographs of the plume from Shot BANEBERRY, which vented to the atmosphere about 4 km west of ground zero of Shot SMOKY, clearly indicate movement of the lower portion toward the northwest, with the upper portion drifting north and east under the influence of winds aloft.

Second, ground survey teams that approached ground zero from different radial directions within the first 3 hours after detonation of Shot SMOKY, including a monitor who approached from the southwest, encountered conditions that caused erratic instrument readings and contamination of the instruments (REECO, undated; Harris et al., 1981). Those conditions could have been caused only by airborne radionuclides that were not carried in the rapidly rising fireball but were separated from the fireball, probably by reflection of the blast wave from the slopes of surrounding mesas, including a mesa to the north indicated in Figure F.1. If most of the radionuclides had been carried in the fireball, the extensive fallout encountered by the ground survey team would not have occurred for several hours, given the typical height of a cloud of about 10 km (see Section IV.C.2.1.5) and a typical fall velocity of large fallout particles of about 10 cm s^{-1}

FIGURE F.2   Photograph of plume from underground Shot BANEBERRY showing separation of lower and upper portions due to different directions of winds near the ground
and aloft.

(Sehmel, 1984). That assertion is supported by a report that stable measurements
by the survey team were possible by 0900 hours, or about 4 hours after detonation
(Harris et al., 1981). Airborne radionuclides at low elevations that were separated
from the fireball were available for transport to the northwest toward Quartzite
Ridge as an updraft at the slopes of mesas farther to the northwest occurred.

Thus, the committee believes that the most likely scenario for exposure of at
least one of the platoons on Quartzite Ridge is that airborne radionuclides that
were separated from the rapidly rising fireball were transported from the area
around ground zero by surface winds generated by an updraft at the slopes of
mesas to the north and west of Quartzite Ridge as the slopes were heated by the
morning sun. The credibility of that scenario is supported by a similar occurrence
at underground Shot BANEBERRY near ground zero of Shot SMOKY and by
the evident presence of substantial amounts of airborne radionuclides at low
elevations in the vicinity of ground zero for more than 3 hours after detonation. In
the scenario, the unexpected radiological conditions experienced by at least one
of the platoons presumably were due to fallout at planned locations of the assaults
on Objectives 2A and 2B.

F.7 EVALUATION OF APPROACH TO DOSE RECONSTRUCTION

The committee's primary concern about the unit dose reconstruction for Task Force WARRIOR (Goetz et al., 1979) involves the assumed exposure scenario for the two platoons that assaulted Quartzite Ridge and encountered unexpected radiological conditions that resulted in premature termination of the exercise. It is the committee's opinion that by forcing the scenario to be consistent with the expected radiation environment, the analysts developed a scenario that required implausible and unsupported assumptions about the actions of one of the platoons and thus lacked credibility. The analysts apparently assumed that available survey data after Shot SMOKY defined the radiation environment in areas near Quartzite Ridge during the assault and thus precluded high radiation levels at those locations and times. However, radiation levels were not measured near Quartzite Ridge to support the assumed scenario (REECO, undated). Rather, radiation levels to the northwest of ground zero were simply assumed on the basis of an extrapolation of measurements elsewhere and the belief that the entire plume traveled south and east away from Quartzite Ridge. The committee believes that there is a plausible scenario that is consistent with all available information and would explain the unexpected radiological conditions encountered during the assaults on Quartzite Ridge, without the need to invoke an assumption that one of the platoons inadvertently or deliberately disobeyed orders and marched a considerable distance in a different direction and into a high-radiation area near ground zero.

The committee's concern about the exposure scenario assumed by Goetz et al. (1979) goes beyond the reconstruction of external dose for Task Force WARRIOR. An assumption of an implausible scenario does not have important consequences for estimating external doses to members of the platoons on Quartzite Ridge, because doses could be estimated from film-badge readings. Rather, the implausible scenario is important, in part, because it ignores possible doses due to inhalation of descending fallout, which probably contained important longer-lived radionuclides that were deposited in the area after previous shots and were resuspended by the Shot SMOKY blast wave.

Of greater concern to the committee is that the assumption of an implausible exposure scenario in this case is not an isolated occurrence. In its review of 99 randomly selected individual dose reconstructions, as discussed in Section V.A, the committee encountered several cases in which an analyst developed an exposure scenario based on prior expectations of exposure conditions or a scenario that conformed to a plan of operation or operational radiation protection guidelines but in doing so ignored evidence, including statements by participants, that indicated that the actual conditions of exposure did not conform to the assumptions. Plausible scenarios that could have resulted in substantially higher doses than were obtained in a dose reconstruction were not considered. Indeed, in contrast to the unit dose reconstruction for Task Force WARRIOR, in which the analysts assumed that troops must have disobeyed orders to receive external doses indicated by film-badge readings, analysts have argued in other dose reconstructions that a plausible scenario could not have occurred because it was against orders, did not conform to a plan of operation, or resulted in doses exceeding operational guidelines (see case #3, 58, 77, and 92).

The committee believes that development of exposure scenarios for use in dose reconstructions for atomic veterans should not be dictated by plans of operation. Rather, plausible alternatives involving unexpected occurrences should be considered and evaluated when they are supported by available information, as is sometimes the case. The goal should be to develop plausible scenarios that are consistent with the body of available information and result in the highest estimates of dose to give veterans the benefit of the doubt as required by regulations governing the NTPR program.

REFERENCES IN APPENDIX F

Goetz, J. L., Kaul, D., Klemm, J., McGahan, J. T. 1979. Analysis of Radiation Exposure for Task Force WARRIOR—Shot SMOKY—Exercise Desert Rock VII-VIII, Operation PLUMBBOB. McLean, VA: Science Applications, Inc.; Report DNA 4747F.

Harris, P. S., Lowery, C., Nelson, A. G., Obermiller, S., Ozeroff, W. J., Weary, E. 1981. Shot SMOKY, a Test of the PLUMBBOB Series, 31 August 1957. Alexandria, VA: JAYCOR; Report DNA 6004F.

Hawthorne, H. A. (Ed). 1979. Compilation of Local Fallout Data from Test Detonations 1945–1962 Extracted from DASA 1251. Volume I. Continental U.S. Tests. Santa Barbara, CA: General Electric Company; Report DNA 1251-1-FX.

Jensen, W. A. 1957. Report of Test, Infantry Troop Test, Exercise Desert Rock VII and VIII. San Francisco, CA: Headquarters, Sixth US Army; Report AMCDR-S-3 (December 11).

REECO (Reynolds Electrical & Engineering Company, Inc.). Undated. PLUMBBOB On-Site Rad-Safety Report. Las Vegas, NV: U.S. Atomic Energy Commission; Report OTO-57-2.

Sehmel, G. A. 1984. Deposition and Resuspension. In: Atmospheric Science and Power Production, ed. by D. Randerson. Washington, DC: U.S. Department of Energy; Report DOE/TIC-27601, pp. 533–583.

Glossary

absorbed dose: The energy imparted by ionizing radiation per unit mass of material irradiated. For purposes of radiation protection and assessing risks to human health, the quantity normally calculated is the **average absorbed dose** in an organ or tissue, given by the total energy imparted to that organ or tissue divided by the total mass. The SI unit of absorbed dose is the joule per kilogram ($J kg^{-1}$), and its special name is the gray (Gy). In conventional units used in this report, absorbed dose is given in rads; 1 rad = 0.01 Gy.

absorption: The passage of a substance across an exchange barrier in the respiratory tract, gastrointestinal tract, or skin into blood.

accuracy: The extent of agreement between a measurement or prediction of a quantity and its actual value. An accurate measurement or prediction should be precise and unbiased. See also **bias** and **precision**.

activation: The production of radionuclides by capture of radiation (for example, neutrons) in atomic nuclei.

activity: The rate of transformation (or distintegration or decay) of radioactive material. The SI unit of activity is the reciprocal second (s^{-1}), and its special name is the becquerel (Bq). In conventional units used in this report, activity is given in curies (Ci); $1 \text{ Ci} = 3.7 \times 10^{10}$ Bq.

activity median aerodynamic diameter: The diameter in an aerodynamic particle-size distribution for which the total activities on particles above and below this size are equal. A lognormal distribution of particles sizes usually is assumed.

agent: An active force (such as ionizing radiation) or substance that produces or is capable of producing an effect.

alpha particle: An energetic nucleus of a helium atom, consisting of two protons and two neutrons, that is emitted spontaneously from nuclei in decay of some radionuclides; also called **alpha radiation** and sometimes shortened to **alpha** (for example, alpha-emitting radionuclide). Alpha particles are weakly penetrating and can be stopped by a sheet of paper or the outer dead layer of skin.

atmospheric testing: Detonation of nuclear weapons or devices in the atmosphere or close to the earth's surface as part of the nuclear-weapons testing program.

atom: The smallest particle of a chemical element that cannot be divided or broken up by chemical means. An atom consists of a central nucleus of protons and neutrons and orbital electrons surrounding the nucleus.

atomic bomb: A nuclear weapon that relies on fission only, in contrast to a thermonuclear ("hydrogen") bomb that uses fission and fusion.

atomic nucleus: The dense core of an atom, composed of protons and neutrons.

atomic veteran: A person who, while serving as a member of the armed forces, was a participant at one or more atmospheric nuclear-weapons tests, served in occupation forces in Hiroshima or Nagasaki, Japan, or was a prisoner of war in Japan at the time of the bombings of Hiroshima and Nagasaki.

attenuation: The reduction in intensities of radiation in passing through matter by a combination of scattering and other interactions with electrons and atomic nuclei.

background radiation: Ionizing radiation that occurs naturally in the environment including: cosmic radiation; radiation emitted by naturally occurring radionuclides in air, water, soil, and rock; radiation emitted by naturally occurring radionuclides in tissues of humans and other organisms; and radiation emitted by human-made materials containing incidental amounts of naturally occurring radionuclides (such as building materials). Background radiation may also include radiation emitted by residual fallout from nuclear-weapons tests that has been dispersed throughout the world. The average annual effective dose due to natural background radiation in the United States is about 0.1 rem, excluding the dose due to indoor radon, and the average annual effective dose due to indoor radon is about 0.2 rem.

badged dose: An estimate of a person's external radiation dose, specifically the **deep equivalent dose** from external exposure to photons, as derived from readings of exposure by one or more film badges assigned to the person.

basal cells: Cells in the epidermis that give rise to more specialized cells and act as stem cells.

basal cell carcinoma: A malignant growth originating from basal cells that is most common in fair-skinned or sun-exposed areas; the most common form of skin cancer.

becquerel: The special name for the SI unit of activity; $1 \text{ Bq} = 1 \text{ s}^{-1}$.

beta-to-gamma dose ratio: An estimated ratio of the equivalent dose to the skin or lens of the eye from external exposure to beta particles to the associated equivalent dose to the whole body from external exposure to photons.

beta particle: An energetic electron emitted spontaneously from nuclei in decay of some radionuclides and produced by transmutation of a neutron into a proton; also called **beta radiation** and sometimes shortened to **beta** (for example, beta-emitting radionuclide). Beta particles are not highly penetrating, and the highest-energy beta radiation can be stopped by a few centimeters of plastic or aluminum.

bias: The systematic tendency of a measurement or prediction of a quantity to overestimate or underestimate the actual value, on average. See also **accuracy** and **precision**.

bias factor: In external radiation dosimetry, an estimated ratio of the **exposure** recorded by a film badge to the corresponding **deep equivalent dose** in humans. The bias factor normally is greater than 1.

bioassay: The determination of the kinds, quantities or concentrations, and, in some cases, locations of radioactive material in the human body, either by direct measurement (*in vivo* counting) or by analysis of materials excreted or removed from the body.

biokinetic model: A model describing the time course of absorption, translocation, distribution in organs or tissues, metabolism, and excretion of a substance (such as a radionuclide) introduced into the body by ingestion, inhalation, or absorption through the skin or an open wound.

biological effectiveness: The ability of ionizing radiation to induce biological responses in tissues of humans. The biological effectiveness of a particular type of ionizing radiation may be represented by the **quality factor, radiation effectiveness factor, radiation weighting factor,** or **relative biological effectiveness.**

biological half-time: The time required for half the quantity of a material taken into the body to be eliminated from the body by biological processes. For radionuclides, the biological half-time does not include elimination by radioactive decay.

biological response: A significant adverse effect in an organism resulting from exposure to a hazardous agent. The determination of whether an effect is significant or adverse sometimes involves subjective judgment. Often called a biological endpoint or biological effect in the literature.

calibration: The check or correction of a measuring instrument by comparing an instrument reading with a standard of known accuracy, in order to ensure acceptable operational characteristics.

cancer: A malignant tumor of potentially unlimited growth that expands locally by invasion and systemically by metastasis.

carcinogen: An agent capable of inducing cancer.

carcinoma: A malignant tumor that occurs in epithelial tissues, which cover the body or body parts and serve to enclose and protect those parts, to produce secretions and excretions, and to function in absorption.

cataract: A clouding of the lens of the eye, or its capsule, that obstructs the passage of light.

central estimate: A "best" estimate of the dose received by an individual, as distinct from an upper bound of the dose that accounts for uncertainty in that estimate.

chronic lymphocytic leukemia: A slowly progressing form of leukemia, characterized by an increase in the number of white blood cells known as lymphocytes, which studies have not shown to be caused by radiation in humans.

ciliary action: Locomotion produced by the movement of minute hair-like cells on the surface of tissues, especially in the respiratory tract.

clearance: The removal of inhaled substances from the respiratory tract by mechanical processes or absorption.

Code of Federal Regulations: Codification of general and permanent rules published in the *Federal Register* by executive departments and agencies of the federal government and published annually by the US Government Printing Office.

coefficient of variation: Ratio of the **standard deviation** of a set of values to the **mean**.

cohort: A group of individuals having a common association or factor.

committed dose: The dose (that is, the **absorbed dose**, **equivalent dose**, **effective dose**, or **effective dose equivalent**) delivered to specified organs or tissues over a specified period after an acute intake of a radionuclide by ingestion, inhalation, or absorption through the skin or an open wound. For adults, the period over which committed doses are calculated normally is 50 years.

composite dose coefficient: A **dose coefficient** that applies to an assumed mixture of radionuclides of specified relative activities.

confidence interval: An estimate of the range within which the true value of an uncertain quantity is expected to occur in a specified percentage of measurements or predictions. For example, a 90% confidence interval of (x, y) means that, on the basis of available information, the probability is 0.9 that the true value lies between x and y. See also **lower confidence limit** and **upper confidence limit**.

correlation: The degree of linear association between two variables, normally described by a unitless correlation coefficient that lies between -1 and $+1$. A correlation coefficient of 0 implies no linear association, whereas a value of -1 or $+1$ implies a perfect linear association, one variable increasing while the other decreases in the first instance, and both increasing in the second.

cosmic radiation: Particulate and electromagnetic radiation that originates in space, including secondary radiation produced by interactions with the constituents of the earth's atmosphere.

curie: The conventional unit of activity, equal to 3.7×10^{10} disintegrations per second (Bq).

database: An organized set of data or collection of files that can be used for a specific purpose.

deck log: A document that records the daily activities of Navy and Coast Guard ships, including a list of officers on board.

deep equivalent dose: The equivalent dose from external exposure of the whole body estimated at a depth of 1 cm in tissue and intended to represent an upper bound of the equivalent dose to the major organs and tissues of the body other than skin and lens of the eye.

deposition: (A) The transfer of airborne materials to the ground or other surface. (B) The accumulation of materials in organs or tissues of the body after intake by inhalation, ingestion, or absorption through the skin or an open wound.

deposition velocity: Ratio of the flux of a contaminant from the atmosphere to the ground or other surface to the concentration in air above the surface.

diffusion: The spreading of a material in a medium due to thermal or mechanical agitation in response to a concentration gradient.

dispersion: The spreading of a flowing substance in a medium due to random variations in the structure of the medium or in the speed and direction of flow.

dose: A quantification of exposure to ionizing radiation, especially in humans. In this report, the term is used to denote **average absorbed dose** in an organ or tissue, **equivalent dose, effective dose**, or **effective dose equivalent**, and to denote dose received or **committed dose**. The particular meaning should be clear from the context in which the term is used.

dose assessment: Estimation of radiation doses received by specified individuals or populations under specified conditions of exposure. See also **dose reconstruction**.

dose coefficient: (A) For intakes of a specific radionuclide by inhalation, ingestion, or absorption through the skin or an open wound, the committed dose per unit activity intake. (B) For external exposure to a specific radionuclide in air, water, or soil or on the body surface, the dose rate per unit concentration in the specified source region. Usually referred to as a dose conversion factor in earlier literature.

dose conversion factor: See **dose coefficient**.

dose equivalent: See **equivalent dose**.

dose rate: Dose received per unit time, often expressed as an average over some period (such as an hour or a day).

dose reconstruction: The process of estimating doses to individuals or populations at some time in the past from exposure to ionizing radiation (or other hazardous agents) on the basis of assumed exposure scenarios.

dose-response analysis: A statistical analysis to estimate values of parameters that describe the relationship between the dose of a hazardous agent (such as ionizing radiation) and an increase in a specified biological response (such as a cancer or other health effect) above the normal (background) incidence. In assessing cancer risks in humans from exposure to ionizing radiation, for example, linear or linear-quadratic dose-response relationships are used most commonly.

dosimeter: A portable instrument for measuring and registering the total accumulated exposure to ionizing radiation.

dosimetric model: (A) For internal exposure to radionuclides, a model that estimates the dose in specific organs or tissues per disintegration of a specific radionuclide in a specified source organ (site of deposition or transit in the body). (B) For external exposure, a model that estimates the dose rate in specific organs or tissues per unit activity concentration of a specific radionuclide in a specified source region, or the dose rate in specific organs or tissues per unit fluence of radiations at the body surface.

dosimetry: The measurement and recording or estimation by calculation of radiation doses or dose rates.

effective dose: The sum over specified organs or tissues of the **equivalent dose** in each tissue modified by the **tissue weighting factor**, as defined in ICRP (1991a). Supersedes **effective dose equivalent**.

effective dose equivalent: The sum over specified organs or tissues of the **average dose equivalent** in each tissue modified by the **tissue weighting factor**, as defined in ICRP (1977). Now superseded by **effective dose**.

effective resuspension factor: A factor used to estimate airborne concentrations of contaminants (assumed to be uniformly distributed with height) that resulted in known concentrations deposited on the ground or other surface, given by the reciprocal of the assumed height of the atmospheric cloud from which deposition occurred. See also **resuspension factor**.

electron: An elementary particle with a unit charge of about 1.602×10^{-19} coulomb and rest mass of 1/1837 that of a proton. Electrons that orbit the nucleus of an atom determine its chemical properties.

electron paramagnetic resonance: The process of resonant absorption of radiation by ions or molecules that are mildly attracted to a magnetic field with at least one unpaired electron spin and in the presence of a static magnetic field.

electron volt: The kinetic energy attained when a particle of unit electronic charge is accelerated through a difference in electric potential of 1 volt (V).

element: A substance that cannot be separated by ordinary chemical methods. Elements are distinguished by the numbers of protons in the nuclei of their atoms.

epidemiology: The study of the incidence, distribution, and causes of health conditions and events in populations.

epidermis: The outer layer of the skin.

equivalent dose: A quantity developed for purposes of radiation protection and assessing risks to human health in general terms, defined as the **average absorbed dose** in an organ or tissue modified by the **radiation weighting factor** for the type, and sometimes energy, of the radiation causing the dose, as defined in ICRP (1991a). Supersedes **average dose equivalent**, as defined in ICRP (1977). The SI unit of equivalent dose is the joule per kilogram (J kg^{-1}), and its special name is the sievert (Sv). In conventional units used in this report, equivalent dose is given in rem; 1 rem = 0.01 Sv.

error: The difference between an estimated value of a quantity and its actual value.

estimate: A measure of or statement about the value of a quantity that is known, believed, or suspected to incorporate some degree of error.

exposure: (A) A general term indicating human contact with ionizing radiation, radionuclides, or other hazardous agents. (B) For the purpose of measuring levels of ionizing photon radiation, the absolute value of the total charge of ions of one sign produced per unit mass of air when all electrons and positrons liberated or created by photons in air are completely stopped in air. Exposure is the quantity measured, for example, by a film badge. The SI unit of exposure is the coulomb per kilogram (C kg^{-1}). In conventional units used in this report, exposure is given in roentgens (R); 1 R = 2.58 $\times$ 10^{-4} C kg^{-1}.

exposure pathway: The physical course of a radionuclide or other hazardous agent from its source to an exposed person.

exposure route: The means of intake of a radionuclide or other hazardous agent by a person (such as ingestion, inhalation, or absorption through the skin or an open wound).

exposure scenario: In this report, the set of circumstances in which an individual or group was exposed to ionizing radiation. Characterization of an exposure scenario usually relies on assumptions about the activities of an individual or group, the times and locations of the activities, and the radiation environment in which the activities took place.

external dose: The dose to organs or tissues of the body due to sources of ionizing radiation located outside the body, including sources deposited on the body surface.

extrapolation: Use of a dataset or model under conditions different from those for which it was established.

fallout: Deposition of radioactive particles produced by detonation of a nuclear weapon.

film badge: Photographic film shielded from light and worn by a person or placed in a specific location to measure and record external exposure to ionizing radiation.

fireball: The highly luminous cloud of vaporized fission and activation products, device constituents, and surrounding support material created by a nuclear detonation.

fissile: Capable of undergoing fission by interaction with neutrons. Fissile isotopes used in nuclear weapons include uranium-235 and plutonium-239.

fission: The splitting of an atomic nucleus into two or more atomic nuclei accompanied by release of neutrons, photons, and energy in the form of kinetic energy of the fission products. In nuclear weapons, fission occurs mainly as a result of capture of neutrons by nuclei of uranium-235 or plutonium-239.

fission product: An atomic nucleus, either stable or radioactive, produced in fission or by decay of a radionuclide produced in fission.

fission yield: See **yield**.

fluence: The number of radiations incident on a sphere per unit cross-sectional area.

flux: The volume of material crossing or impinging on a given cross-sectional area of a surface per unit time divided by the area of the cross section.

fractionation: The chemical and physical separation of radionuclides produced in a nuclear detonation caused by differences in condensation rates as the fireball cools.

free-in-air exposure: The amount of ionization in air produced by incident photons in the absence of any other medium, such as the human body or a structure that might result in attenuation of the radiation.

fusion: The joining together of two atomic nuclei to form heavier nuclei accompanied by release of energy caused by the smaller mass of the heavier nucleus compared with the combined masses of the original nuclei.

gamma radiation: Electromagnetic radiation emitted in de-excitation of atomic nuclei, frequently occurring as a result of decay of radionuclides; also called **gamma rays** and sometimes shortened to **gamma** (for example, gamma-emitting radionuclide). High-energy gamma radiation is highly penetrating and requires thick shielding, such as up to 1 m of concrete or a few tens of centimeters of steel. See also **photon** and **x radiation**.

gastrointestinal tract: Organs of the digestive system, including the esophagus, stomach, small intestine, and upper and lower large intestine (colon).

Geiger counter: An instrument, consisting of a gas-filled tube containing electrodes between which an electric voltage is maintained, used to detect ionizing radiation. When radiation passes through the tube, short pulses of current are generated, which are measured and related to the intensity of the radiation.

generic: Of, applied to, or referring to a whole kind, class, or group. In this report, the term refers to assumptions intended to be broadly applicable to a defined group of participants in the atmospheric nuclear-weapons testing program, especially assumptions concerned with the group's exposure to ionizing radiation.

geometric mean: The nth root of the product of n observations or predictions of a quantity.

geometric standard deviation: The exponential of the standard deviation of the natural logarithms of a set of values.

gray: The special name for the SI unit of absorbed dose; $1 \text{ Gy} = 1 \text{ J kg}^{-1}$.

ground zero: The point on the surface of land or water at or vertically below or above the center of the burst of a nuclear weapon.

half-life: The time required for half the atoms of a particular radionuclide to decay to another nuclear state.

Hodgkin's disease: A type of lymphoma that appears to originate in a particular lymph node and to spread to the spleen, liver, and bone marrow and is characterized by progressive enlargement of the lymph nodes, spleen, and general lymph tissue.

incidence: The rate of occurrence of new cases of a specific disease, calculated as the number of new cases during a specified period divided by the number of individuals at risk of the disease during that period.

internal dose: The dose to organs or tissues of the body due to sources of ionizing radiation in the body.

International System of Units (SI): A modern version of the meter-kilogram-second-ampere system of units, which is published and controlled by an international treaty organization (International Bureau of Weights and Measures).

ion: An atom or molecule that carries a positive or negative electric charge as a result of having lost or gained one or more electrons.

ionizing radiation: Any radiation capable of displacing electrons from atoms or molecules, thereby producing ions. Examples include alpha particles, beta particles, gamma rays or x rays, and cosmic rays. The minimum energy of ionizing radiation is a few electron volts (eV); $1 \text{ eV} = 1.6 \times 10^{-19}$ joules (J).

isopleth: A line on a map connecting points at which a given variable is assumed to have a specified constant value.

isotope: A form of a particular chemical element determined by the number of neutrons in the atomic nucleus. An element may have many stable or unstable (radioactive) isotopes.

isotropic: Exhibiting properties with the same values in all spatial directions.

kiloton: A measure of explosive force equivalent to that of 1,000 tons of trinitrotoluene (TNT).

latent period: The earliest time after exposure to a carcinogenic agent when a cancer caused by that exposure can be manifested; also called **latency period**.

Latin hypercube sampling method: A technique of stratified random sampling from specified probability distributions of variables in which the distributions are divided into intervals of equal probability and one sample is taken at random from each interval. See also **Monte Carlo analysis**.

leukemia: A group of malignant, commonly fatal blood diseases with common characteristics, including progressive anemia, internal bleeding, exhaustion, and a marked increase in the number of white cells (generally their immature forms) in circulating blood.

linear energy transfer: The energy lost by a charged particle per unit distance traversed in a material. The SI unit of linear energy transfer (LET) is the joule per meter ($J\ m^{-1}$). For purposes of radiation protection, LET normally is specified in water and is given in units of keV μm^{-1}.

lognormal distribution: A set of values whose logarithms have a normal distribution.

lower bound: See **lower confidence limit**.

lower confidence limit: The lowest value in a confidence interval. For example, if (x, y) denotes a 90% confidence interval of an uncertain quantity, the lower confidence limit is x, and since confidence intervals generally are specified symmetrically, the true value is expected to be greater than x in 95% of measurements or predictions (and less than the upper confidence limit y in 95% of cases). See also **confidence interval** and **upper confidence limit**.

lymphoma: Malignant tumors originating in cells of lymphatic tissues.

macular degeneration: A blurring of vision in the central visual field.

malignant: Tending to infiltrate, metastasize, and terminate fatally.

mean: The arithmetic average of a set of values, given by the sum of the values divided by the number of values. The mean of a distribution of values is the weighted average of possible values, each value weighted by its probability of occurrence in the distribution.

mechanical clearance: The remove of inhaled substances from the respiratory tract by processes other than absorption into blood (for example, by ciliary action, coughing, sneezing, or nose-blowing).

median: The value in a set of values for which there is an equal probability of a greater or smaller value; the 50th percentile.

megaton: A measure of explosive force equivalent to that of 1 million tons of trinitrotoluene (TNT).

melanoma: A malignant, and often fatal, tumor in cells of the skin that synthesize dark pigments.

metastasis: The transfer of disease from one organ or part to another not directly connected with it due to transfer of cells in malignant tumors.

mission badge: A film badge issued to a participant in the atmospheric nuclear-weapons testing program on a particular occasion when unusual potential for exposure to ionizing radiation was expected and intended to be worn only on that occasion.

mode: The value in a distribution of values that has the highest probability of occurrence.

model: A mathematical, or sometimes physical, representation of an environmental or biological system, sometimes including specific values for the parameters of the system.

monitoring: The measurement of radiation levels or quantities of radionuclides in environmental media.

Monte Carlo analysis: The computation of a probability distribution of an output of a model based on repeated calculations using random samplings of the model's input parameters (variables) from specified probability distributions. See also **Latin hypercube sampling**.

morning report: A document that records the previous day's activities of a military unit, including the number of people in the unit and a list of the unit's officers.

mortality: A measure of the number of people who die from a specific disease or condition in a specified population during a specified period.

multiple myeloma: The proliferation of plasma cells that often replace all other cells within bone marrow, leading to immune deficiency and, frequently, destruction of the outer layer of bone.

neutron: An elementary uncharged particle, of mass slightly greater than that of a proton, that is a constituent of atomic nuclei.

noble gas: Any of a group of rare gases (helium, neon, argon, krypton, xenon, and radon) that exhibit great stability and very low chemical reaction rates.

nonpresumptive disease: Any radiogenic disease in an atomic veteran that is not presumed in law to have been caused by participation in the atmospheric nuclear-weapons testing program but requires an evaluation of the likelihood that the disease was caused by radiation exposure during such participation.

nonrespirable: Incapable of being transported in substantial amounts to regions of the respiratory tract beyond the nose and throat when inhaled, because of the large size of the inhaled materials.

normal distribution: A symmetrical and unbounded distribution, often referred to as a "bell-shaped curve," in which the frequency of occurrence, f, of a distributed quantity, x, is given by

$$f(x) = \frac{1}{\sqrt{2\pi}\sigma} \exp\left[-\frac{1}{2}\left(\frac{x-\mu}{\sigma} \right)^2 \right],$$

where μ is the mean of the distribution and σ^2 is the variance. In a normal distribution, the mean, median, and mode are the same. The probability that a value in the distribution lies between any two numbers, a and b, is equal to the area under the curve $f(x)$ between a and b.

nuclear weapon: A weapon that derives its explosive force from nuclear fusion or nuclear fission reactions.

nucleus: See **atomic nucleus**.

operation: See **series**.

parameter: Any of a set of variables in a model whose values determine the characteristics or behavior of the model, especially the model output. An example of a parameter is the resuspension factor in a model to estimate airborne concentrations of radionuclides on the basis of estimated concentrations on a surface.

parathyroid adenoma: A benign tumor of glands that are adjacent to or embedded in the thyroid and produce a hormone involved in calcium metabolism.

pathway: See **exposure pathway**.

percentile: The value between 0 and 100 that indicates the percent of values in a distribution that are equal to or below it. For example, the 95th percentile is the value that equals or exceeds 95% of the values in a distribution.

permanent badge: A film badge issued to a participant in the atmospheric nuclear-weapons testing program that is intended to be worn at all times of potential radiation exposure until the time of turn-in.

photon: A quantum of electromagnetic radiation, having no charge or mass, that exhibits both particle and wave behavior, especially a **gamma ray** or an **X ray**.

positron: A positively charged electron.

potential: In the context of dose reconstruction, a term denoting exposure to ionizing radiation (or other hazardous agents) of uncertain occurrence.

precision: The degree of reproducibility of a measurement or prediction of a quantity. A measurement or prediction can be precise without being accurate and unbiased. See also **accuracy** and **bias**.

presumptive disease: Any disease in an atomic veteran that is presumed in law to have been caused by exposure to ionizing radiation during participation in the atmospheric nuclear-weapons testing program without regard for whether radiation exposure occurred during such participation and for the magnitude of any such exposure.

probability: The likelihood (chance) that a specified event will occur. Probability can range from 0, indicating that the event is certain not to occur, to 1, indicating that the event is certain to occur.

probability distribution: A representation of the likelihood of occurrence of possible values of an uncertain quantity, such as a model parameter or model output. See, for example, **lognormal distribution** and **normal distribution**.

probability of causation: The probability that a specific disease in a person was caused by their exposure to a particular hazardous agent (such as ionizing radiation). Probability of causation (PC) is estimated as a quotient of two risks: $PC = R/(R+B)$, where R is the estimated risk of the disease in that person due to exposure to the particular hazardous agent and B is the estimated background (baseline) risk of the disease in that person from all other causes (that is, the risk in the absence of exposure to that agent).

Probability of causation differs from risk in that it is conditional on the occurrence of a disease. See also **risk**.

proton: An elementary particle, with a positive charge numerically equal to the charge of an electron and a mass of 1.672×10^{-27} kg, that is a constituent of atomic nuclei.

quadrature: The process of estimating the variance of a quantity calculated as the sum of independent variables by adding the variances of each variable.

quality assurance: A process of ensuring proper documentation of data, interpretations of data that are embodied in assumptions, and computer codes.

quality control: A process of ensuring that measurements and calculations are free of significant error.

quality factor: A dimensionless quantity developed for purposes of radiation protection and assessing risks to human health in general terms that accounts for differences in biological effectiveness between different types of ionizing radiation in producing stochastic effects (such as cancer), which is used to modify the **average absorbed dose** in an organ or tissue to obtain the **average dose equivalent**; see ICRP (1977). The quality factor is assumed to be 1 for photons and electrons of any energy, and higher values are assumed for alpha particles and neutrons. Now superseded in radiation protection by the **radiation weighting factor**.

rad: The special name for the conventional unit of absorbed dose; 1 rad = 100 ergs g^{-1} = 0.01 Gy.

radiation: Energy emitted in the form of waves or particles. See also **ionizing radiation**.

radiation effectiveness factor: A dimensionless quantity that represents the **relative biological effectiveness** of a specific type, and sometimes energy, of ionizing radiation for purposes of estimating cancer risks and probability of causation of specific cancers in persons on the basis of estimates of average absorbed dose from each radiation type in specific organs or tissues in which a cancer has occurred.

radiation exposure: See **exposure**.

radiation protection: The control of exposure to ionizing radiation by use of principles, standards, measurements, models, and such other means as restrictions on access to radiation areas or use of radioactive materials, restrictions on releases of radioactive effluents to the environment, and warning signs. Sometimes referred to in the literature as radiological protection.

radiation weighting factor: A dimensionless factor developed for purposes of radiation protection and assessing risks to human health in general terms, which is intended to replace the **quality factor**; see ICRP (1991a).

radioactive: Exhibiting radioactivity.

radioactive decay: The spontaneous transformation of the nucleus of an atom to a state of lower energy.

radioactivity: The property or characteristic of an unstable atomic nucleus to spontaneously transform with the emission of energy in the form of radiation.

radioepidemiological tables: A tabulation of estimated probabilities of causation of specific cancers in a person at various doses of ionizing radiation. See also **probability of causation** and **risk**.

radiogenic: Causally linked to or possibly associated with exposure to ionizing radiation.

radiological: Of or related to ionizing radiation.

radionuclide: A naturally occurring or artificially produced radioactive element or isotope.

radiosensitive: Susceptible to the injurious action of ionizing radiation.

reconstructed dose: A dose to a person estimated by any means other than a reading of external photon exposure by a film badge worn by the person.

Reference Man: A hypothetical aggregation of human physical, anatomical, and physiological characteristics arrived at by international consensus and used in radiation protection for purposes of calculating radiation doses to organs and tissues of the body from external and internal exposure.

refractory: Capable of enduring relatively high temperatures (for example, about 3000°C or higher) without boiling.

relative biological effectiveness: For a specific radiation (A), the ratio of the absorbed dose of a reference radiation required to produce a specific level of a response in a biological system to the absorbed dose of radiation A required to produce an equal response, with all physical and biological variables, except radiation quality (LET), being held as constant as possible. The reference radiation normally is gamma rays produced in decay of ^{60}Co or ^{137}Cs or X rays produced in an electron tube in which the highest potential difference is about 180–250 kV. RBE is specific to each study and generally depends on the dose, dose per fraction if the dose is fractionated, dose rate, reference radiation, and biological response.

rem: The special name for the conventional unit of equivalent dose; 1 rem = 100 ergs g^{-1} = 0.01 Sv.

respirable: Capable of being transported in substantial amounts to regions of the respiratory tract beyond the nose and throat when inhaled, because of the small size of the inhaled materials.

respiratory protection: An apparatus, such as a respirator or a mask, used to reduce a person's intake of airborne materials.

respiratory tract: A system of organs subserving breathing and associated functions, consisting of the lungs (bronchial and pulmonary regions), respiratory lymph nodes, and the channels by which these are continuous with the outer air (nose and throat).

respiratory-tract model: A biokinetic model describing the deposition, translocation, and absorption of inhaled materials in different regions of the respiratory tract.

resuspension: The transfer of material that has been deposited on the ground or other surface to the atmosphere; also commonly used to mean suspension of material on the ground or other surface that was not deposited from the atmosphere.

resuspension factor: Ratio of the concentration of a resuspended or suspended contaminant in air above the ground or other surface to its concentration on the surface. Resuspension factors usually are determined at a height of 1 m and are given in units of m^{-1}.

retention: The act of remaining at a site of deposition in the body.

retention half-time: The time required for half the quantity of a radionuclide taken into the body to be eliminated from the body by biological processes and radioactive decay combined; also may be referred to as residence half-time in the literature.

risk: The probability of an adverse event. In regard to adverse effects of ionizing radiation on humans, the term usually refers to the probability that a given radiation dose to a person will produce a health effect (such as cancer) or the frequency of health effects produced by given radiation doses to a specified population within a specified period. The risk of cancer due to a given radiation dose generally depends on the cancer type, sex, age at exposure, and time since exposure (attained age), and it may depend on dose rate; the risk of lung cancer also depends on a person's smoking history, and the risk of melanoma or basal cell carcinoma also depends on a person's race.

roentgen: The special name for the conventional unit of exposure; 1 R = 2.58 × 10^{-4} coulomb per kilogram (C kg^{-1}).

safety shot: A test of a nuclear device in which only conventional explosives are detonated intentionally. Safety shots were carried out to ensure that an accidental triggering of the conventional explosive in a nuclear device would not result in significant nuclear fission.

scenario: See **exposure scenario**.

screening: A process of rapidly identifying potentially important radionuclides or exposure pathways by eliminating those of known negligible importance.

screening code: A code assigned to military units indicating exposure conditions under which the committed equivalent dose to bone from inhalation of radionuclides should be less than 0.15 rem for personnel in those units.

screening dose: A dose to a specific organ or tissue that is assumed to correspond to a 99% upper bound (upper confidence limit) of a probability of causation of 50%.

screening model: A model that incorporates pessimistic assumptions about potential exposures to ionizing radiation and a criterion intended to correspond to a negligible dose for the purpose of identifying radionuclides or exposure pathways of negligible importance.

series: An official grouping of nuclear-weapons tests that were carried out by a military task group over a particular interval and in a particular area (the Nevada Test Site or the Pacific), for example, the PLUMBBOB test series at the NTS in 1957.

shielding factor: Ratio of the external exposure rate or dose rate indoors to that outdoors due to attenuation of photon or neutron radiations outside the structure by the structure components.

shot: A detonation of a nuclear device; used synonymously with **test** in the nuclear-weapons testing program.

sievert: The special name for the SI unit of equivalent dose; $1 \text{ Sv} = 1 \text{ J kg}^{-1}$.

SI units: See **International System of Units**.

solar index: A whole number between 1 and 10 that represents the intensity of solar ultraviolet radiation at the earth's surface and is associated with the likelihood of damage to the skin or eye and the time it takes for damage to occur.

spectrum: An array of energies of radiation (such as radiation emitted in decay of radionuclides or detonation of nuclear weapons) and their associated intensities.

squamous cell carcinoma: A malignant growth originating from plate-like cells found in the outer layer of the skin and usually occurring on the skin, lips, inside of the mouth, throat, or esophagus, which studies have not shown to be radiogenic in humans.

standard deviation: Square root of the **variance**.

stochastic: Of, pertaining to, or arising from chance; involving probability; random.

stratification: The process or result of separating a sample into subsamples according to specified criteria such as, for example, age, sex, or dose received.

surface porosity: The presence of openings on the surface of a material.

suspension: See **resuspension**.

systemic: Present in the body after absorption from the respiratory tract, gastrointestinal tract, or skin into blood.

test: See **shot**.

tissue weighting factor: A dimensionless factor that represents the ratio of the stochastic risk attributable to a specific organ or tissue to the total stochastic risk attributable to all organs and tissues when the whole body receives a uniform exposure to ionizing radiation; see ICRP (1977; 1991a).

toxicant: A poisonous agent.

translocation: The movement of a substance from one part of the body to another, especially in blood.

transuranium: Of, related to, or being an element with an atomic number greater than that of uranium (92). Examples of transuranium elements are plutonium and americium.

uncertainty: The lack of sureness or confidence in results of measurements or predictions of quantities owing to stochastic variation or to a lack of knowledge founded on an incomplete characterization, understanding, or measurement of a system.

uncertainty analysis: An analysis of the variability in model predictions due to uncertainty in the input parameters or other assumptions.

uncertainty factor: In this report, the ratio of an upper confidence limit of an estimated dose to a central estimate.

uniform distribution: A probability distribution in which the frequency of occurrence of any value between the lowest and highest possible values is the same.

unit dose reconstruction: A reconstruction of radiation doses to an average member of a military unit.

upper bound: See **upper confidence limit**.

upper-bound factor: See **uncertainty factor**.

upper confidence limit: The highest value in a confidence interval. For example, if (x, y) denotes a 90% confidence interval of an uncertain quantity, the upper confidence limit is y, and since confidence intervals generally are specified symmetrically, the true value is expected to be less than y in 95% of measurements or predictions (and greater than the lower confidence limit x in 95% of cases). See also **confidence interval** and **lower confidence limit**.

urinalysis: A method of bioassay involving measurement of quantities of contaminants excreted in urine for the purpose of estimating intake.

variability: The variation of a property or quantity among members of a population. Variability is often assumed to be random and can be represented by a probability distribution.

variance: A measure of the spread of a distribution of values, denoted by σ^2 and given by the mean of the squares of the differences between the individual values and their mean:

$$\sigma^2 = \frac{1}{N} \sum_i (x_i - \mu)^2,$$

where μ is the mean of the distribution of values x_i, and N is the number of values. See also **standard deviation**.

volatile: Capable of boiling at relatively low temperatures (for example, about 1500°C or lower).

weathering: (A) A reduction in availability of material deposited on the ground or other surfaces for resuspension, owing to physical, chemical, or biological processes other than radioactive decay. (B) Actions of the weather (wind, precipitation, or temperature changes) to reduce the size of particles deposited on the ground or other surfaces or to cause radionuclides present in the uppermost soil layer to migrate to greater depth.

whole body: For purposes of estimating radiation dose, especially from external exposure, the head, trunk (including male gonads), arms above the elbow, and legs above the knee.

whole-body counting: The measurement of radioactivity in the body by use of radiation detectors outside the body to detect penetrating radiation (gamma rays and x rays) emitted by the sources.

X radiation: (A) Electromagnetic radiation emitted in de-excitation of bound atomic electrons, frequently occurring in decay of radionuclides, referred to as **characteristic X rays**, or (B) electromagnetic radiation produced in deceleration of energetic charged particles (such as beta radiation) in passing through matter, referred to as **continuous X rays** or **bremsstrahlung**; also called **X rays**. See **gamma radiation** and **photon**.

yield: The total energy released in a nuclear detonation, usually expressed in kilotons (kT) or megatons (MT). The announced yield includes the energy released by fission and fusion.

LIST OF ABBREVIATIONS

AMAD	Activity median aerodynamic diameter
ANSI	American National Standards Institute
ASME	American Society of Mechanical Engineers
ASQC	American Society for Quality Control
BCT	Battalion Combat Team
BRER	Board on Radiation Effects Research
C and P	Compensation and Pension
CDR	Camp Desert Rock
CFR	Code of Federal Regulations
CIC	Coordination and Information Center
CIRRPC	Committee on Interagency Radiation Research and Policy Coordination
CV	Coefficient of variation
DHHS	U.S. Department of Health and Human Services
DNA	Defense Nuclear Agency
DOD	U.S. Department of Defense
DOE	U.S. Department of Energy
DRI	Desert Research Institute
DTRA	Defense Threat Reduction Agency
EPA	U.S. Environmental Protection Agency
EPR	Electron paramagnetic resonance
FBE	Film-badge equivalent
FOIA	Freedom of Information Act
GAO	General Accounting Office
GI	Gastrointestinal
GSD	Geometric standard deviation
GZ	Ground zero
ICRP	International Commission on Radiological Protection
ICRU	International Commission on Radiation Units and Measurements
IOM	Institute of Medicine
IREP	Interactive RadioEpidemiological Program
LAN	Local area network
LET	Linear energy transfer
MC	Monte Carlo
MP	Military police
NA	Not applicable
NAAV	National Association of Atomic Veterans
NAS	National Academy of Sciences
NCI	National Cancer Institute
NCRP	National Council on Radiation Protection and Measurements

NIH	National Institutes of Health
NIOSH	National Institute for Occupational Safety and Health
NQA	Nuclear Quality Assurance
NRC	National Research Council
NTPR	Nuclear Test Personnel Review
NTS	Nevada Test Site
NuTRIS	Nuclear Test Review Information System
OPHEH	Office of Public Health and Environmental Hazards
ORNL	Oak Ridge National Laboratory
PC	Probability of causation
POW	Prisoner of war
PPG	Pacific Proving Ground
QA	Quality assurance
QC	Quality control
QF	Quality factor
Rad-safe	Radiological safety
RBM	Red bone marrow
RCT	Regimental Combat Team
REECO	Reynolds Electrical and Engineering Company
REF	Radiation Effectiveness Factor
SAIC	Science Applications International Corporation
SD	Standard deviation
SF	Shielding factor
SI	Système International (International System)
SOP	Standard operating procedure
TGLD	Task Group on Lung Dynamics
TNT	Trinitrotoluene
TU	Task unit
VA	U.S. Department of Veterans Affairs
VACO	Department of Veterans Affairs Central Office
VARO	Department of Veterans Affairs Regional Office
VITAL	Veterans Issue Tracking Adjudication Log
UB	Upper bound
USC	U.S. Code
USMC	U.S. Marine Corps
USS	U.S. Ship
UV	Ultraviolet

Committee Biographies

John Till, PhD (Chair), president, Risk Assessment Corporation. Dr. Till is a recognized authority in dose reconstruction and communication efforts in radiological assessment, dose reconstruction, and risk analysis. Dr. Till was a member of the IOM committee that produced the year 2000 National Academies report *The Five Series Study: Mortality of Military Participants in U.S. Nuclear Weapons Tests*. He is the 1995 recipient of the E. O. Lawrence Award in Environmental Science and Technology. He has chaired a number of committees and task groups on issues related to radiation dosimetry. He was responsible for the dosimetry estimates in a major University of Utah epidemiological dose reconstruction project and was chairman of the technical steering panel that directed the Hanford Environmental Dose Reconstruction Project. Dr. Till's scientific achievements include more than 150 publications.

Harold L. Beck, BS, retired director of the Environmental Sciences Division of the Department of Energy (DOE) Environmental Measurements Laboratory (EML) in New York City. Mr. Beck previously served as director of the EML Instrumentation Division and as acting deputy director of the laboratory. Mr. Beck received his BS from the University of Miami summa cum laude and did graduate work in physics and mathematics at Cornell University from 1960 to 1962. He is the author or coauthor of over 100 publications in radiation physics, radiation protection, environmental radiation, dosimetry, and instrumentation. His development of the scientific approach to reconstructing fallout doses to the US population from aboveground nuclear-weapons tests in Nevada earned him the DOE Meritorious Service Award in 1988, the second-highest award in the department. He is a member of the American Association for the Advancement of Science, the American Nuclear Society, and the National Council on Radiation Protection and Measurements (NCRP), and he is a fellow of the Health Physics Society. Mr. Beck served as the scientific vice president of the NCRP for radiation measurement and dosimetry in 1996-2002. He also served as a member of BRER's Committee on Dosimetry for the Radiation Effects Research Foundation.

William J. Brady, BS, was former Principal Health Physicist, Reynolds Electrical and Engineering Co., Inc. (REECo), a subsidiary of EG&G Inc., at the Nevada Test Site (NTS). For 35 years he held health physics and technical positions and was director of the Rad Safe Division Reactor Branch and director of the Laboratory Branch. He wrote the first NTS monitoring manual (1956) and emergency monitoring manual (1957), and developed the personnel radiation dosimeter worn at the NTS for 26 years. In 1957 he started collecting US dosimetry records for nuclear testing and put them into a computerized Master File during 1966–1969. Those records were the basis of research used by the Nuclear Test Personnel Review (NTPR) program of the Defense Nuclear Agency (DNA). He continued assisting DNA as a dosimetry expert until retirement in 1991 and wrote several histories of underground testing. Mr. Brady is a past scientific advisor of the National Association of Atomic Veterans. In 1988, he received the Department of Energy's highest department commendation, the Award of Excellence, "In recognition of significant contributions to radiological safety in support of the weapons testing program at the Nevada Test Site." He has served on two prior NRC committees, one on ionizing-radiation dosimetry and the other on film-badge dosimetry in atmospheric tests.

Thomas Gesell, PhD, is professor of health physics, director of the Technical Safety Office, and director of the Environmental Monitoring Program at Idaho State University. He has worked in multiple capacities for the DOE Idaho Operations Office, including deputy assistant manager for Nuclear Programs, and director of the Radiological and Environmental Sciences Laboratory on the Idaho National Engineering Laboratory site. Dr. Gesell was a faculty member of the University of Texas School of Public Health in Houston for 10 years. Dr. Gesell is a member of several committees and professional organizations, including the EPA Science Advisory Board's Radiation Advisory Committee and the National Council on Radiation Protection and Measurements (NCRP). He chaired committees whose work led to three publications of the NCRP and served on one previous committee of the National Research Council. He is a fellow of the Health Physics Society and was recently elected to its Board of Directors. Dr. Gesell was also a consultant to the President's Commission on the Accident at Three Mile Island. Dr. Gesell was coauthor of *Environmental Radioactivity from Natural, Industrial and Military Sources* (1997) with Merril Eisenbud.

David G. Hoel, PhD, Distinguished University Professor, Department of Biometry and Epidemiology at the Medical University of South Carolina, Charleston. Dr. Hoel earned his PhD from the University of North Carolina at Chapel Hill. His research specialties include environmental causes of cancer, especially ionizing radiation; risk-assessment models; statistical inference, particularly sequential procedures; statistical and mathematical applications in biology and medicine; epidemiology; and radiation health effects. Dr. Hoel was formerly an associate director of the Radiation Effects Research Foundation, Hiroshima, Japan. He is a member of the Institute of Medicine, and has served on numerous national committees, including committees of the National Research Council.

David G. Kocher, PhD, senior scientist at SENES Oak Ridge, Inc., Center for Risk Analysis, Oak Ridge, Tennessee. He earned his PhD in physics from the University of Wisconsin-Madison in 1970. He joined the research staff at Oak Ridge National Laboratory (ORNL) in 1971, where he worked as an environmental health physicist from 1976 to 2000. His principal research activities at ORNL involved the development and implementation of models and databases for estimating radiation doses to the public from exposure to radionuclides in the environment; the results of this work have been widely used in assessing radiological effects of releases from operating nuclear facilities and from disposal of radioactive waste. An important focus of his work has involved evaluations of environmental dose-assessment models for regulatory and decision-making purposes. He served as a member of several technical advisory groups for the Department of Energy, the Science Advisory Board of the Environmental Protection Agency, the Nuclear Regulatory Commission, and the International Atomic Energy Agency in environmental radiological assessment and management of radioactive and hazardous chemical wastes, and he served on the National Research Council's Committee on Evaluation of EPA Guidelines for Exposure to Naturally Occurring Radioactive Materials. In 1999, he was elected to membership in the National Council on Radiation Protection and Measurements (NCRP), and he has served on NCRP scientific committees on risk-based classification of radioactive and hazardous chemical wastes, performance assessment for disposal of low-level radioactive waste, and risk management analysis for decommissioned sites. He has lectured widely in external and internal dosimetry, environmental radiological assessments, radioactive-waste management, and laws and regulations addressing public exposures to radionuclides and hazardous chemicals in the environment. He joined SENES Oak Ridge, Inc., in December 2000. His activities at SENES have included development of the radiation effectiveness factors (REFs) used in the compensation program for sick workers at Department of Energy facilities and participation in the dose reconstruction for historical releases at Idaho National Engineering Laboratory.

Jonathan D. Moreno, PhD, Emily Davie and Joseph S. Kornfeld Professor of Biomedical Ethics at the University of Virginia, where he is also director of the Center for Biomedical Ethics. Dr. Moreno received his bachelor's degree from Hofstra University in 1973, with highest honors in philosophy and psychology. He was a University Fellow at Washington University in St. Louis, receiving his doctorate in philosophy in 1977, and was a Mellon Postdoctoral Fellow in cooperation with the Aspen Institute for Humanistic Studies. In 1998, he received an honorary doctorate from Hofstra. Dr. Moreno is a member of the Board on Health Sciences Policy of the Institute of Medicine and of the Council on Accreditation of the Association for the Accreditation of Human Research Protection Programs. He is president-elect of the American Society for Bioethics and Humanities. He was a member of the National Human Research Protection Advisory Committee, a senior consultant for the former National Bioethics Advisory Commission, and has advised the White House Office of Science and Technology Policy. During 1994-1995, he was senior policy and research analyst for the President's Advisory Committee on Human Radiation Experiments. He is also a bioethics adviser for the Howard Hughes Medical Institute and Genomics Collaborative, Inc., a faculty affiliate at the Kennedy Institute of Ethics at Georgetown University, and a Fellow of the Hastings Center and the New York Academy of Medicine. His book *Undue Risk: Secret State Experiments on Humans* was published by Routledge in 2001. Dr. Moreno has also published around 200 papers and book chapters and is a member of several editorial boards.

Clarice Weinberg, PhD, chief, Biostatistics Branch, National Institute of Environmental Health Sciences, and expert in biostatistics. Dr. Weinberg served on the IOM committee that produced the year 2000 National Academies report *The Five Series Study: Mortality of Military Participants in U.S. Nuclear Weapons Tests.* She is a fellow of the American Statistical Association and former editor of the *American Journal of Epidemiology*, and she serves on the editorial boards of *Epidemiology* and *Environmental Health Perspectives*.

Index